T0319349

Evolution of Wireless Communication Ecosystems

IEEE Press
445 Hoes Lane
Piscataway, NJ 08854

Evolution of Wireless Communication Ecosystems

Dr. Suat Seçgin

The ComSoc Guides to Communications Technologies
Nim K. Cheung, Series Editor

Published by John Wiley & Sons, Inc., Hoboken, New Jersey.
Published simultaneously in Canada.

For general information on our other products and services or for technical support, please contact our Customer Care Department within the United States at (800) 762-2974, outside the United States at (317) 572-3993 or fax (317) 572-4002.

Wiley also publishes its books in a variety of electronic formats. Some content that appears in print may not be available in electronic formats. For more information about Wiley products, visit our web site at www.wiley.com.

Library of Congress Cataloging-in-Publication Data applied for:
Hardback ISBN: 9781394182312

Cover Design: Wiley
Cover Image: © Dr Pixel/Getty Images

Set in 9.5/12.5pt STIXTwoText by Straive, Pondicherry, India

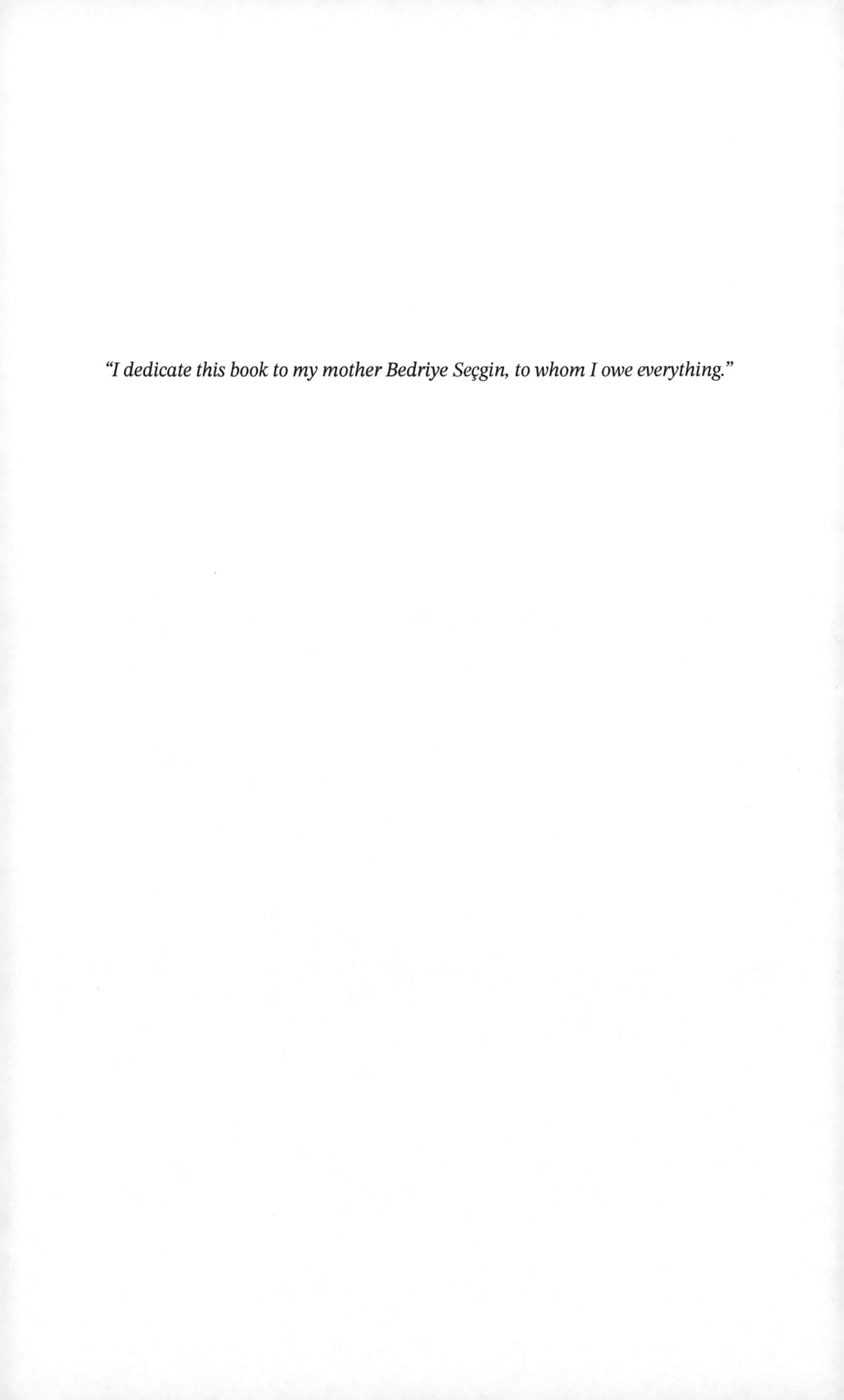

"I dedicate this book to my mother Bedriye Seçgin, to whom I owe everything."

Contents

About the Author

Dr. Suat Seçgin has worked in the telecommunication sector for nearly 30 years in access systems, core networks, IT, and customer management. After completing his Electrical and Electronics Engineering undergraduate education, Dr. Seçgin received his master's degree in Computer Engineering on mobile networks and data access strategies. Subsequently, he earned his Ph.D. in Computer Engineering in data science and decision support systems.

He has publications in peer-reviewed and indexed journals on customer analytics in the telecommunications industry, and this is his second book on telecommunication systems.

Preface

Since Alexander Graham Bell's first "hello" (1876), communication systems have witnessed revolutionary developments. These communication systems, which used to work entirely in circuit-switching and wired transmission environments, are now turning into wireless systems, especially in access techniques, apart from the main backbone. Initially, copper wire circuits were used in communication, while fiber optic cables began in the 1990s. Today, fiber services are put into use in the home. In parallel with these developments in wired communication systems, wireless communication systems continue to develop rapidly. With the developments in multiplexing and modulation techniques, the bandwidth provided to each user is progressing at an increasing speed.

With the introduction of the Internet into our lives in the 1990s, Information Technologies and Communication Technologies began to converge. After this point, we started talking about Information Communication Technologies. Today, we live in scenarios where everything is connected to the vast Internet cloud with 4G, 5G, and 6G wireless communication systems. However, it is still in the fictional stage. Communication systems, previously used as discrete structures, are almost becoming living organisms thanks to this substantial interconnected communication infrastructure. It would be much more appropriate to call this cyber organism structure "Wireless Communication Ecosystem," which extends from wireless sensor networks operating at the most extreme points to the edges. This is a gigantic ecosystem that extends from big data, artificial intelligence, blockchain, and machine learning to quantum communication on the one hand. The book is designed to explain the main elements of this ecosystem. If we liken this ecosystem to a human body, each body organ is introduced in the book.

Before going into the details of these systems, modulation and multiplexing techniques are also explained to understand the communication generations and to visualize the big picture in our minds. In addition to increasing the efficiency of the frequency spectrum, users were introduced to high-frequency and high-bandwidth by using multiplexing techniques together. Systems that used to work with classical circuit switching have evolved into packet switching-based systems that provide flexible and scalable solutions.

With the convergence of communication and information systems, information and communications technology (ICT) systems were developed that radically changed our daily lives and created super-intelligent societies. The support of software systems has led to paradigm shifts in network metrics such as management, security, configuration, scaling, resilience, and more.

Starting with 1G systems, we will talk about communication systems at terahertz levels that have developed due to spectrum-efficient modulation techniques and advances in electronic circuits. In addition, users can receive services at high bandwidths using three-dimensional multiplexing techniques.

Wireless communication systems, which continue to progress without slowing down, have evolved into the fifth-generation communication systems as of the 2020s. Communication speeds up to 20 GHz with 5G, and thanks to these speeds, the concepts of ultrareliable low-latency communication (uRLLC), enhanced mobile broadband (eMBB), and massive machine-type communication (mMTC) entered our lives. The transition occurred from the Internet of Things (IoT) to the Internet of Everything (IoE). Communication systems, applications, and services have become more intelligent using topics such as artificial intelligence, blockchain, and big data in the software field. Smart homes, smart cities, innovative health systems, and autonomous vehicles are now inseparable parts of our lives.

In the years when the book was written (2021/2022), 5G applications entered our lives, and the sixth-generation communication systems, which will be put into use starting from the 2030s, became talked about and fictionalized. Along with 6G, concepts such as 3D networks, intelligent networks, quantum communication, blockchain technologies, deep learning, and programmable surfaces are designed together with communication infrastructures.

The book's primary purpose is to describe the big picture of wireless communication generations and applications running on this communication medium. The book describes the infrastructure of 4G, 5G, and 6G systems, this all-connected communication ecosystem, the subcomponents of this ecosystem, and the relationship among them.

Since IoT systems are an integral part of wireless communication infrastructure in parallel with 4G, 5G, and 6G systems, access techniques, protocols, and security issues for these systems are also explained in detail in our book. In addition to these access techniques, the methods used in M2M and IoT connections at the endpoints are given. Security is also a significant challenge in this ecosystem where everything is connected with everything. Vital security breaches, especially for terminals/users at the endpoint, are also described in the book.

In the first part, the basic concepts of communication systems are explained. In the following, the events seen in the capillaries of the communication echo system are described by explaining the switching techniques, modulation, and multiplexing techniques. Thus, it is aimed at understanding the applications running at higher levels. With this aspect, the book has been designed to guide the reader who wants to advance in each subject. The book, which has a pervasive literature review for each section, is also an essential resource for researchers.

Dr. Suat Seçgin
Electrical & Electronics Engineer (BSc)
Computer Engineer (MSc & PhD)

List of Abbreviations

3GPP	Third-Generation Partnership Project
6LoWPAN	IPv6 over low-power WPANs
A-CSCF	Access Session Border Controller
ADSL	Asymmetric Digital Subscriber Line
AM	Amplitude Modulation
AMPS	Advanced Mobile Phone System
AMQP	Advanced Message Queuing Protocol
ANN	Artificial Neural Network
ASK	Amplitude Shift Keying
ATM	Asynchronous Transfer Mode
AuC	Authentication Center
BBU	Baseband Unit
BER	Bit Error Rate
BGP	Border Gateway Protocol
BICN	Bearer-Independent Core Network
BSC	Base Station Controller
BSS	Base Station Subsystem
BTS	Base Transceiver Station
CA	Carrier Aggregation
CDM	Code Division Multiplexing
CDMA	Code Division Multiple Access
CDR	Call Detail Record
CINR	Carrier to Interference + Noise Ratio
CIR	Channel Impulse Responses
CNN	Convolutional Neural Network
CoAP	Constrained Application Protocol
CPC	Continuous Packet Connectivity
C-RAN	Cloud/Centralized Radio Access Network
CSCP	Call Session Control Function

DAS	Distributed Antenna System
DC	Dual Connectivity
DDoS	Distributed Denial of Service
DDS	Data Distribution Service
DNS	Domain Name System
DPWS	Devices Profile for Web Services
D-RAN	Distributed Radio Access Network
DSCP	Differentiated Service Code Point
DTLS	Datagram Transport Layer Security
EDGE	Enhanced Data Rates for GSM Evolution
EIR	Equipment Identity Register
eMBB	Enhanced Mobile Broadband
EPC	Evolved Packet Core
EV-DO	Evolution-Data Optimized
FDD	Frequency Division Duplexing
FDM	Frequency Division Multiplexing
FDMA	Frequency Division Multiple Access
FM	Frequency Modulation
FSK	Frequency Shift Keying
FTAM	File Transfer, Access, and Management
FTP	File Transfer Protocol
GAN	Generative Adversarial Network
GEO	Geosynchronous Equatorial Orbit
GERAN	GSM Edge Radio Access Network
GGSN	Gateway GPRS Support Node
GMSC	Gateway Mobile Switching Center
GPOS	General Purpose Operating System
GPRS	General Packet Radio Service
GSM	Global System for Mobile Communication
HART	Wireless Highway Addressable Remote Transducer Protocol (HART)
HLR	Home Location Register
HMIMOS	Holographic MIMO Surfaces
HSDPA	High-Speed Downlink Packet Access
HSPA	High-Speed Packet Access
HSUPA	High-Speed Uplink Packet Access
IFFT	Inverse Fast Fourier Transformation
IMEI	International Mobile Equipment Identity Number
IMS	IP Multimedia Subsystem
IMSI	International Mobile Subscriber Identity Number
IM-SSF	IP Multimedia Services Switching Function

IMT-2000	International Mobile Telecommunication Standard-2000
IN	Intelligent Network
IP	Internet Protocol
IrDA	Infrared Data Association
ISDN	Integrated Service Digital Network
ISI	Intersymbol Interference
ISM	Industrial, Scientific, and Medical
ISO	International Organization for Standardization
ITU	International Telecommunication Union
IWSN	Industrial Wireless Sensor Network
LEO	Low Earth Orbit
LiFi	Light Fidelity
LLC	Logical Link Control
LOS	Line of Sight
LPWAN	Low-Power Wide-Area Networks
LTE	Long-Term Evolution
LTE-M	Long-Term Evolution for Machines
LTP	Lean Transport Protocol
M2M	Machine to Machine
MAC	Media Access Control
MAEC	Multi-Access Edge Computing
MCC	Mobile Cloud Computing
MCS	Mobile Switching Center
MEC	Mobile Edge Computing
MEO	Medium Earth Orbit
MGW	Media Gateway
MIMO	Multiple Input Multiple Output
MIOT	Massive IoT
MLP	Multilayer Perceptrons
MME	Mobile Management Entity
MMS	Multimedia Messaging Service
mMTC	Massive Machine-Type Communication
mmWave	Milimeter Wave
MOM	Message-Oriented Middleware
MQTT	Message Queue Telemetry Transport Protocol
MSISDN	Mobile Subscriber Integrated Services Digital Network Number
NAMPS	Narrowband Advanced Mobile Phone
NB-IoT	Narrowband IoT
NFC	Near Field Communication
NFV	Network Function Virtualization
NGN	Next-Generation Networks

NMT	Nordic Mobile Telephone
NOMA	Non-Orthogonal Multiple Access
NSS	Network Subsystem
OAMM	Orbital Angular Momentum Multiplexing
OFDM	Orthogonal Frequency Division Multiplexing
OSA-GW	Open Service Access Gateway
OSI	Open System Interconnection
OTT	Over the Top
OWC	Optical Wireless Communication
PAN	Personal Area Network
PAPR	Peak to Average Power Ratio
PCRF	Policy and Charging Rule Functions
P-CSCF	Proxy Call Session Control Function
PCU	Packet Control Unit
PDM	Polarization Division Multiplexing
PDN	Public Data Network
PDU	Protocol Data Unit
PER	Packet Error Rate
PGW	PDN Gateway
PLC	Power Line Communication
PLMN	Public Land Mobile Network
PM	Phase Modulation
POTS	Plain Old Telephone System
PSK	Phase Shift Keying
PSTN	Public Switched Telephone Network
QAM	Quadrature Amplitude Modulation
qDC	Quantum-assisted Data Center
qEdge	Quantum-assisted Edge Network
QKD	Quantum Key Distribution
QML	Quantum Machine Learning
QoE	Quality of Experience
QoS	Quality of Service
qRAN	Quantum-assisted RAN
qSIN	Quantum Space Information Network
qWAI	Quantum-assisted Wireless Artificial Intelligence
RAN	Radio Access Network
RAT	Radio Access Technology
REST	Representational State Transfer
RF	Radio Frequency
RFID	Radio Frequency Identification
RIS	Reconfigurable Intelligent Surfaces

RNC Radio Network Controller
RNN Recurrent Neural Networks
RPMA Random Phase Multiple Access
RRH Remote Radio Head
RRU Remote Radio Unit
RTOS Real-Time Operating System
RTP Real-Time Transport Protocol
SCE Service Creation Environment
SCP Service Control Point
SC-PTM Single Cell-Point to Multipoint
SDM Spatial Division Multiplexing
SDN Software Defined Network
SE Spectral Efficiency
SGSN Serving GPRS Support Node
S-GW Serving Gateway
SIC Successive Interference Cancellation
SIM Subscriber Identity Module
SINR Signal Interference + Noise Ratio
SIP Session Initiation Protocol
SLA Service-Level Agreement
SMS Short Message Service
SMTP Simple Mail Transfer Protocol
SNR Signal to Noise Ratio
SOAP Simple Object Access Protocol
SON Self-Organizing Network
SS7 Signaling System Number 7
STOMP Simple Text-Oriented Messaging Protocol
TACS Total Access Communication System
TAS Telephony Application Server
TCP Transmission Control Protocol
TDD Time Division Duplexing
TDM Time Division Multiplexing
TDMA Time Division Multiple Access
TFTP Trivial File Transfer Protocol
TSCH Time Slotted Channel Hopping
UCA Uniform Circular Array
UCDC Unconventional Data Communication
UMTS Universal Mobile Telecommunication Standard
uRLLC Ultra-Reliable Low Latency Communication
UTRAN UMTS Radio Access Network
VLC Visible Light Communication

VLR	Visitor Location Register
VoLTE	Voice over LTE
VoWiFi	Voice over WiFi
VPN	Virtual Private Network
WAP	Wireless Application Protocol
WCDMA	Wideband Code Division Multiple Access
WDM	Wavelength Division Multiplexing
WIA-PA	Wireless Network for Industrial Automation-Process Automation
WiMAX	Worldwide Interoperability for Microwave Access
WPAN	Wireless Personal Area Network
WSDL	Web Service Description Language
WSN	Wireless Sensor Network
WWAN	Wireless Wide Area Network
XMPP	Extensible Messaging and Presence Protocol

1

Basic Concepts

1.1 Introduction

The input of a communication system is a sound, image, or text file to be transmitted to the other end. The output is naturally this original information signal, which goes through many processes (modulation, coding, multiplexing, etc.) until it reaches the end. This section explains the layers through which the information passes from where it enters the system to where it leaves.

1.2 Main Components of Communication Systems

The main components of an end-to-end communication system are the transmitter, transmission medium, and receiver (Figure 1.1). Any factor that negatively affects the operation of the system is called noise.

- **Information source:** The first step in sending a message is to convert it into an electronic form suitable for transmission. For voice messages, a microphone is used to convert the sound into an electronic audio signal. For TV, the camera converts the light information in the scene into a video signal. In computer systems, the message is typed on the keyboard and converted into binary codes that can be stored in memory or transmitted in serial. Transducers convert physical properties (temperature, pressure, light intensity, etc.) into electrical signals.
- **Transmitter:** The transmitter is a collection of electronic components and circuits designed to convert the electrical signal into a signal suitable for transmission over a given communication medium. Transmitters consist of oscillators, amplifiers, tuned circuits and filters, modulators, mixers, frequency synthesizers, and

Evolution of Wireless Communication Ecosystems, First Edition. Suat Seçgin.

Published 2023 by John Wiley & Sons, Inc.

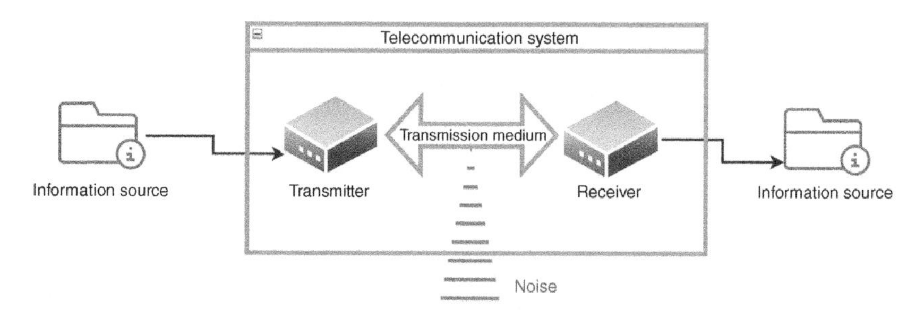

Figure 1.1 Block schema of a communication system.

other circuits. The original signal is usually modulated with a higher frequency carrier sine wave produced by the transmitter and amplified by power amplifiers. Thus, the information signal is rendered transmittable in the transmission medium.

- **Communication channel:** The communication channel is the medium in which the electronic signal is sent from one place to another. Many media types are used in communication systems, including wire conductors, fiber optic cable, and free space. Of these, electrical conductors can be a pair of wires that carry an audio signal from the microphone to the headphone. It could be a coaxial cable similar to that used to have signals. Or it could be a twisted-pair cable used in a local area network (LAN). The communication medium may also be a fiber optic cable or "light pipe" that carries the message on a light wave. These are used today to carry out long-distance calls and all Internet communications. The information is converted into a digital form that will be used to turn a laser diode on and off at high speeds. Alternatively, audio or video analog signals can be used to vary the amplitude of the light. When space is media, the resulting system is known as radio. Radio, also known as wireless, is the general term applied to any form of wireless communication from one point to another. Radio makes use of the electromagnetic spectrum. Information signals are converted into electric and magnetic fields that propagate almost instantly in space over long distances.
- **Receiver:** The receiver is a collection of electronic components and circuits that accepts the message transmitted through the channel and converts it back into a form that can be understood. Receivers include amplifiers, oscillators, mixers, tuned circuits and filters, and a demodulator or detector that retrieves the original information signal from the modulated carrier. The output is the initial signal that is then read or displayed. It can be an audio signal sent to a speaker, a video signal fed to an LCD screen for display, or binary data received by a computer and then printed or displayed on a video monitor.

- **Transceiver:** Most electronic communications are two-way. Therefore, both parties must have a transmitter and a receiver. As a result, most communications equipment contains both sending and receiving circuits. These units are often called transceivers. All transmitter and receiver circuits are packaged in a single enclosure and often share some common circuitry, such as the power supply. Telephones, walkie-talkies, mobile phones, and computer modems are examples of transceivers.
- **Attenuation:** Regardless of the transmission medium, signal attenuation or degradation is inevitable. The attenuation is proportional to the square of the distance between the transmitter and receiver. Media is also frequency selective because a particular medium acts as a low-pass filter for a transmitted signal. Thus, digital pulses will be distorted, and the signal amplitude will significantly reduce over long distances. Therefore, a significant amount of signal amplification is required at both the transmitter and receiver for successful transmission. Any medium also slows signal propagation to a slower-than-light speed.
- **Noise:** Noise is mentioned here because it is one of the most important problems of all electronic communication. Its effect is experienced in the receiving part of any communication system. Therefore, we consider noise in Chapter 9 as a more appropriate time. While some noise can be filtered out, the general way to minimize noise is to use components that contribute less noise and lower their temperature. The measure of noise is usually expressed in terms of the signal-to-noise ratio (SNR), which is the signal power divided by the noise power and can be expressed numerically or in decibels (dB). A very high SNR is preferred for the best performance.

1.3 Circuit, Packet, and Cell Switching

A circuit, packet, or cell switching technique is used on the communication line established to communicate two terminals at two opposite endpoints.

1.3.1 Circuit Switching

Circuit switching is the first method used in communication systems. When you somehow pull a cable (or establish a wireless link) between the two opposite ends that will communicate, we establish a circuit between the two terminals. A one-to-one connection between the terminals in the matrix structure and connected to the switching center (switchboard) with a circuit is established between the terminals that require connection by the switching center. Thus, a circuit is established (switched) that can only be used by those two terminals at the communication time. Since packet-switched communication is widely used in today's

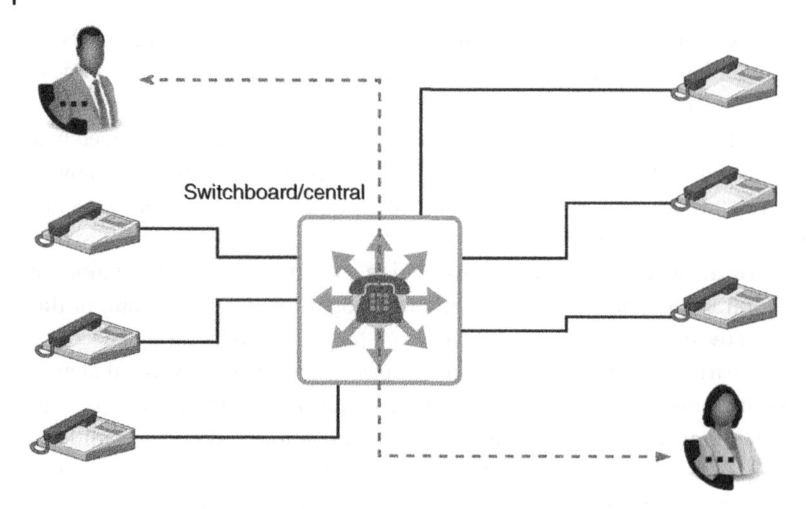

Figure 1.2 Circuit switching.

communication, virtual circuits specific to end terminals can be established by defining virtual paths on packet-based circuits (Figure 1.2).

In circuit switching, a link is established between both terminals, which is used only by these terminals. As long as the link connection is used, other terminals cannot use this line. As we mentioned earlier, only two terminals can use the virtual circuits established on the packet-switched circuits (for example, an IP network). A virtual private network (VPN) can be given as an application example. Unlike packet-switched circuits, the capacities of unused circuits cannot be transferred to currently used circuits. In this sense, circuit switching is insufficient for the efficient use of transmission lines.

1.3.2 Packet Switching

We have mentioned that in the circuit switching technique, a "dedicated" circuit is installed on the terminals at the opposite ends, which is used only by these two terminals at the time of communication. The circuit switching technique is insufficient due to limited bandwidths and increasing communication speed needs. Even if the connected terminals do not exchange information over the circuit, other terminals cannot use this circuit. The packet switching technique divides the data to be transmitted into packets. Each of these packets contains the address of the sender (IP) and the receiver's addresses. These packets are left to the transmission medium and delivered to their destination via packet switching devices (switch, router, etc.). Thus, a transmission medium can be used by hundreds of terminals (millions if we consider the Internet environment) instead of being divided into only two terminals (Figure 1.3).

We can compare the packet switching circuit to highways where hundreds of vehicles (packages) are present simultaneously. Each vehicle proceeds on the

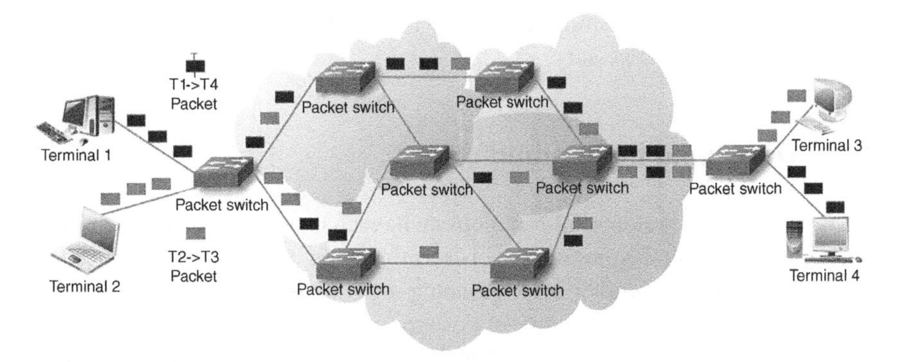

Figure 1.3 Packet switching.

same road (backbone) and reaches its destination by entering secondary roads when necessary. The critical limitation is the slowdowns due to increased vehicle (package) traffic. In this case, traffic engineering methods come into play and make essential optimizations on the network to prevent jams.

1.3.3 Cell Switching

We can describe cell-switched systems as a mixture of the circuit and packet-switched systems. What is decisive here is that the packet lengths are divided into tiny packets of 53 bytes in size. A circuit is then virtually allocated between opposing terminals (physically on a single line). These small packets are exchanged extremely quickly over these dedicated virtual circuits (Figure 1.4).

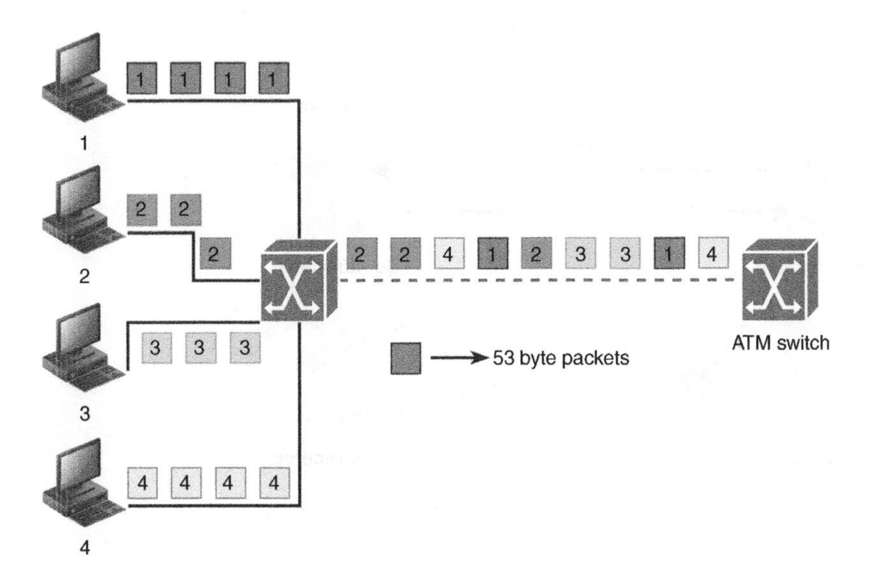

Figure 1.4 Cell switching.

Virtual circuits not transmitting packets for a certain period are closed and re-established when necessary.

1.4 Duplexing in Communication

In communication systems, information can be exchanged in three different ways between two mutual communication terminals. In simplex communication (Figure 1.5a), the transmitter is broadcasting continuously. Classical radio broadcasting can be given as an example of this type of communication.

On the other hand, simultaneous telephone conversations are a good example of full-duplex communication (Figure 1.5b). In this type of communication, the terminals perform both the receiving and transmitting functions at the same time.

Finally, the type of communication in which one of the terminals acts as a receiver and the other as a transmitter at a given time interval is called half-duplex communication (Figure 1.5c). While one terminal transmits information, the other is in a listening state, and these roles change according to the need during the conversation. Conversations made from police radio devices can be given as an example of this type of communication.

In wireless communication systems, one channel should be reserved for upload/transmit and one for download (receive) for the terminal in connection with the base station. Two doubling techniques create this simultaneous transmission

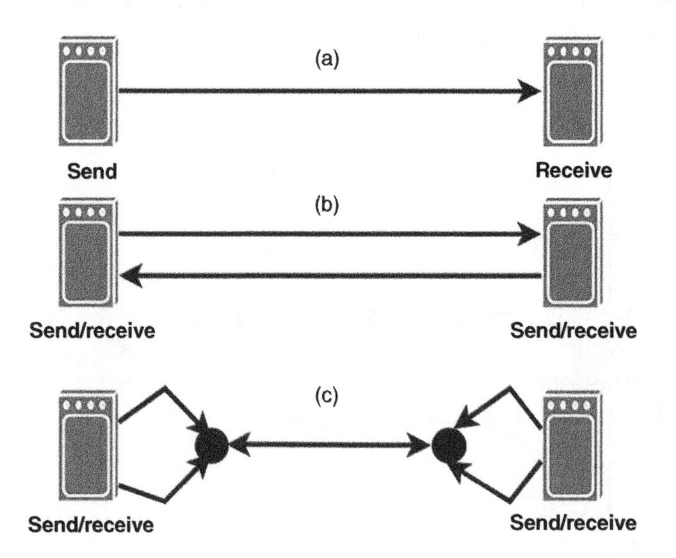

Figure 1.5 Duplexing methods. (a) Simplex; (b) full-duplex; (c) half-duplex.

environment: frequency division duplexing (FDD) and time division duplexing (TDM). In the FDD mechanism, two-way communication is carried out by defining different frequency ranges (carriers) for each of the transmit/receive channels. In the TDM mechanism, two-way communication is provided by sending at a given moment of t_1 and receiving at a consecutive moment of t_2 [1].

1.5 Historical Developments of Wireless Communication Systems

Starting with 1G systems (1980), we will talk about communication systems at tera hertz levels with spectrum efficient modulation techniques and advances in electronic circuits. Additionally, users can receive services at high bandwidths using three-dimensional multiplexing techniques.

Wireless mobile communication systems, which started with only voice calls (1G) in the 1980s, were introduced into our lives with the 2G short message service (SMS) in the 1990s. In both generations, communication was carried out using circuit switching techniques. On the other hand, the third-generation (3G) systems have been a turning point. With this generation, packet-switched (data) services have been used in the wireless communication ecosystem. With 3G, multimedia content started to be used among users in the 2000s. With 4G, communication was carried out entirely with packet switching; thus, users could operate 24/7 Internet access. Although machine-to-machine communication exists, we have now met the Internet of things (IoT) concept with 4G (Figure 1.6).

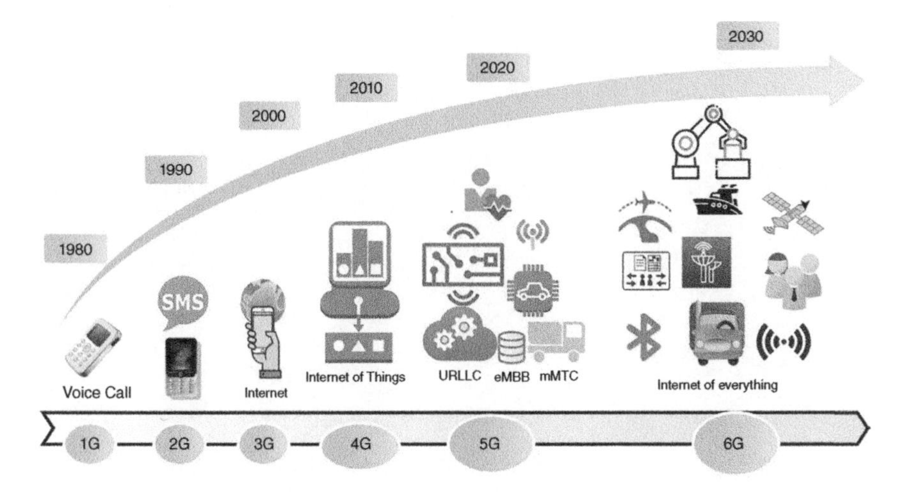

Figure 1.6 Evolution of wireless communication systems.

Wireless communication systems, which continue to progress without slowing down, have evolved into the fifth-generation communication systems as of the 2020s. Communication speeds up to 20 GHz with 5G, and thanks to these speeds, the concepts of ultra-reliable low latency communication (uRLLC), enhanced mobile broadband (eMBB), and massive machine type communication (mMTC) entered our lives. The transition phase from the IoT to the Internet of everything occurred at this stage. Communication systems, applications, and services have become much more intelligent using topics such as artificial intelligence, blockchain, and big data in the software field. Smart homes, smart cities, intelligent health systems, and autonomous vehicles are now inseparable parts of our lives.

In the years when the book was written (2021/2022), 5G applications entered our lives, and the sixth-generation communication systems, which will be put into use starting from the 2030s, became talked about and fictionalized. Concepts such as 6G and 3D networks, intelligent networks, quantum communication, blockchain technologies, deep learning, and programmable surfaces are designed together with communication infrastructures.

With the wireless communication systems enabling high-speed connection anytime and anywhere, the concept of IoT has started to take more place in our lives.

Reference

1 Frenzel, L.E. (2016). *Principles of Electronic Communication Systems*. New York: McGraw-Hill Education.

2

Modulation and Demodulation

2.1 Introduction

Before explaining the communication generations in the book, it is helpful to mention the concepts that will form the basis of communication. When the reader has an idea about modulation and demodulation, the view of the systems that will be explained in steps will become more meaningful. Without modulation and demodulation techniques, there would be no telecommunication systems. In this regard, the reader should have a good understanding of these concepts before moving on to more advanced topics.

2.2 What Are Modulation and Demodulation?

Before moving on to the details of modulation techniques, let us try to explain the subject with an analogy. Imagine you have a piece of paper with information on it. We aim to transmit this information over a long distance (for example, to a friend 100 m away). If we try to throw the information sheet to our friend by arm strength, the paper will not exceed a few meters. But if we wrap (modulate) this paper in a small piece of rock (carrier), we can easily send the data to our friend. Our friend, who receives the data wrapped in the rock, will be able to read the information we send by scraping the paper from the stone (demodulation) (Figure 2.1).

Moving from the preceding example, we can explain the modulation process as sending the information signal at a lower frequency and power by superimposing the information signal on the carrier signal at a much higher power intensity and frequency. The electrical/electromagnetic signals in the communication environment are in sinusoidal form. We can talk about three variable sizes of this sine sign: amplitude, frequency, and phase. We perform the modulation process when

Evolution of Wireless Communication Ecosystems, First Edition. Suat Seçgin.

Published 2023 by John Wiley & Sons, Inc.

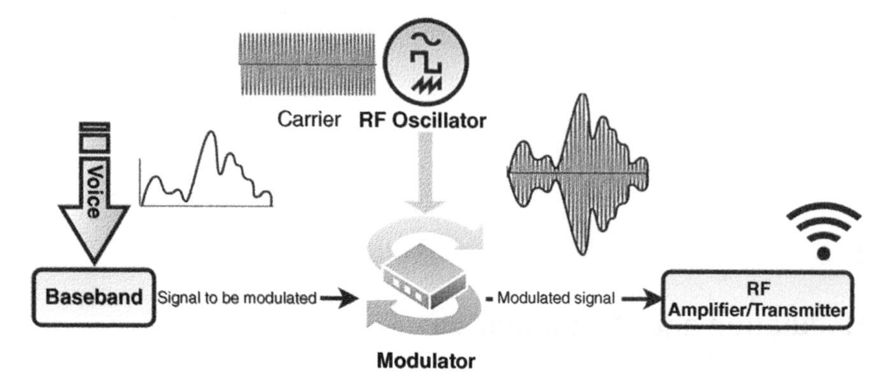

Figure 2.1 Modulation process.

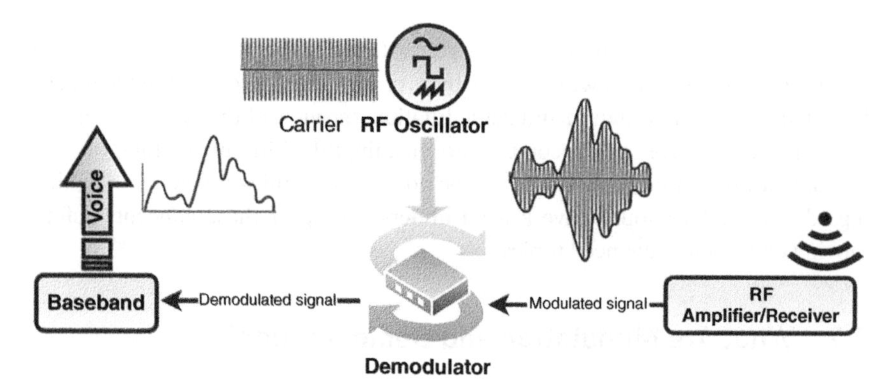

Figure 2.2 Demodulation process.

we change the amplitude, frequency, or phase of this sinusoidal signal depending on the amplitude of the information signal we will carry. If the signal is received from the opposite side, the information signal is obtained by reversing this process. This process is called demodulation. Modulation and demodulation processes are performed by devices called modems (modulator-demodulator) (Figure 2.2).

The following sections will describe three basic modulation techniques using these parameters (i.e. amplitude, frequency, and phase).

2.3 Analog Modulation Methods

In analog modulation methods, the information signal in a sinusoidal form is transmitted by "overlaying" the carrier's amplitude, frequency, or phase change, which is also in sinusoidal form. The system is defined as analog, as the information signal is in analog (sine) form.

2.3.1 Amplitude Modulation

In amplitude modulation, the amplitude of the carrier signal is a function of the information signal. In other words, the amplitude of the carrier signal is shaped by the form of the information signal. The amplitude of the carrier signal changes depending on the variation of the information signal. These amplitude changes are detected in the modem located at the opposite end. The information signal is re-obtained in its original form (Figure 2.3).

2.3.2 Frequency Modulation

In frequency modulation, the frequency of the carrier signal is a function of the information signal. In other words, the frequency of the carrier signal is shaped by the form of the information signal. The frequency of the carrier signal changes depending on the variation of the information signal. The information signal is recovered in its original form by detecting these frequency changes in the modem at the opposite end (Figure 2.4).

2.3.3 Phase Modulation

In phase modulation, the phase of the carrier signal is a function of the information signal. In other words, the phase of the carrier signal is shaped according to the form of the information signal. The phase of the carrier signal changes depending on the shift in the information signal. These phase changes are detected in the modem at the opposite end, and the information signal is recovered in its original form (Figure 2.5).

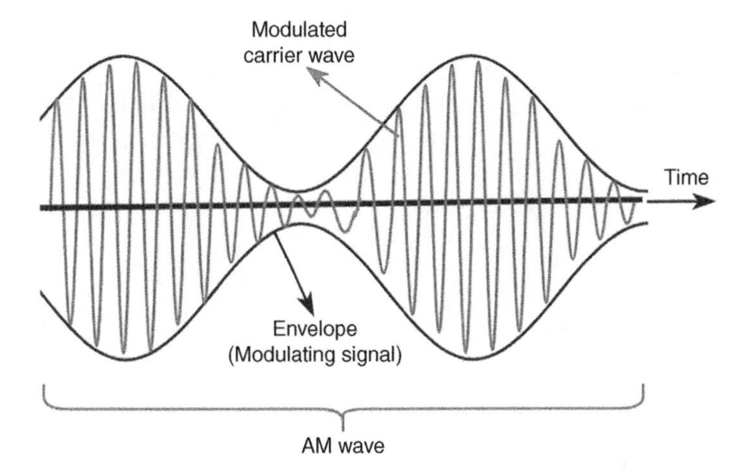

Figure 2.3 Amplitude modulation.

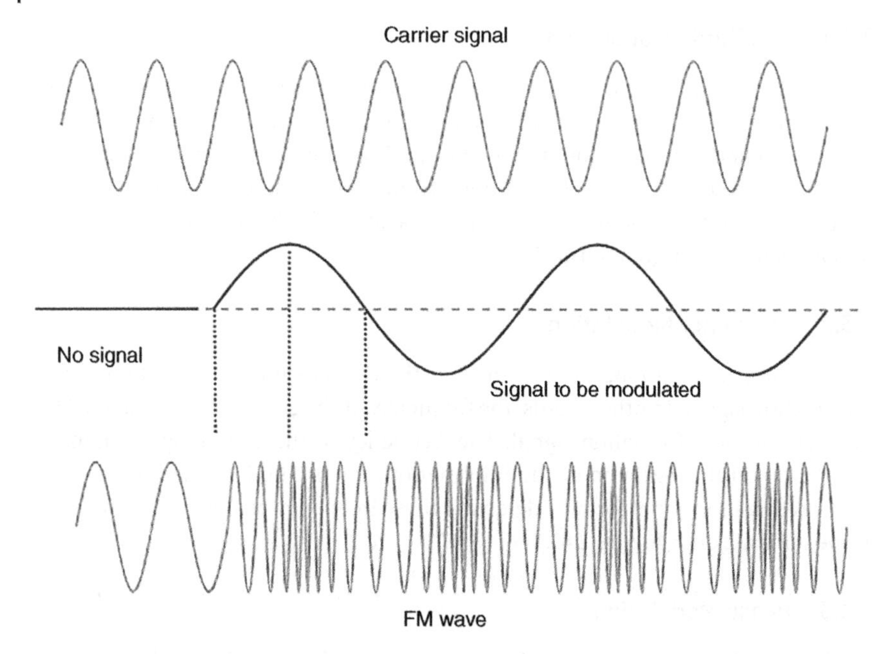

Figure 2.4 Frequency modulation (FM).

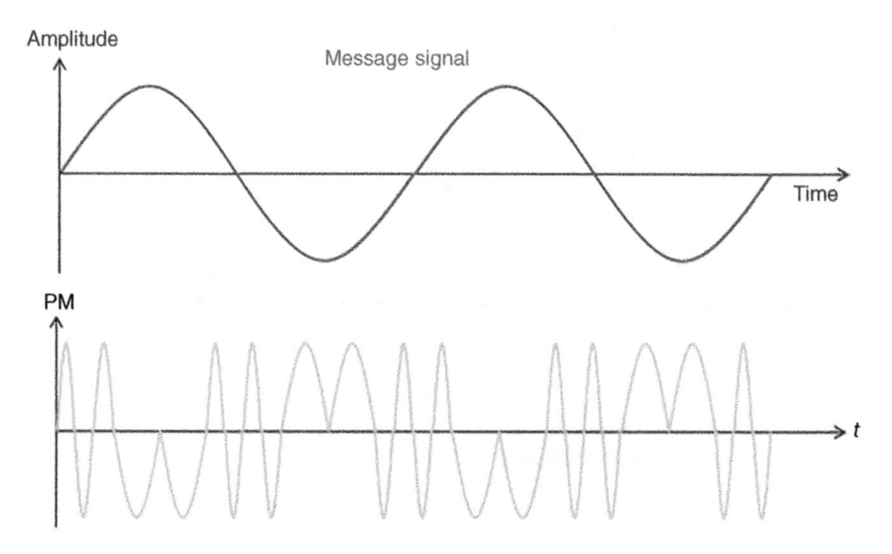

Figure 2.5 Phase modulation.

2.4 Digital Modulation Methods

In analog modulation systems, the information signal is in analog form. On the other hand, in digital modulation systems, the information signal is in the form of transitions or squared amplitude changes expressing 1s and 0s. This digital information form is obtained (especially in audio and video transmission) by applying the analog-digital conversion of the information signal originally produced as analog. That is, the analog information signal is converted into 1 and 0 forms by going through digital transformation. At the opposite end, the original analog information signal is obtained by digital-analog conversion.

2.4.1 Amplitude Shift Keying (ASK) Modulation

As shown in Figure 2.6, there is the carrier signal amplitude at the points where the amplitude of the information signal is positive. On the other hand, in negative values of information signal amplitude, carrier signal amplitude decreases to 0 levels (100% modulation). In 50% modulation, 50% of the amplitude of the carrier signal is taken at the points where the information signal is zero (Figure 2.6).

2.4.2 Frequency Shift Keying (FSK) Modulation

The information signal has two distinctive amplitude values that are positive and negative. In this case, the carrier signal carries information by entering two

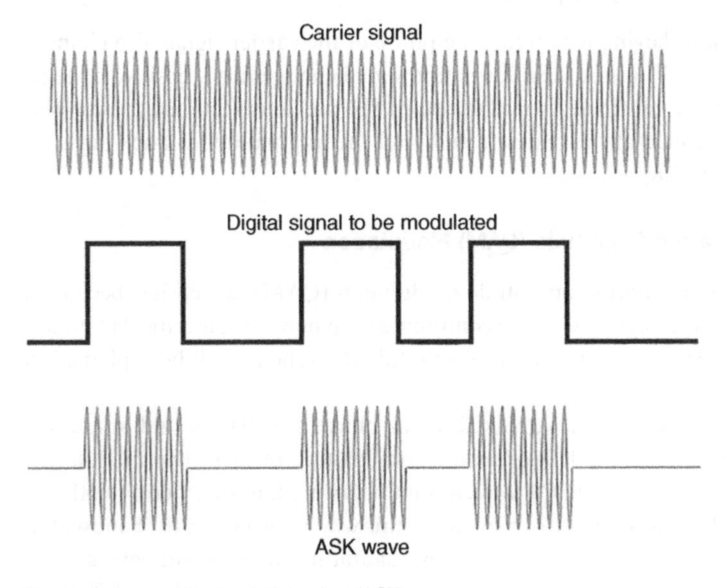

Figure 2.6 Amplitude shift keying (ASK).

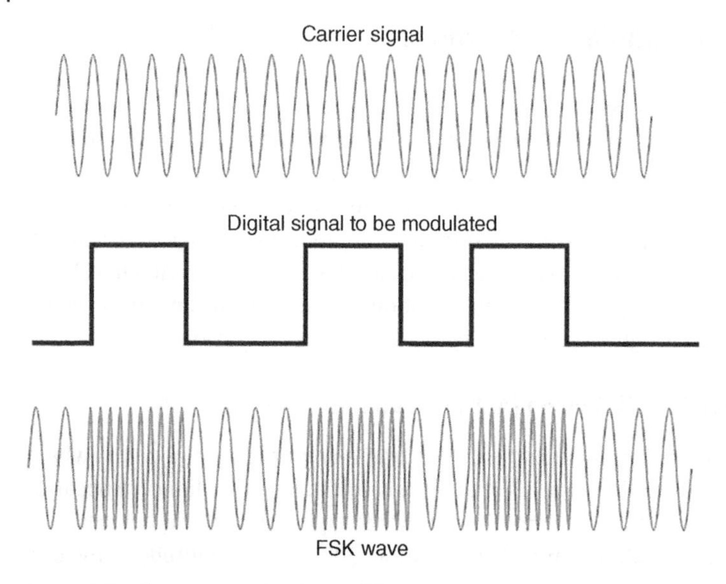

Figure 2.7 Frequency shift keying (FSK).

different frequency forms. For matters of 0, the x frequency is used; for 1, the y frequency is used ($x < y$) (Figure 2.7).

2.4.3 Phase Shift Keying (PSK) Modulation

In the phase shift keying technique, the phase of the carrier signal also changes whenever the amplitude of the information signal changes from positive to negative or negative to positive. As mentioned in the previous sections, these phase changes are detected at the opposite end, and the original information signal is obtained (Figure 2.8).

2.4.4 Quadrature Amplitude (QAM) Modulation

Understanding quadrature amplitude modulation (QAM), which has become a standard in 4G and beyond wireless communication networks, is critical to understanding these systems. Therefore, this modulation scheme will be explained in more detail.

In the previous chapters, we talked about phase modulation and amplitude modulation. QAM can be defined as the combination of these two modulation types. In other words, in the QAM technique, information is carried in both the amplitude and phase of the carrier signal. In this way, we get twice the bandwidth. Based on this definition, QAM is known as quadrature carrier multiplexing [1].

In QAM, two carrier signals of the same frequency but with a phase difference of 90° are used in their simplest form. The signal starting from the 0° phase is called the "I," and the other signal in the 90° phase is called the "Q" signal.

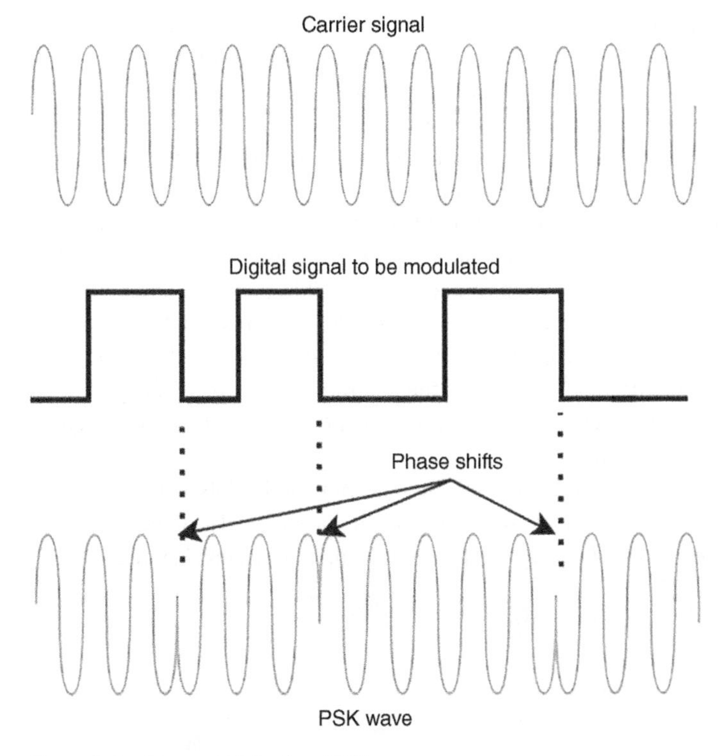

Figure 2.8 Phase shift keying (PSK).

Considering that there is a phase difference of 90° between the sine and cosine signals, we can simultaneously transmit both carrier signals at the same frequency from the transmission medium (Figure 2.9). As a basic concept, the method we give in Figure 2.9 is analog QAM and is used to send color information in analog video television broadcasts.

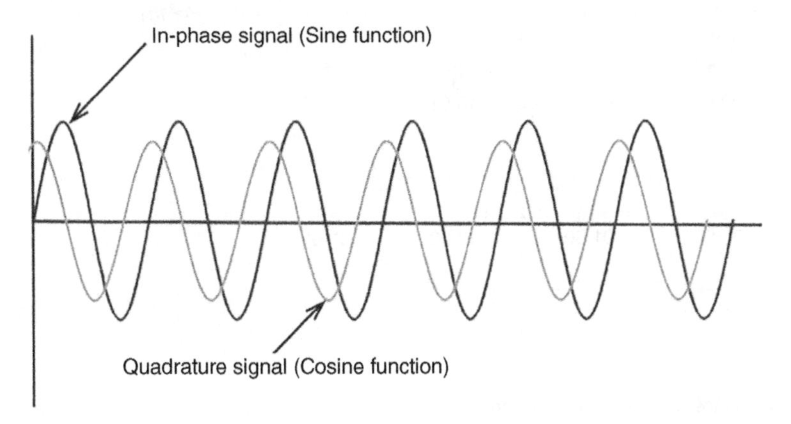

Figure 2.9 In-phase signal and quadrature signal component.

QAM is known as quantized/digital QAM in cellular communication and Wi-Fi fields. With digital QAM, much higher data rates are achieved compared to amplitude and phase modulations. In addition to high-speed data transmission, QAM is also used for noise resistance, low error values, etc.

In digital QAM schemes, the number of bits transmitted simultaneously can be increased using different phase and amplitude values. This constellation diagram structure can mark possible message points (polar coordinates). In its simplest form, we can transmit one bit of information simultaneously (first level QAM) [2]. When multilevel schemes are used, it becomes possible to send incremental bit numbers. In the constellation diagram, for example, we can send 4 bits of information simultaneously with a QAM (16-QAM) scheme with 16 levels (16 different points in the constellation, $2^4 = 16$). This means transmitting 16 other symbols is possible. Similarly, 32-QAM (25), 64-QAM (26), 128-QAM (27), and 256-QAM (28) schemes are used in practice.

Similarly, considering the 64-QAM scheme, we can mark 6 bits per symbol and 64 symbols. In terms of modulation, we can say that in the 64-QAM scheme, the carrier signal is modulated in one of 64 different phase and amplitude states (Figure 2.10).

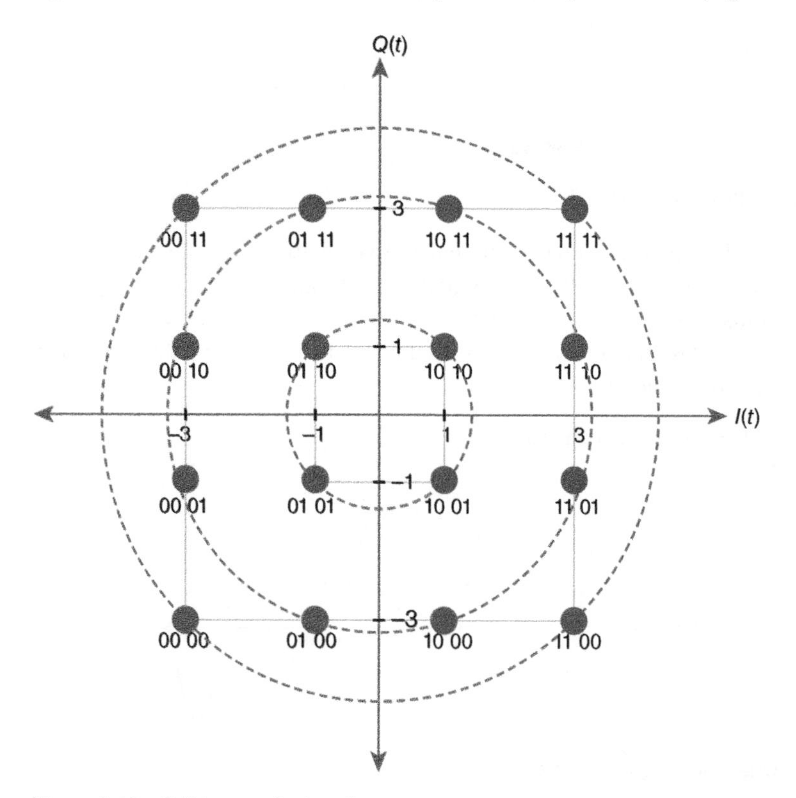

Figure 2.10 QAM constellation diagram.

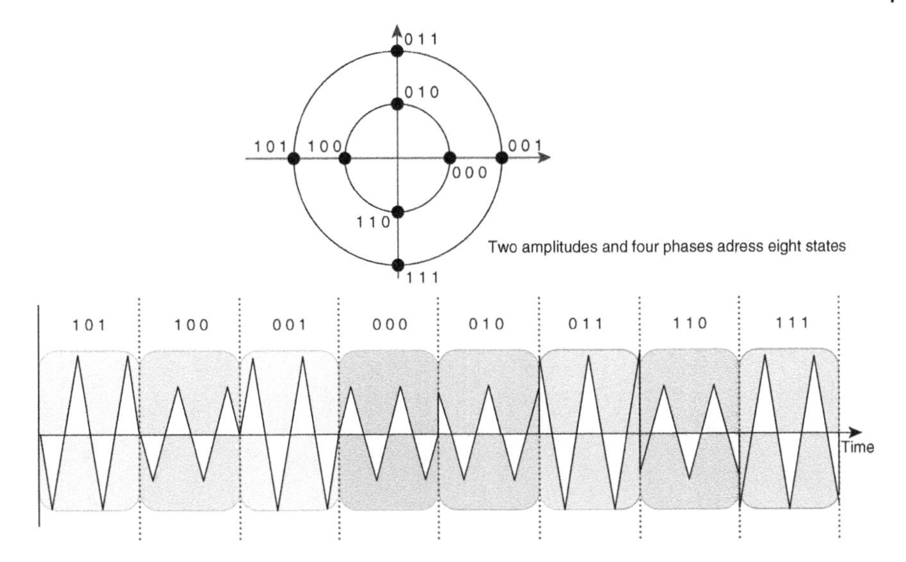

Figure 2.11 8-QAM signal.

We can say that there are 3 different amplitudes (Euclidean distance from each point to the origin) and 12 different phases (the angle formed by each end) in the 16-QAM diagram. In other words, when we look at the chart, we can say that the distances of each polar coordinate to the origin are in 1 of three different values. Looking at the chart again, we can say that each polar coordinate is in 1 of a total of 12 phases (angles). For example, we can say that the "1111" symbol will be sent with a carrier with an amplitude of $\sqrt{2}$ and phase of 45°. An example of an 8-QAM signal image is in the Figure 2.11 [3].

References

1. Quadrature Amplitude Modulation (QAM) (2023). What is it? *Electrical4U*. https://www.electrical4u.com/quadrature-amplitude-modulation-qam/ (accessed 14 February 2023).
2. Beard, C., Stallings, W., and Tahiliani, M.P. (2015). *Wireless Communication Networks and Systems, Global Edition*. Pearson.
3. Salvi, S. and Geetha, V. (2019). From light to Li-Fi: research challenges in modulation, MIMO, deployment strategies and handover. *International Conference on Data Science and Engineering (ICDSE)*, Patna, India (26–28 September 2019), pp. 107–19.

3

Multiplexing Methods

3.1 Introduction

Countless user information is carried between two different (e.g. switchboard or backbone router) ends in other locations. Therefore, efficient use of the transmission line (copper, fiber, radio link, and satellite) is essential when considering non-technical aspects such as installation and operating costs and technical factors such as line length, possible losses on the line, interference, bandwidth, etc. [1].

It is an effective method that has been used for a long time to divide the transmission line capacity into channels by using the time and frequency variables of the electrical signals running on the line. The allocation makes this capacity increase different frequencies to different channels or creates other channels in different time zones. With the introduction of fiber optic cables in the transmission area, different light wavelengths began to create additional channels. This "channeling" process using frequency, time, and light wavelengths is called multiplexing. The method of recovering the information signal is called demultiplexing (Figure 3.1).

Frequency division multiplexing (FDM) and wavelength division multiplexing (WDM) techniques are analog and time division multiplexing (TDM) are digital multiplexing methods. Knowledge of multiplexing methods is critical for understanding wireless access techniques in the following sections [2] (Figure 3.2).

We can explain the multiplexing process as sharing a transmission medium by using components such as frequency, time, or code of carrier signals in a way that allows the use of more than one user.

We can call the structure formed by putting multiplexing methods at users' service multiple access ecosystems. In other words, for example, in a system that serves users with frequency multiplexing, users are provided with a frequency division multiple access (FDMA) environment. It is essential to understand multiplexing and the access techniques running on these multiplexing systems to

Evolution of Wireless Communication Ecosystems, First Edition. Suat Seçgin.

Published 2023 by John Wiley & Sons, Inc.

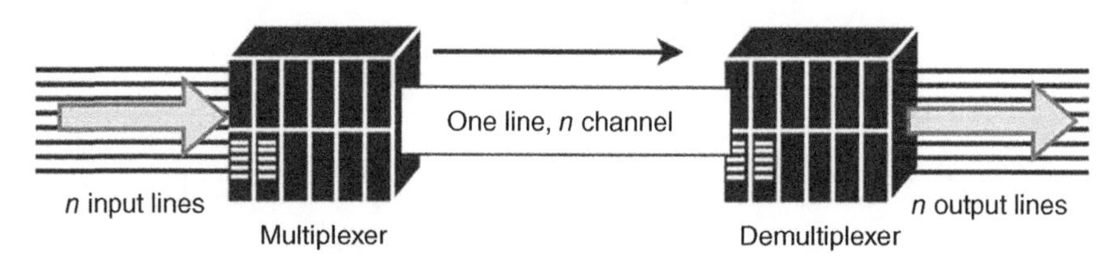

Figure 3.1 Multiplexing-demultiplexing.

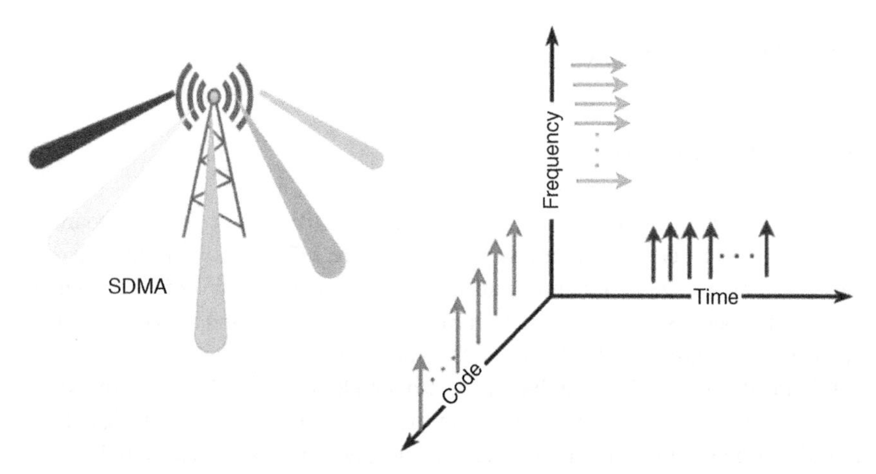

Figure 3.2 Multiple accessing methods.

understand the communication generations starting from 1G to 6G. Therefore, these access methods will be explained before moving on to xG communication systems.

3.2 Frequency Division Multiplexing

In FDM, frequency ranges with a certain width of guard bands are constructed as channels. If we give FM radio broadcasts for clarity, we can access and listen to hundreds of radio channels in the same time zone on an FM radio. Since a specific frequency range is determined for each radio broadcaster, hundreds of radio broadcasts can be listened to simultaneously without interfering. To prevent the channel frequencies from mixing, guard bands are reserved between each channel. Thus, using different frequency bands establishes more than one channel in a single transmission environment (Figure 3.3).

For example, let us observe the European ITU-T (Telecommunication Standardization Sector of the International Telecommunication Union)

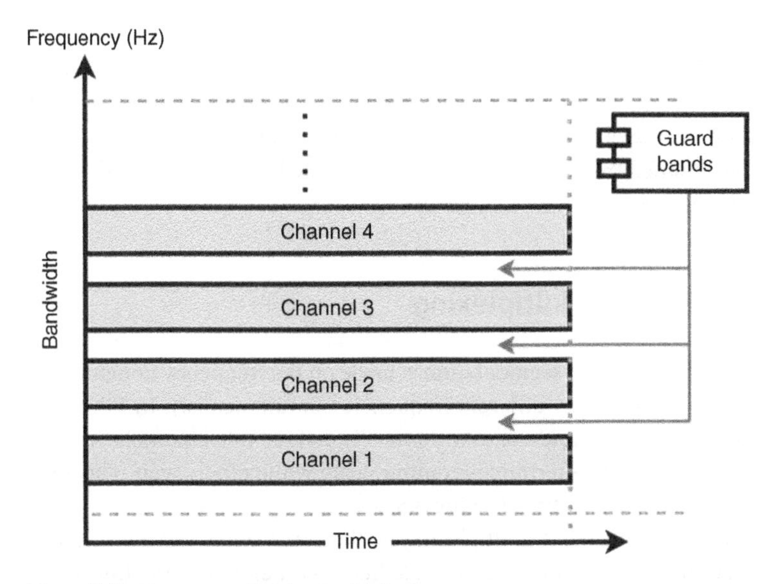

Figure 3.3 Frequency division multiplexing.

application used in analog telephone systems. In first-level frequency multiplexing, 12 channels are used between the 60 and 108 kHz band. This structure is called a group. The bandwidth of each channel is 4 kHz. Five groups of 12 channels are regrouped in the 312–552 kHz band range. As a result, a supergroup structure with 60 audio channels is formed. Similarly, a master group hierarchy is established in the 812–244 kHz band range, so five supergroups will create 300 different voice channels.

Since different carrier frequencies are defined for each channel, communication opportunity is provided for more than one terminal simultaneously. The multiplexing/demultiplexing application scheme for FDM is given in Figure 3.4 [3].

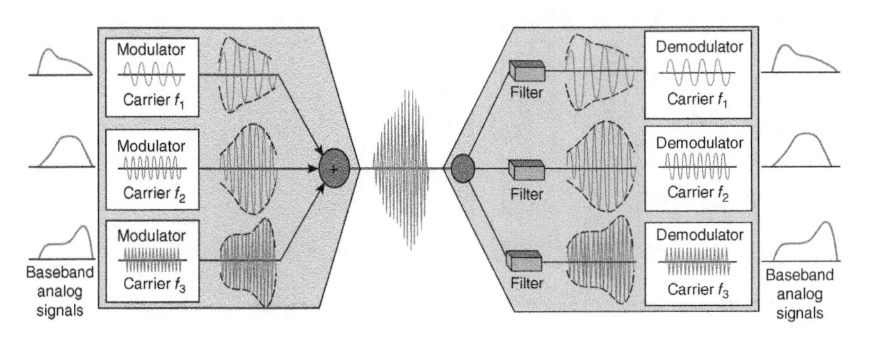

Figure 3.4 FDM mux-demux.

Modulation, filtering, and demodulation processes are applied to the baseband signals in the FDM process. Information signals are transmitted through different channels on the sender, modulating with carriers of different frequencies, and passed to the transmission medium. On the receiving side, the relevant information signal is obtained by being sorted and demodulated with "band-pass" filters that give only the desired carrier frequency.

3.3 Time Division Multiplexing

We have stated that in FDM, segmentation is made on the frequency domain, and channeling is performed over different frequencies simultaneously. In TDM, the frequency is kept constant, and the time domain is divided into channels. In each different time slot, different terminals communicate while others wait. This process proceeds circularly over time (Figure 3.5).

TDM technology is a digital multiplexing method that sends different signals (inputs) at certain time intervals over a single line. In other words, the time environment is divided into different time slots, and information signals are sent in these slots. Multiplexers and splitters (mux-demux) on opposite ends work synchronously in terms of time by making simultaneous switching (Figure 3.6).

TDM is divided into four types: synchronous, asynchronous, interleaving, and statistical TDM.

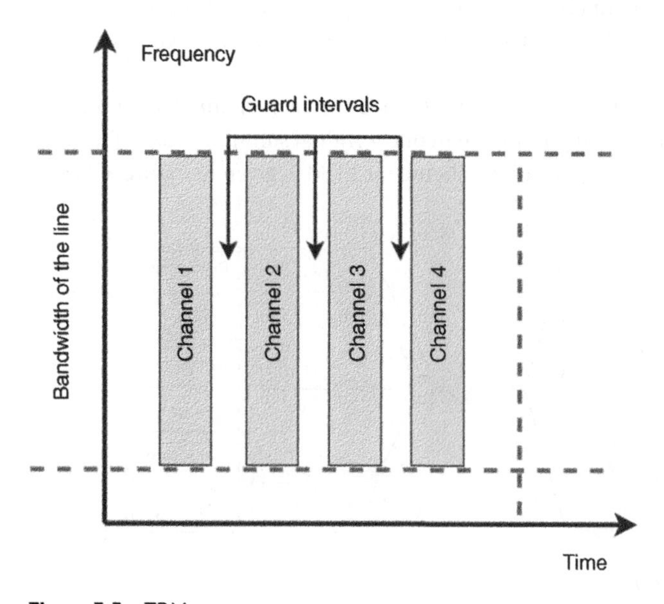

Figure 3.5 TDM.

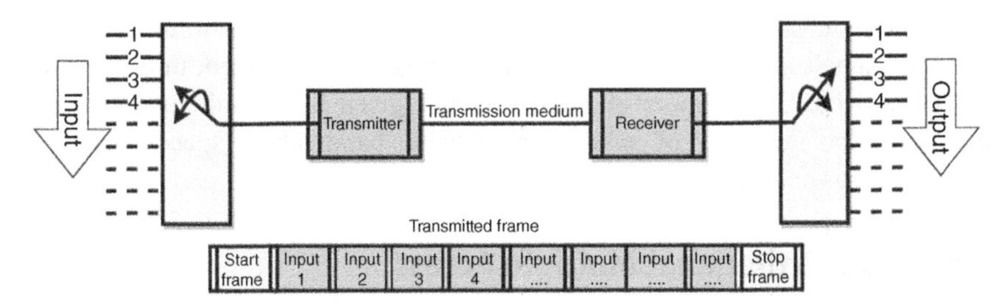

Figure 3.6 TDM mux-demux.

3.4 Orthogonal Frequency Division Multiplexing

As can be seen in Figure 3.3, the frequency spectrum is used in conventional FDM, which will be inefficient in terms of both guard bands and allocated frequencies. Due to the increasing number of users and the increasing need for speed for each user due to the introduction of 4G and 5G systems into our lives, it has become inevitable to use the frequency spectrum as efficiently as possible. Because of this need, orthogonal frequency division multiplexing (OFDM) has been in practice. Today OFDM is used in Wi-Fi 802.11ac, 4G, and 5G communication infrastructures.

OFDM can be thought of as a multicarrier modulation scheme. Due to the elimination of guard bands and the carrier signals orthogonal to each other, multiple users can exchange data simultaneously in the available band range, thanks to OFDM technology and subcarriers (Figure 3.7). According to the OFDM scheme,

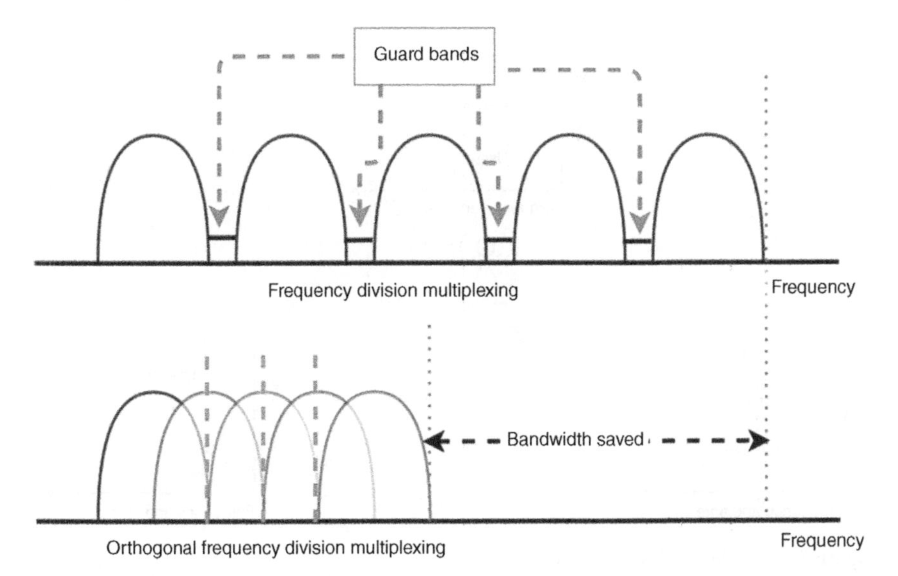

Figure 3.7 FDM and OFDM.

when the peak value of a subcarrier at any frequency is at its maximum, the two adjacent signals are at the zero level (orthogonal). As can be seen, unlike FDM, OFDM provides much more efficient use in a given bandwidth. In addition, thanks to the flexibility of subcarriers, flexible bandwidths (speeds) can be allocated to users/terminals as needed [4].

3.5 Non-Orthogonal Multiple Access

Non-orthogonal multiple access (NOMA) is one of the most talked about access techniques in new-generation wireless communication. We can also use the secondary name power division multiplexing for NOMA, most commonly encountered in practice. Thanks to NOMA, where multiple users are supported in a single resource, the performance and capacity of the user and the entire system can be increased [9]. NOMA ignores signal interference, unlike other access methods (TDMA, FDMA). The critical point here is that even if there is interference between the signals, NOMA allows simultaneous communication (multiplexing) by adjusting/adapting the power levels of different signals.

If we explain the subject from the two-user scenario given in Figure 3.8, suppose user one is closer to the base station (higher channel gain) and user two is farther from the base station. To provide the same quality service, the base station sends a higher power signal to the channel of user two. Similarly, the strength of the outgoing call is lower in channel one. In this way, the power levels differ in both user signals sent using the same frequencies.

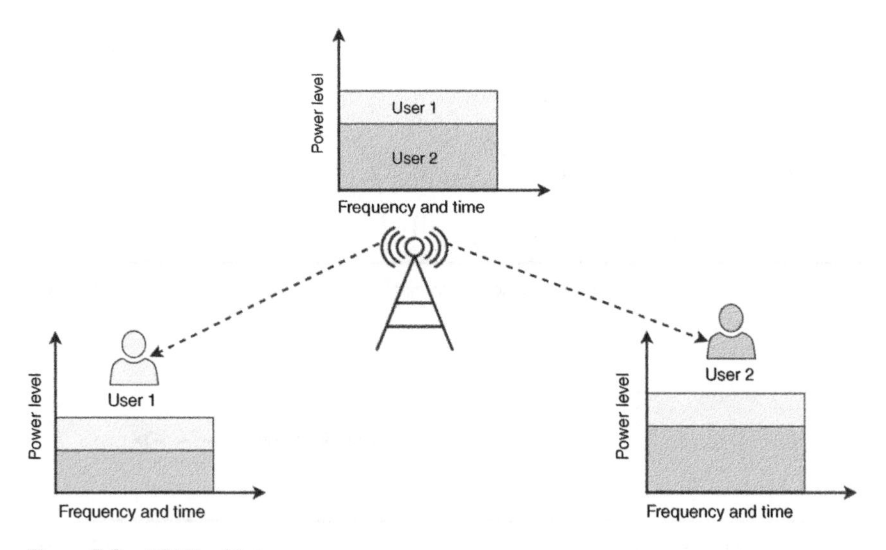

Figure 3.8 NOMA with two users.

Since the power level of user one's signal is low in the signal received on user two's side, this signal is detected as noise and eliminated. On user one's side, this signal cannot be extracted by noise detection since the power level of the signal belonging to user two is higher. Therefore, the "successive interference cancellation" (SIC) mechanism comes into play. With this mechanism, the user two signal is not eliminated when it is not detected as noise because the power level is high on the user one side [5].

The advantages of the NOMA access scheme, which is expected to be a standard access method in 5G and beyond systems, over OFDMA are given as follows:

- High spectral efficiency with multiple users using the same frequency source.
- Dens connectivity (IoT, M2M) due to its ability to provide simultaneous service to users with higher density.
- Lower latency times as it provides simultaneous transmission instead of dedicated scheduled time slots.
- Much better quality of service (QoS) thanks to flexible power management.
- Improved communication performance thanks to the multiple input multiple output (MIMO) antenna structure.

Some of the disadvantages of NOMA, such as the need for receivers to be more complex and the need for high energy consumption, can be counted.

As 5G and beyond transmission systems occur in our lives, we will often start seeing NOMA-based machine-to-machine (M2M) communication and Internet of things (IoT) applications. As we will discuss in the relevant section, NOMA solutions will be integrated with applications such as MIMO, beamforming, and space-time coding, which have already been used in practice.

3.6 Wavelength Division Multiplexing

Wavelength division multiplexing (WDM) transmission systems are used in fiber backbones. They are not used in user-based access in multiple access systems. The WDM multiplexing method aims to create different channels by using different wavelengths (in other words, colors) of light (Figure 3.9).

While it may seem complicated at first glance, the underlying idea is quite simple. As it is known, the light prism can gather the light beams (according to the angle and wavelength) that come to it in a single light beam. In the opposite case, when a single light is passed through a light prism, multiple (multicolored, different wavelength) light beams are formed on the prism's other side, called the light spectrum. WDM multiplexing and deduplication processes are done on this principle. In this way, high-bandwidth transmission lines can be created.

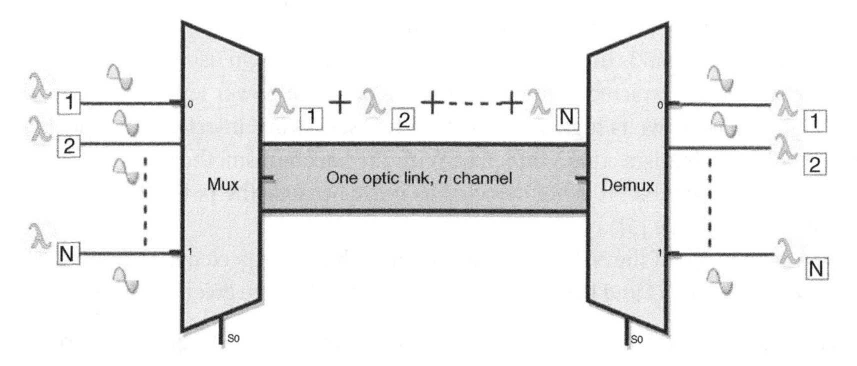

Figure 3.9 WDM.

3.7 Code Division Multiplexing

In code division multiplexing, a unique code is generated for each channel, keeping the time and frequency constant. Using these propagation codes, the relevant terminal/user filters the channel of their code and evaluates the channels carrying other codes as noise. This method provides advantages in spectrum efficiency, QoS, power consumption, and security (Figure 3.10).

The following Figure 3.11 shows the block diagram of code division multiplexing. As can be seen, a different code is generated for each channel. These pieces of information with other codes are sent to the transmission line in a mixed way. Since the code of each information is known on the receiving side, the original data is obtained using this code.

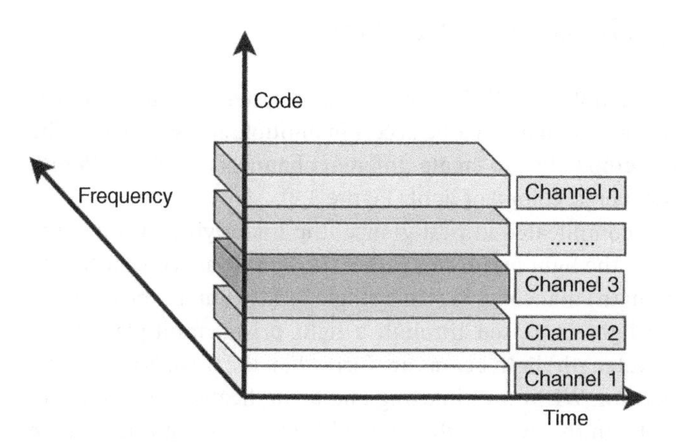

Figure 3.10 CDM.

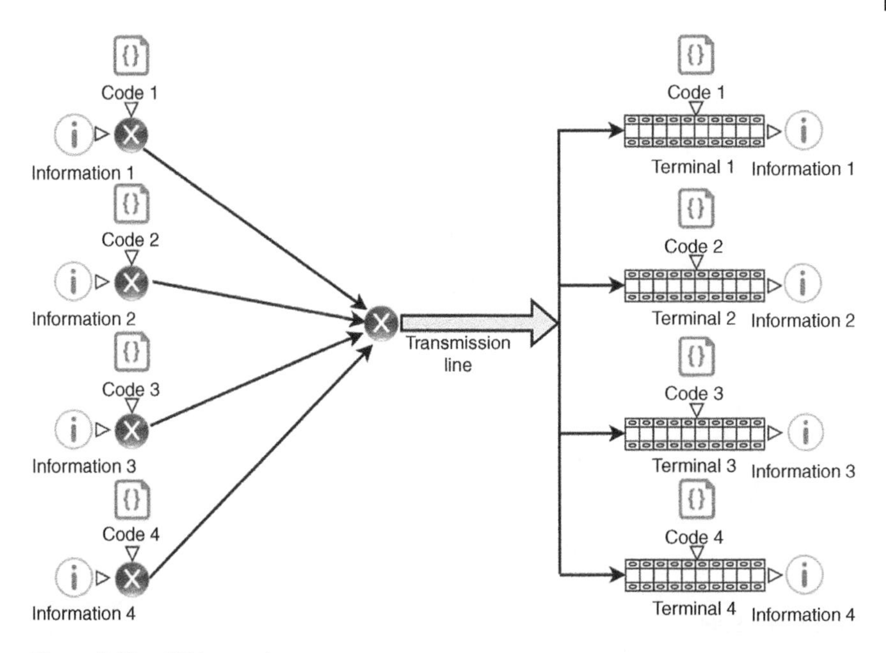

Figure 3.11 CDM mux-demux.

3.8 Spatial Division Multiplexing

One of the most critical developments in 5G and later generation wireless communication systems is the design of MIMO antenna systems. This way, base stations equipped with antenna arrays have developed an environment for each user/terminal to emit electromagnetic emissions at the same frequency but with different beams. With this directed signaling, spatial division multiplexing (SDM) was created without the frequency interferences [6] (Figure 3.12).

3.9 Orbital Angular Momentum Multiplexing

In recent years, wireless communication using orbital angular momentum (OAM) has become an important application area for post-5G systems due to its potential to enable high-speed wireless transmission. OAM is implemented by utilizing the physical characteristic of electromagnetic waves characterized by a helical

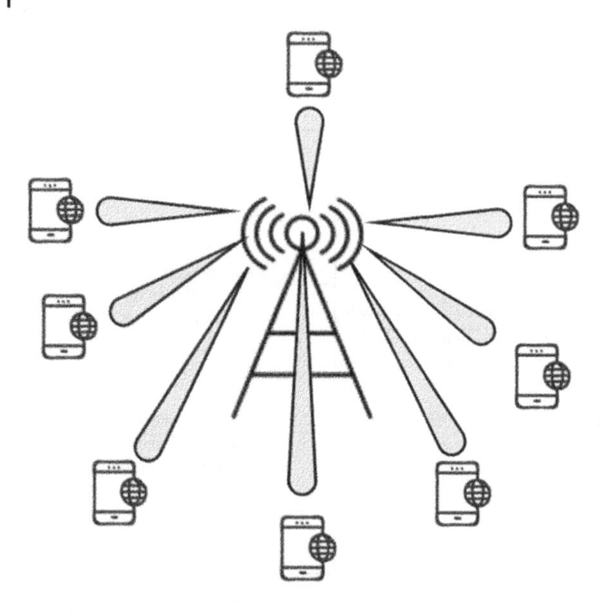

Figure 3.12 SDM.

phase front in the direction of propagation. Since this characteristic can create multiple independent channels, wireless OAM multiplexing can effectively increase the transmission rate in a point-to-point link such as a wireless backhaul and fronthaul by creating multiple orthogonal channels [2, 7].

An essential advantage of this method is that it can produce multiple orthogonal channels in line of sight (LOS) channel environments without requiring complex signal processing techniques such as channel diagonalization. Table 3.1 compares mMIMO, LOS MIMO, and OAM [7] (Figure 3.13).

The following steps are applied in the OAM process. The transmitter generates the data and performs typical baseband operations such as modulation, interleaving, channel coding, and power precompensation. The output stream is split into multiple streams by serial-parallel conversion. Each stream is then digital-to-analog converted before being sent to the radio frequency (RF) stage. Each RF stream generates beams carrying different OAM modes using single or multiple uniform circular arrays (UCA). In the final stage, all OAM signals are transmitted simultaneously in the same frequency band. The point that should not be overlooked here is that OAM beam generation and multiplexing are done by devices such as phase shifters and combiners that do not require digital signal processing. We can also see OAM as a specific application of MIMO technologies.

Table 3.1 Comparison of mMIMO, LOS MIMO, and OAM for fixed wireless links.

	Massive MIMO	**LOS MIMO**	**OAM multiplexing**
Type	*P2MP*	*P2P*	*P2P*
Channel	*LOS/NLOS*	*LOS*	*LOS*
Antenna configuration	*Linear array*	*Linear array*	*Circular array*
Antenna size	*Large/medium*	*Large*	*Large*
Stream per user	*Single*	*Multiple*	*Multiple*
Circuit complexity	*High/medium*	*Medium*	*Medium*
Mobility tracking	*Eligible*	*Not eligible*	*Not eligible*
Robustness for coaxial distance variations	*Good*	*Bad*	*Good*
Robustness for axis misalignments	*Good*	*Bad*	*Good*

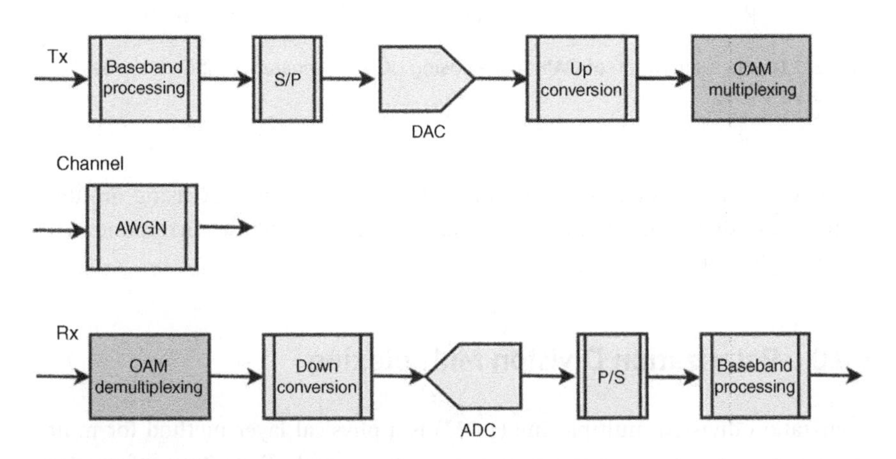

Figure 3.13 Block diagram of OAM.

In Figure 3.14, modes 0, 1, and 2 OAM beamforming using regular circular arrays of eight antenna components (elements) are given. The separation of the beams carrying the OAM modes can be performed similarly to the generation of these beams, with antenna elements connected with phase shifters that make opposite rotation directions. As long as the number of antenna elements exceeds $2n$, $n * 360°$ turns are orthogonal to each other. Therefore, each OAM mode can be

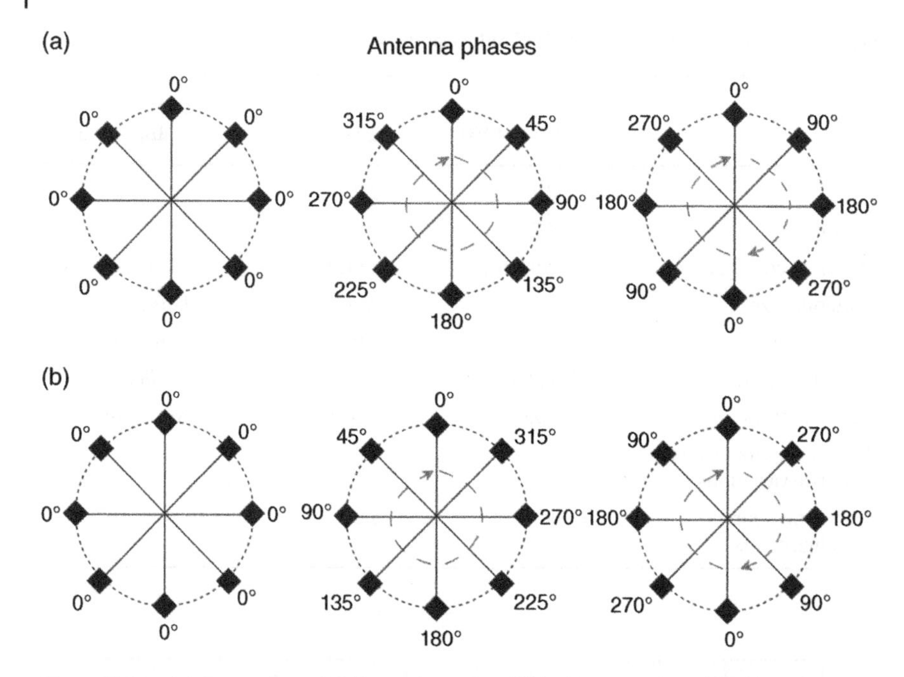

Figure 3.14 (a) Generation of OAM modes using UCA; (b) separation of OAM modes with UCA.

separated from the signals of the mixed OAM modes without using an alias. Figure 3.14a shows an example of each antenna element phase corresponding to the example Figure 3.14b [7].

3.10 Polarization Division Multiplexing

Polarization division multiplexing (PDM) is a physical layer method for multiplexing signals carried on electromagnetic waves and allows two information channels to be transmitted on the same carrier frequency using waves of two orthogonal polarization states. It is used in microwave links such as satellite television downlinks to double the bandwidth using two orthogonally polarized feed antennas in satellite antennas. It is also used in fiber optic communication by transmitting separate left and right circularly polarized light beams over the same optical fiber [8] (Figure 3.15).

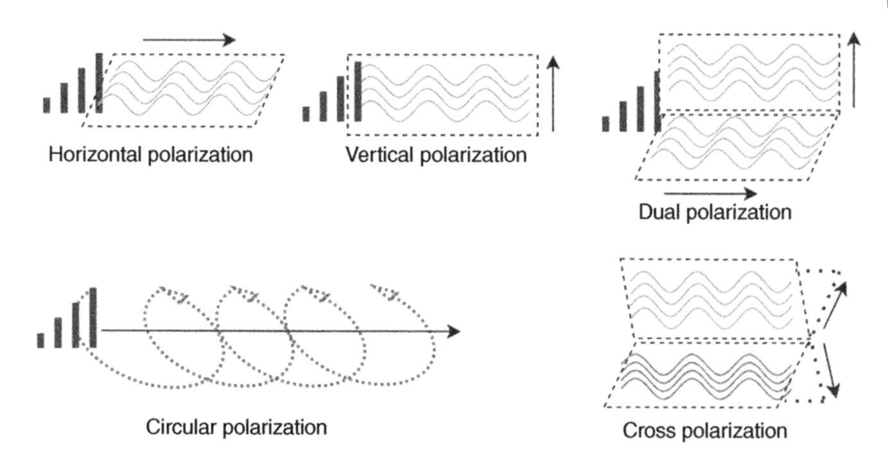

Figure 3.15 Polarization types.

References

1 Grami, A. (2015). *Introduction to Digital Communications*. Academic Press.

2 Yan, Y., Xie, G., Lavery, M.P.J. et al. (2014). High-capacity millimetre-wave communications with orbital angular momentum multiplexing. *Nature Communications* 5 (1): 4876.

3 Forouzan, B.A. and Fegan, S.C. (2007). *Data Communications and Networking*. Tata McGraw-Hill Education.

4 Penttinen, J. (2015). *The Telecommunications Handbook*. Wiley.

5 Venkataraman, H. and Trestian, R. (2017). *5G Radio Access Networks*. CRC Press.

6 Cheng, F., Jiang, P., Jin, S., and Huang, Q. (2015). Dual-polarized spatial division multiple access transmission for 3D multiuser massive MIMO. *2015 International Conference on Wireless Communications & Signal Processing (WCSP)*, Nanjing, China (5–17 October 2015), pp. 1–5.

7 Lee, D., Sasaki, H., Fukumoto, H. et al. (2017). Orbital angular momentum (OAM) multiplexing: an enabler of a new era of wireless communications. *IEICE Transactions on Communications* 100 (7): 1044–1063.

8 RF-Design (2023). Antenna polarization options via Pasternack. *RF-Design*. https://rf-design.co.za/2018/12/11/antenna-polarization-options-via-pasternack/ (accessed 14 February 2023).

9 Vaezi, M., Ding, Z., and Poor, H.V. (ed.) (2019). *Multiple Access Techniques for 5G Wireless Networks and Beyond*, vol. 159. Berlin: Springer.

4

Network Performance Metrics

4.1 Introduction

With the introduction of 3G, 4G, and 5G technologies into our lives, streaming multimedia, interactive games, mobile TV, video games, 3D services, and video sharing applications have become frequently used in daily life. With the spread of these applications, the need for bandwidth per user has also increased. Adequate access and core network usage have become more critical than ever. At this point, controlling and maximizing the network metrics is necessary to maintain and maximize the network metrics to optimize the network performance. The following sections describe these network performance metrics.

4.2 Spectral Efficiency

Spectral efficiency, defined as the number of bits carried per Hertz, aims to increase the number of bits transmitted on the subchannel. For example, in 4G (IMT-Advanced) wireless networks, downlink spectrum efficiency is 15 bps/Hz and uplink spectrum efficiency is 6.75 bps/Hz. Another definition of the spectrum or modulation efficiency (Erlangs/MHz/km^2) about modulation is given in the following formula [1].

$$\mathrm{SE} = \frac{\text{Total traffic carried by the system}}{(\text{Bandwidth})(\text{Total coverage})}$$

The graph of spectral efficiency vs. signal-to-noise ratio is given in Figure 4.1 [2].

Figure 4.1 shows R throughput, B bandwidth, and E represents the energy of a single bit in joules C channel capacity in bits per second. The dotted line in the chart represents Shannon Limit that converges ln(2) = −1.6. Various modulation

Evolution of Wireless Communication Ecosystems, First Edition. Suat Seçgin.

Published 2023 by John Wiley & Sons, Inc.

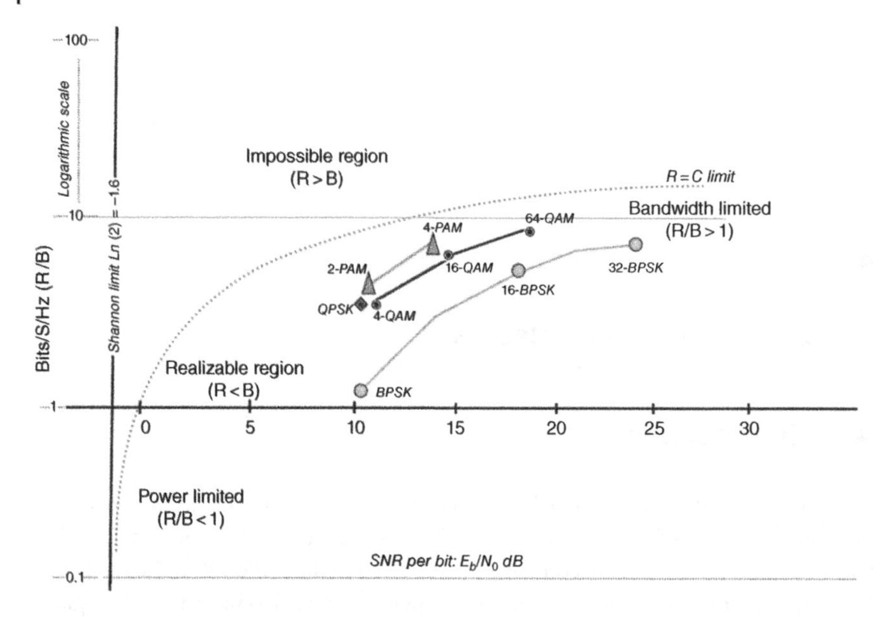

Figure 4.1 Spectral efficiency vs. SNR.

schemes are shown under Shannon Limit with a typical range of 4G-long-term evolution (LTE) signals.

Areas of interest in the figure include the R > B "Impossible Zone." This region is above the Shannon boundary of the curve. It indicates that no form of reliable information exchange can be above the borderline. The area below the Shannon boundary is called the "Executable Region," where R < B. Every protocol and modulation technique in any form of communication tries to get as close as possible to the Shannon limit. By looking at this graph, we can see where the typical 4G-LTE using various forms of modulation is.

As it can be understood from the explanations given above, spectrum efficiency is optimizing the bandwidth and spectrum with the maximum amount of data with minimal error. Since the frequency spectrum is a limited resource, it is critical to use it effectively and efficiently. We can examine efficiency from economic, technical, and functional aspects. From a financial point of view, we see licensing and pricing models. In technical efficiency, the maximum amount of data to be sent in a given spectrum range is considered. In functional efficiency, the intent is related to how well the allocated spectrum resource meets the user's needs as desired. One of the biggest reasons for using quadrature amplitude modulation (QAM; 4–8–16 etc.) systems in wireless communication is to achieve effective spectral efficiency. Of course, regardless of modulation, as the distance increases, the data transmission rate will decrease due to losses, and the spectrum efficiency will also decrease [3] (Figure 4.2).

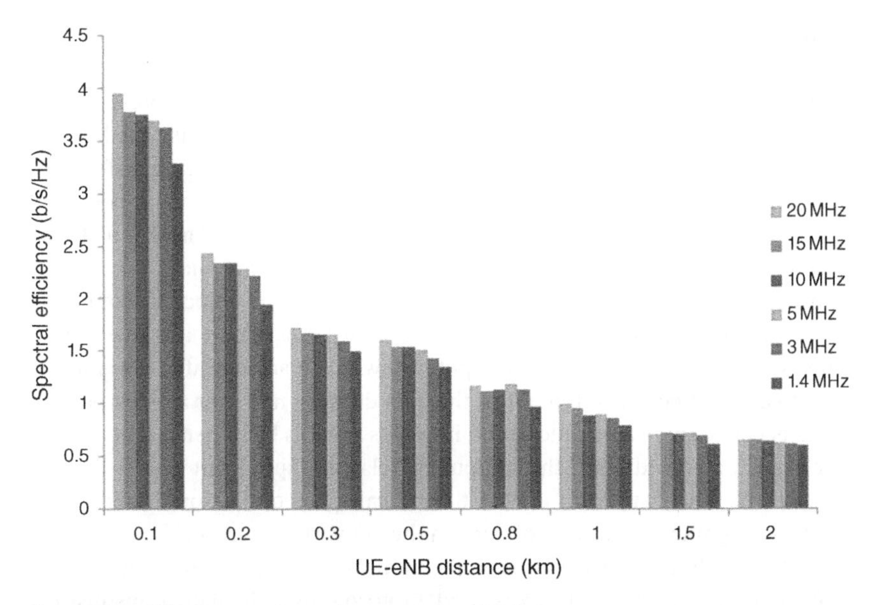

Figure 4.2 LTE distance dependent spectrum efficiency.

4.3 Important Network Performance Metrics

In this section, network performance metrics will be explained. Optimizing these parameters, which directly affect network performance, is extremely important for the efficiency of applications running on the network.

- **Packet loss**: When accessing the Internet or any network, small units of data called packets are sent and received. It is called packet loss when one or more packets cannot reach the desired destination. For users, packet loss manifests as a network outage, slow service, or even a complete loss of network connectivity. In other words, packet loss describes lost data packets that do not reach their destination after being transmitted over a network. Packet loss occurs when network congestion, hardware problems, software errors, and other factors cause packets to drop during data transmission. Any application can be interrupted by packet loss, but the most affected applications rely on real-time packet processing, such as video, audio, and game programs. Packet loss is involved in the trio of two other major network performance complications: latency and jitter.

  Acceptable packet loss depends on the type of data sent. For example, packet loss between 5% and 10% in voice over IP networks affects the quality significantly. Values between 1% and 2.5% are acceptable, while values below 1% are regarded as "good."

- **Bandwidth**: Network bandwidth is a measurement that indicates the maximum capacity of a wired or wireless communication link to transmit data over a network connection over a given period. Typically, bandwidth is represented by the number of bits, kilobits, megabits, or gigabits that can be transmitted in one second. Bandwidth, synonymous with capacity, defines the data transfer rate. Bandwidth is not a measure of network speed, contrary to a common misconception. But it is a concept related to network speed. There is a strong relation between bandwidth and throughput that is explained next.
- **Throughput**: The actual amount of traffic flowing from a specific source or group of resources to a particular destination or target group at a given time. This is an important point: Throughput is how much actual traffic flows when you do a real-time measurement or the data delivery rate over a given period. Packets per second can be measured in values such as bytes per second or bits per second. Bandwidth and throughput are related to speed, but what is the difference? In short, bandwidth is the theoretical speed of data on the network, while throughput is the speed of data on the network (Figure 4.3).

  Factors such as the transmission medium limitation, network congestion, latency, packet losses and errors, and protocols used in data transmission directly affect throughput. The closer the throughput is to the bandwidth, the higher is the performance of the communication network.
- **Delay:** Network latency refers to the time it takes for a packet to get from point A to point B. If Point A is the source and point B is the destination, the delay is the time passed in end-to-end communication. It can be explained under four headings:
  - Propagation delay: Propagation delay is the time it takes for a bit to reach one end of a link to the other. The delay depends on the distance between the sender and the receiver and the propagation speed of the wave signal.
  - Transmission delay: Transmission delay refers to the time it takes for a data packet to be transmitted to the outgoing connection. The packet's size and the outgoing connection's capacity determine the delay.

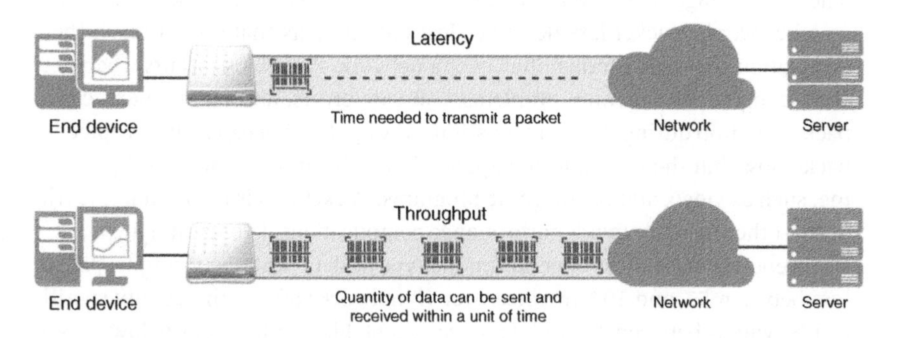

Figure 4.3 Latency and throughput.

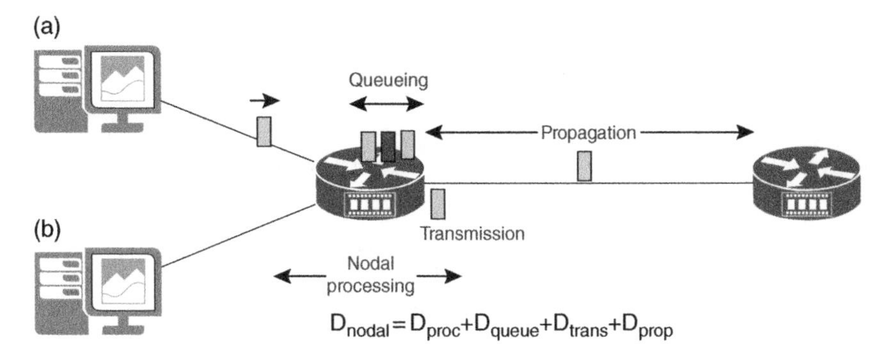

Figure 4.4 Four types of delay.

- Queuing delay: Queuing delay refers to the time a packet waits for processing in a switch's buffer. The delay depends on the arrival rate of incoming packets, the transmission capacity of the outgoing link, and the nature of the network's traffic.
- Processing delay: Transaction latency is the time it takes a switch to process the packet header. The latency depends on the processing speed of the key (Figure 4.4).

- **Latency**: Latency is a measure of the delay metric. In a network, latency is measured in milliseconds as the time it takes for a given data to reach its destination over the network. It is usually a measure in the form of the data reaching its destination and returning to the source of the response (round trip delay). Especially in TCP/IP networks, the round trip delay is an important parameter.

  The term delay is often used interchangeably with latency, but there is a subtle difference between the two. Propagation delay refers to the time it takes for the first bit to travel over a link between the sender and receiver, while network latency refers to the total time it takes to send the entire message. While the average latency is 50 ms in 4G networks, this value has decreased to 1 ms with 5G networks.
- **Jitter**: Wave distortion can be defined as the delay experienced in delays. In other words, jitter is the fluctuations in the delay values of received packets. This is usually due to network congestion and sometimes route changes. The high jitter value in real-time applications (VoIP, video, etc.) dramatically affects the quality. Essentially the longer data packets arrive, the more jitter can negatively impact the video and audio quality. As a typical example, jitter is expected not to exceed 30 ms in VoIP networks (Figure 4.5).
- **Peak-to-average power ratio (PAPR):** PAPR is the relationship between the maximum power of a sample in a given orthogonal frequency division multiplexing (OFDM) transmission symbol divided by the average power of that OFDM symbol. In simple terms, PAPR is the ratio of peak power to the average

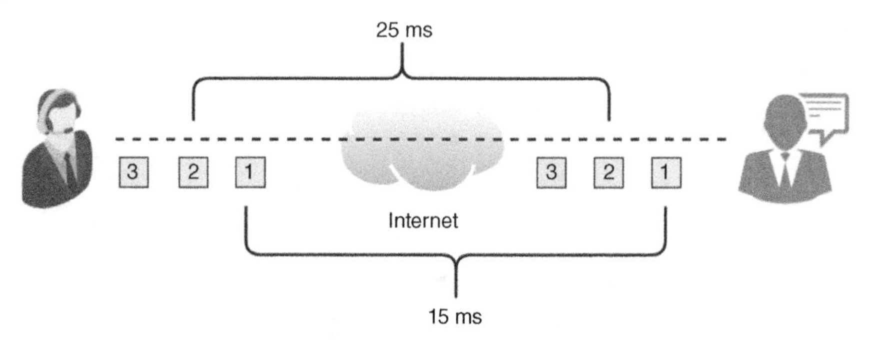

Figure 4.5 Jitter.

power of a signal. It is expressed in dB, and low values are desirable. PAPR occurs when different subcarriers are out of phase in a multicarrier system. OFDM is a multicarrier modulation technique where the available spectrum is divided into subcarriers, each containing a low-rate data stream. Signals transmitted through an OFDM system typically have high peaks in the time domain, and in an OFDM system, all subcarriers are out of phase with each other.

At each phase value, the signals differ from each other, but if, in a given situation, all data points reach the maximum value simultaneously, it will cause the output to spike. This causes a "peak" in the output envelope. Because an OFDM system has independently modulated subcarriers, the peak value of the system can be very high relative to the average of the entire system. This occasional ripple (resulting in a high PAPR) is one of the significant disadvantages of the OFDM system as it reduces the efficiency of the power amplifier in the transmitter. High PAPR can also cause problems such as out-of-band and in-band distortion. In-band distortions include high error vector size (EVM) and degraded receiver performance. Out-of-band distortions have increased adjacent channel leak rates and lessened users' performance on adjacent channels. Considering the mathematical formula of PAPR, it is defined as the square of peak amplitude divided by the root mean square (RMS) value.

- **Bit error rate (BER)**: The bit error rate (BER) is the number of bit errors that occur in a specified time. Bit error rate (BER) is a proportional value obtained by dividing the number of error bits by the total number of bits transferred during the specified time interval. In some cases, the packet error rate (PER) is measured. As the number of lost packets in communication systems increases for the reasons we have explained, the BER rate also increases, which is an undesirable situation. The formulas are given as follows.

$$\text{BER} = \frac{\text{bit error amount in unit time}}{\text{transmitted bit amount in unit time}}$$

$$\text{PER} = \frac{\text{packet error amount in unit time}}{\text{transmitted packet amount in unit time}}$$

9–10 is an acceptable BER value for communication, while the minimum value for data transmission should be 10–13.

- **Signal-to-noise ratio (SNR):** In analog and digital communication, a signal-to-noise ratio, usually written S/N or SNR, measures the strength of the desired signal relative to background noise (unwanted signal). The SNR is calculated using a fixed formula that compares the two levels and gives the ratio indicating whether the noise level affects the desired signal. The rate is typically expressed as a single numerical value in decibels (dB). The SNR can be zero, a positive number, or a negative number. An SNR above 0 dB indicates that the signal level is greater than the noise level. In some sources, SNR is also given as CINR (carrier to interference + noise ratio). The higher the percentage, the better is the signal quality.

$$\text{SNR} = \frac{\text{Signal power}}{\text{Noise power}}$$

In data networks, above 20 dB is recommended. This value is expected to be 25 dB and above in voice applications. In data networks, above 20 dB is recommended. This value is expected to be 25 dB and above in voice-centric applications (Figure 4.6).

- **Signal interference + noise ratio (SINR):** In addition to environmental noise, the receiver may hear signals from other wireless communication

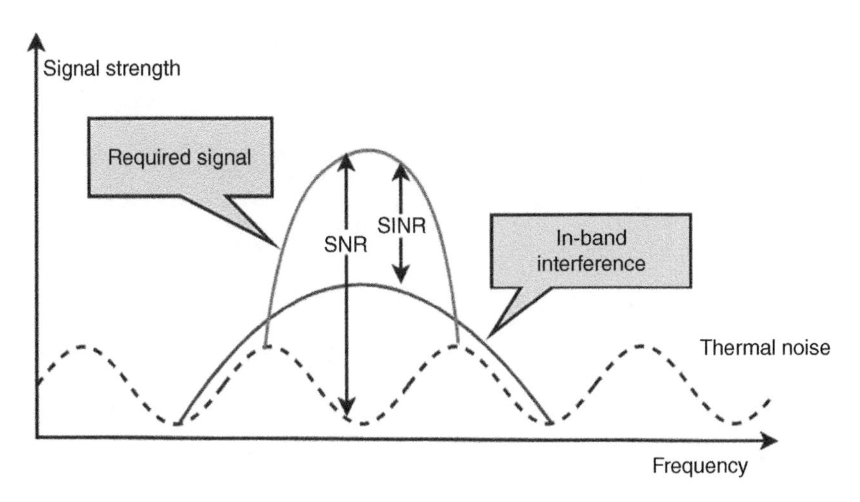

Figure 4.6 SNR and SINR.

systems or other signal sources that may interfere with the voice signal. SINR is a measure of this interference distortion and is a more used metric than SNR. SINR means the strength of the desired signal compared to unwanted interference and noise. Mobile network operators strive to maximize SINR at all sites to deliver the best customer experience by transmitting at a higher power or minimizing interference and noise.

$$\text{SINR} = \frac{\text{Signal power}}{\text{Noise power} + \text{Interference power}}$$

References

1 Beard, C. and Stallings, W. (2015). *Wireless Communication Networks and Systems*. Pearson.

2 Lea, P. (2018). *Internet of Things for Architects: Architecting IoT Solutions By Implementing Sensors, Communication Infrastructure, Edge Computing, Analytics, and Security*. Packt Publishing Ltd.

3 Assefa, T.D. (2015). QoS performance of LTE networks with network coding. Master's thesis. NTNU.

5

Seven Layers of ISO/OSI

5.1 Introduction

To understand the communication issues on a packet-switched network, it is necessary to understand the "Open Systems Interconnection (OSI) model of the International Organization for Standardization (ISO)." The OSI model is a reference model that defines how end-to-end packet communication will be carried out through a seven-layer process. Before the OSI model, each IT company had network infrastructures built according to its standards. With the OSI model, it is ensured that different network topologies and different protocols can work together.

The OSI model is a seven-layer model. In each layer, processes specific to that layer are executed. Conceptually, each layer communicating with its peer is virtually connected (Figure 5.1).

The application first generates the data at layer 7. The data created in each layer is encapsulated with data specific to that layer and transferred to the next layer. Each layer puts information on the data and passes it to a lower layer. The data converted into electrical signals in the last physical layer reaches the physical layer of the opposite host. As data passes upward through each layer, it is decapsulated, and the relevant layer receives data that it will understand on its own [1].

No layer can read data outside of its layer. In other words, layer data other than its peer layer is meaningless for that layer. At each layer, protocols specific to that layer are executed. For example, when we want to access the Internet via a browser, the HTTP (Hypertext transfer protocol) is used at the application layer. Or, when we run an e-mail client, the SMTP (Simple Mail Transfer Protocol) provides mail transferring functions at the application layer.

In a layered system, the devices of a layer exchange data in a different form known as a protocol data unit (PDU). For example, when a user wants to browse a website on the computer, the remote server software first delivers the requested

Evolution of Wireless Communication Ecosystems, First Edition. Suat Seçgin.

Published 2023 by John Wiley & Sons, Inc.

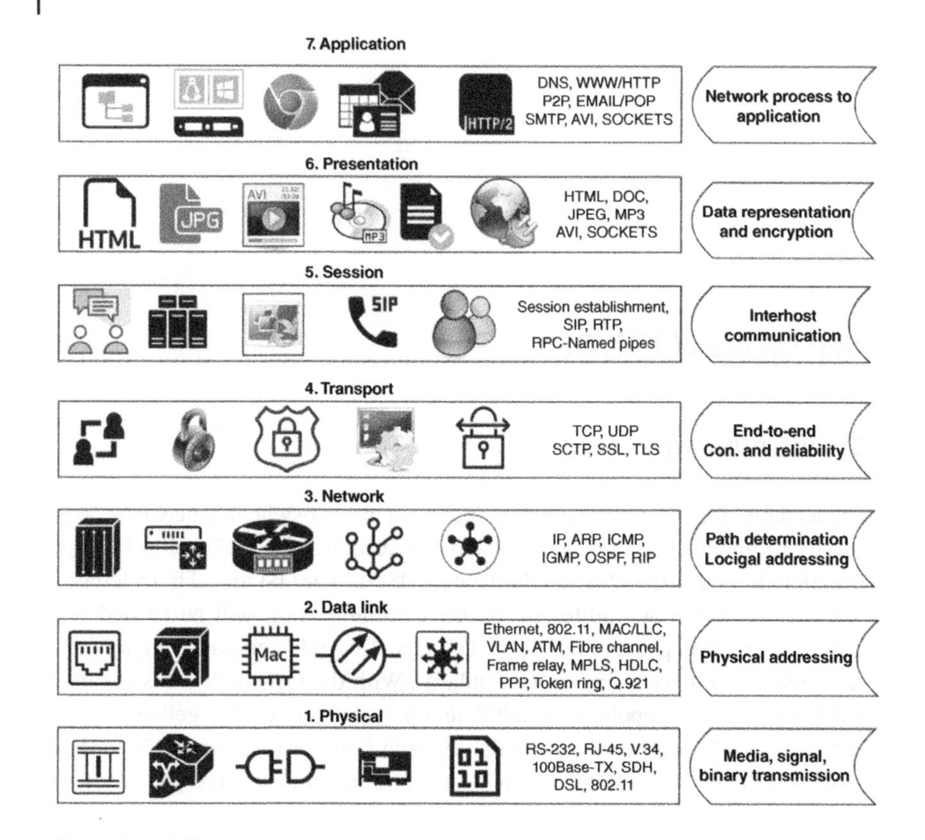

Figure 5.1 OSI layers.

data to the application layer. Each layer passed the information to its adjacent layer encapsulating its data.

During passing the data, each layer adds a header, footer, or both to the incoming PDU from the upper layer, which routes and identifies the packet. This process is called encapsulation. Encapsulation combines the header (and footer) and data from the PDU for the next layer. The process continues until it reaches the lowest level layer (physical layer or network access layer), where data is transmitted to the receiving device. The receiving device reverses the process by de-encapsulating the data in each layer, with the header and footer information driving the operations. Then the application finally uses the data. The process continues until all data has been transmitted and received (Figure 5.2).

Although seven layers are discussed in the initial definitions, in today's applications, a four-layer (TCP/IP) model consisting of an application, transport,

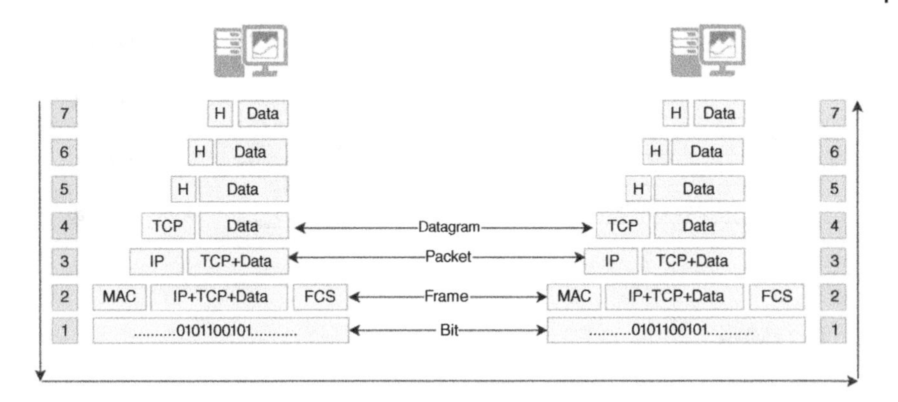

Figure 5.2 Data flow of the OSI model.

Internet, and link layers or a five-layer model comprising an application, transport, network, data link, and physical layers come to the fore. However, to give the reader more detailed information, the seven-layer structure covering other models will be explained.

5.2 Application Layer

It is the top layer of the OSI model. Manipulation of data (information) in various ways is done in this layer, which allows the user or software to access the network. User applications such as email client (e.g. Outlook), web browser (e.g. Chrome) and directory services (e.g. DNS) run in this layer.

The application layer contains various protocols commonly needed by users. One of the widely used application protocols is HTTP, which is the basis of the World Wide Web. When a web page is requested via a browser, the unified resource locater (URL) address of the requested page is sent to the server using HTTP. The relevant server also opens this page to the user (Figure 5.3).

Other application protocols used are File Transfer Protocol (FTP), Trivial File Transfer Protocol (TFTP), SMTP, Teletype Network Protocol (TELNET), Domain Name System (DNS), etc. In terms of functional classification, we can say that four different services are provided in the application layer [2]:

- **Mail services:** This layer provides the base for sending, receiving, and storing e-mail.
- **Network virtual terminal:** This allows the user to connect to a remote site. The application creates terminal emulation software for a terminal on the

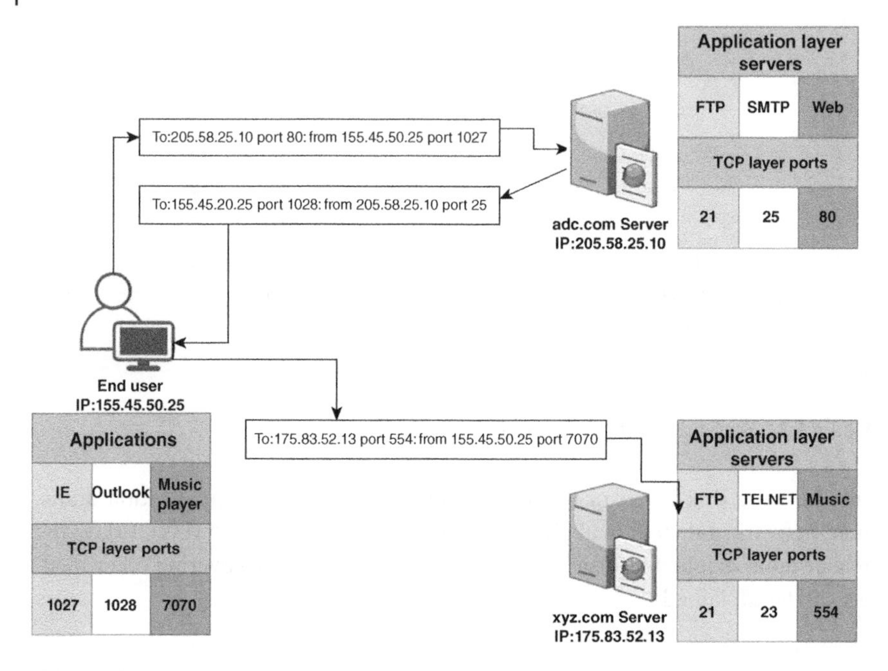

Figure 5.3 Application layer services.

remote host. In this way, the user computer communicates with the remote host through the emulator. The remote host thinks it is communicating with its original terminal.

- **Directory services:** This layer provides access to global information about different services.
- **File transfer, access, and management (FTAM):** It is a standard mechanism for accessing and managing files. Users can access and manage files on the remote computer. They can also receive files from a remote computer.

5.3 Presentation Layer

The primary purpose of this layer is to take care of the syntax and semantics of the information exchanged between two mutual terminal systems. The presentation layer takes care of sending the data in such a way that the receiver can understand the information (data) and use the data. Languages (syntax) may differ in these two terminals. Under this condition, the presentation layer plays the role of translator. The data structures to be exchanged can be defined abstractly to enable computers with different data representations to communicate. The presentation layer manages these conceptual data structures and allows higher-level data structures to be defined and exchanged.

The presentation layer has three main functions:

- **Translation:** Before being transmitted, information in the form of characters and numbers must be changed to bit streams. The presentation layer is responsible for the interoperability between encoding methods as different computers use different encoding methods. It translates data between the formats required by the network and the format of the computer.
- **Encryption:** It performs encryption at the transmitter and decryption at the receiver.
- **Compression:** It performs data compression to reduce the bandwidth consumed by the data to be transmitted. The primary role of data compression is to reduce the number of bits to transmit audio, video, text, etc. It is essential in multimedia (JPEG, MP3, etc.) transmission.

5.4 Session Layer

The session layer controls the dialogs (connections) between computers. It establishes, manages, maintains, and terminates local and remote application connections. Layer 5 software also handles authentication and authorization functions. It verifies that the data has also been delivered. The session layer is often explicitly implemented in applications that use remote procedure calls.

The session layer manages and synchronizes the conversation between two different applications. At the session layer, data streams are flagged and synchronized properly so that messages are not prematurely truncated and data loss is avoided.

Functions of the session layer are:

- **Dialog control:** This function allows the two systems to communicate in half or full duplex.
- **Token management:** This function prevents two parties from simultaneously attempting the same critical transaction.
- **Synchronization:** This function allows a process to add control points to the data stream, which are considered synchronization points. For example, if a system sends an 800-page file, it is recommended to add checkpoints after every 50 pages. This mechanism ensures that the 50-page unit is successfully received and validated. For example, if there is a crash on page 110, there is no need to resubmit the first 100 pages.

5.5 Transport Layer

The transport layer provides the functions and means to transfer data streams from a source to a destination host over one or more networks while maintaining quality of service (QoS) functions and ensuring complete data delivery.

The integrity of the data can be guaranteed by error correction and similar functions. It can also provide an open flow control function [3].

The primary function of the transport layer is to accept the data from the upper layer, divide it into smaller units, transfer these data units to the network layer, and ensure that all the pieces of the data reach the other end correctly. Also, all these processes must be done efficiently and in a way that isolates the upper layers from inevitable changes in hardware technology.

The transport layer also determines what kind of service is provided to the session layer and ultimately to network users. The most popular transport connection is the error-free point-to-point channel connection, which delivers messages or arranges bytes in the order they were sent.

The transport layer is an actual end-to-end layer from source to destination. In other words, a program on the source machine continues to talk to a similar program on the target machine, using message headers and control messages. In this sense, the transport layer is connection oriented.

The functions of the transport layer are given as follows:

- **Service point addressing:** The transport Layer header contains the service point address, which is the port address. This layer forwards each packet to the appropriate process on the opposite computer.
- **Segmentation and reassembling:** A message is divided into sections; each piece contains a sequence number that enables this layer to reassemble the message. The message is correctly reassembled on arrival at the destination, and packets lost in transmission are re-requested.
- **Connection control:** There are two connection types. In a connectionless connection (for example, UDP), each segment is treated as an independent packet and sent to the network. It does not matter whether the packet reaches the destination or not. In the connection-oriented transport layer, a connection is made with the transport layer on the target machine before the package is delivered. After this connection, the packets are sent to the target machine.
- **Flow control:** End-to-end flow control is executed.
- **Error control:** The function checks that the entire message is transmitted to the receiving transport layer without error. Error correction is carried out through retransmission.

5.6 Network Layer

The network layer performs packet routing through logical addressing and switching functions. A network has many nodes connected. Each of these nodes has a logical address (Internet protocol [IP]). When a node wants to forward a message to another node or nodes, the packets are delivered to their destination via one or more routers using the addresses of the source and destination nodes. It also splits outgoing messages into packets and combines incoming packets into messages for upper layers.

The network layer functions include converting logical addresses (IP) to physical addresses, routing packets over routers and gateways, performing flow/sequence control, error control, and converting large packets into smaller packets. There are three different IP definitions on a host connected to a switch device in a local network. The first is the host address; the second is the subnet mask used to determine the local network to which this host is connected and the gateway address (router) required to deliver packets produced for destinations, not in its network.

Communication between two hosts in the same local network is straightforward. When a host sends packets to another host, it checks whether that host is on the same local network. The packet communication is continued through the switch device if it is in its network. However, if the other host is on a different network, it delivers its packets to the router device. The router has two interfaces. One is connected to the local network, while the other is to the wide area network. The router maintains a routing table in which it is recorded which networks are to be accessed via which interfaces.

IP running at this layer is a connectionless protocol. In other words, the sender does not care whether the packet reaches across or not. This function is performed by transmission control protocol (TCP) in the transport layer. Therefore, TCP is a connection-oriented protocol (Figure 5.4).

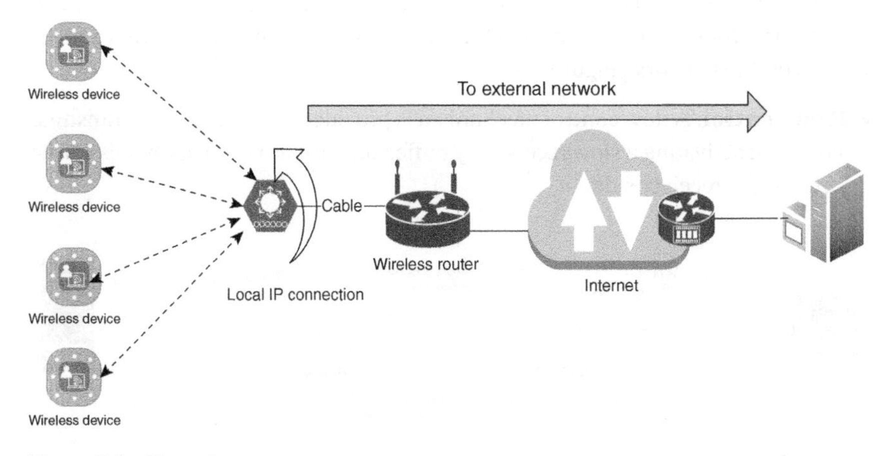

Figure 5.4 IP routing.

5.7 Data Link Layer

The data link layer performs the most reliable node-to-node transmission of data. It creates frames from packets received from the network layer and gives them to the physical layer. It also synchronizes the information transmitted over the data and performs error checking.

The data link layer is generally divided into two sublayers – the media access control (MAC) layer and the logical link control (LLC) layer. The MAC layer controls how devices on a network access media and permission to transmit data. The LLC layer is responsible for defining and encapsulating network layer protocols and controls error checking and frame synchronization.

Its primary functions are given as follows:

- **Framing:** Frames are bitstreams from the network layer. These bits are divided into manageable data units.
- **Physical addressing:** The data link layer adds a MAC header to the frame. These addresses, which we call MAC (or physical) addresses, are hardware addresses uniquely found in every device that can connect with other devices. Even if the IP address of a device varies according to the network, it is connected. The MAC address is unique and does not change. The first 24 bits (3 octets) of the MAC address, which consists of 48 bits and 5 octets, indicate the manufacturer. In comparison, the second 24 bits show information such as hardware model, year, and place of manufacture. While the IP address is visible to all users, the device's MAC address is only visible to users connected to the same network. Even if the hosts' communication starts with IP addresses, it is continued locally via MAC addresses with IP-MAC mappings (Figure 5.5).

Considering the IP (layer 3) and MAC (layer 2) addressing mechanisms, we can say that the routers operate in the OSI layer 3. And it switches packets to other routers on the network (Figure 5.6).

- **Flow control:** A flow control mechanism is provided to prevent a fast transmission from activating a slow receiver by buffering the extra bit. This avoids traffic jams on the receiver side.

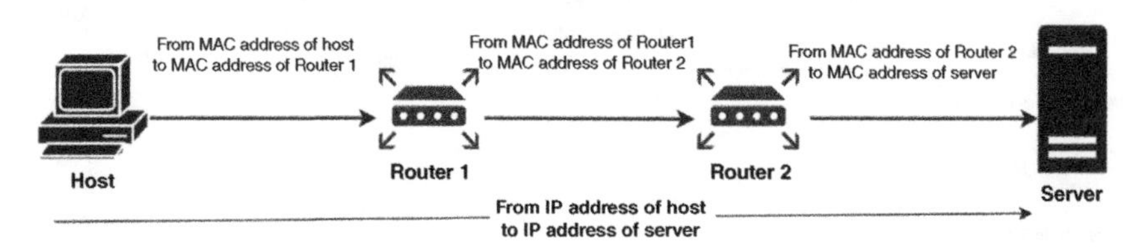

Figure 5.5 MAC address communication.

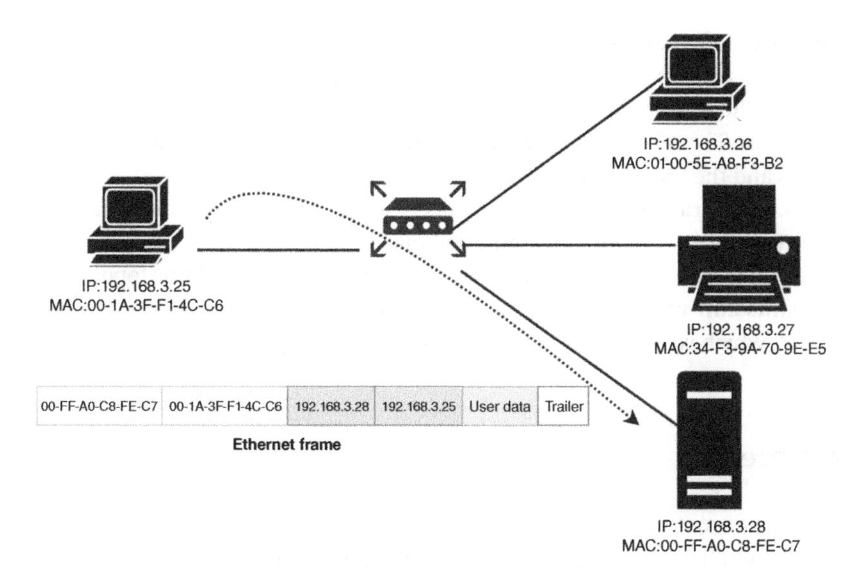

Figure 5.6 IP vs. MAC addresses.

- **Error control:** Error checking is achieved by adding a trailer to the end of the frame. Frames are also prevented from being duplicated using this mechanism.
- **Access control:** The protocols of this layer determine which devices have control over the link at any given time when two or more devices are connected to the same link.

5.8 Physical Layer

The physical layer defines the electrical and physical characteristics of the data link. For example, the layout of the connector's pins, the operating voltages of an electrical cable, fiber optic cable characteristics, and the frequency of wireless devices are specified in this layer. It is responsible for transmitting and receiving unstructured raw data on a physical medium. Bit rate control is done at the physical layer. It is the low-level layer of network equipment and never deals with protocols or other upper-layer elements. It deals with baseband and broadband transmission.

The functions of the physical layer are given as follows:

- **Representation of bits:** The data in this layer consists of a bit stream. Bits must be encoded into signals for transmission. It defines the encoding type, i.e. how the 0's and 1's are changed depending on the signal.

- **Data rate:** This function defines the transmission rate, which is bits per second.
- **Synchronization:** It deals with the synchronization of the transmitter and receiver. The sender and receiver are synchronized at the bit level.
- **Interface:** The physical layer defines the transmission interface between devices and the transmission medium.
- **Line configuration:** This layer connects devices to the environment with point-to-point and multipoint configurations.
- **Topologies:** Devices are connected using Mesh, Star, Ring, and Bus topologies.
- **Transmission modes:** It defines the direction of transmission between two devices, which is unidirectional, half duplex, or full duplex.

References

1 Pérez, A. (2013). *IP, Ethernet and MPLS Networks*. Wiley.
2 Goralski, W. (2017). *The Illustrated Network*. Morgan Kaufmann.
3 Edwards, J. and Bramante, R. (2015). *Networking Self-Teaching Guide*. Wiley.

6

Cellular Communication and 1G Systems

6.1 Introduction

This section explains first-generation wireless communication systems and cellular communication paradigms. The first milestone in the evolution journey to 6G is the cellular communication system using frequency reuse. Therefore, for advanced wireless cellular networks to be understandable, the reader must master cellular communication. Cellular communication systems have been developed to overcome the constraints imposed by the limited frequency source. These systems, established by using the same frequencies on cells that are not adjacent to each other, were an essential step for wireless communication systems.

6.2 A Brief History of Wireless Communication

We can say that the first step in wireless communication was taken with the discovery of infrared. In 1800, Sir William Herschel decided to experiment with sunlight. In this experiment, Sir Herschel observed rainbow colors by passing sunlight through a prism. He then positioned a thermometer under each color of light. In addition, he set a different thermometer at a point just beyond the red light. He saw that the thermometer outside the light spectrum reached the highest temperature in his experiment. Thus, Sir William Herschel discovered infrared light.

A year after this discovery, Johann Wilhelm Ritter, inspired by Herschel's discovery, discovered ultraviolet light. In 1867, James Clerk Maxwell predicted that there should be light with wavelengths even longer than infrared light. In 1887, Heinrich Hertz demonstrated the existence of waves that Maxwell predicted by generating radio waves in his laboratory. Thus, the door of communication on electromagnetic waves was opened.

Evolution of Wireless Communication Ecosystems, First Edition. Suat Seçgin.

Published 2023 by John Wiley & Sons, Inc.

In 1901, Guglielmo Marconi successfully transmitted the first radio signal to America over the Atlantic Ocean. It was believed that radio signals could not travel more than 200 miles. Again, Marconi sent the world's first radio signal from England to Australia in 1918.

Nikola Tesla, working on radio signals simultaneously with Marconi, invented and patented a radio-controlled robot boat in 1898.

As a result of all these efforts, the era of wireless communication started. As will be seen in the following chapters, this development process, which began with the first cellular wireless network established in the 1940s, continues today with the support of the developments in multiplexing and modulation techniques.

6.3 Cellular Communication

One of the most important revolutionary developments in data transmission and communication has been the development of cellular networks. With the development of cellular communication, mobile communication has also become possible. With frequency reuse, the most efficient use of the limited spectrum in communication is ensured.

Wireless communication systems used before cellular communication infrastructures served an area with a radius of 80 km, offering approximately 25 channels. The most practical way to increase the communication capacity in such an infrastructure would be to design smaller, low-power systems. Thus, it will be possible to use many transceivers [1].

At the heart of a cellular network is the placement of multiple, low-power (100 W or less) transmitters and receivers in the middle of what are called cells. This way, communication infrastructure is created by dividing large areas into smaller segments called cells using a transmitter with a low coverage area. The structure in the middle of each cell consisting of transmitter, receiver, and control units is called a base station. While different frequencies are used in neighboring cells to avoid interference and signal degradation, the same carriers can be used in non-adjacent cells (frequency reusing). In this way, maximum utilization is achieved in the frequency domain (Figure 6.1).

In the structure described earlier, a mobile communication service provider has to provide services such as locating the mobile device, tracking the caller, dynamic channel allocation for each call, and continuing the communication without losing the connection of the devices circulating between the cells (roaming).

It is possible to establish a communication infrastructure consisting of hundreds of cells with frequency reuse in cellular communication. Figure 6.2 shows the cell structure for frequency reusing models 4, 7, and 19, respectively [1].

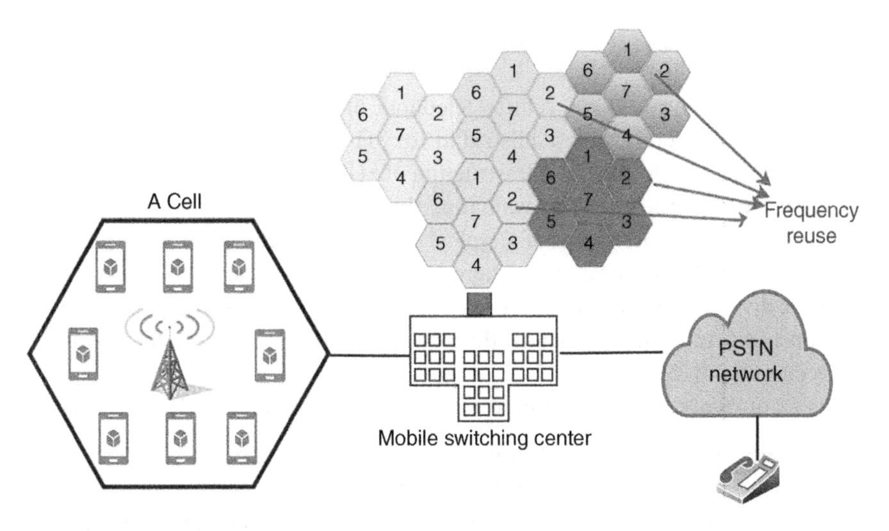

Figure 6.1 Cellular wireless communication system.

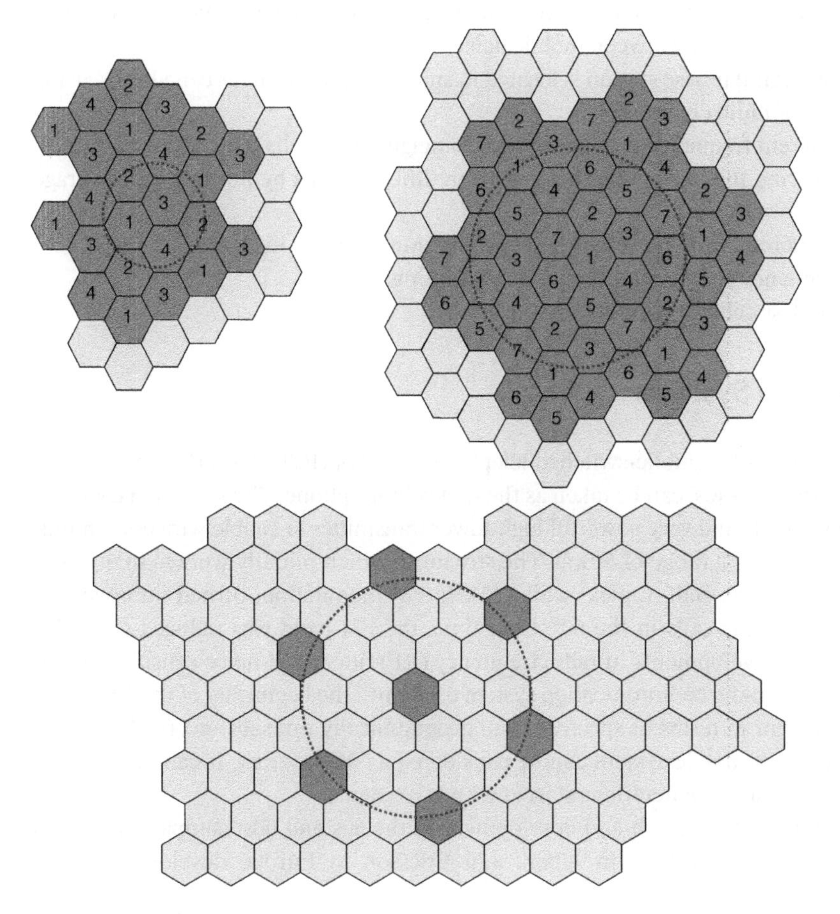

Figure 6.2 Cellular frequency reuse.

Beginning in the late 1970s, cellular telephone or cellular communication usage began. The first application established in Tokyo by the Japanese is the most critical component that forms the infrastructure of xG systems today [2].

Cell widths are optimized by evaluating the population of the geographic area and the call traffic. The radius of a typical cell ranges from 1.5 to 20 km. The cell radius also expands in rural areas where call traffic and population are less dense. Cells are divided into subcells where call traffic is too much to handle. Because of the constraint in frequency bands, the same frequencies can be used repeatedly between non-adjacent cells.

The features of cellular communication systems are summarized as follows [3]:

- High utilization capacity in the limited frequency spectrum.
- Reusing the radio channels in different cells again.
- Effective distribution and use of a certain number of channels among the number of users more than the number of channels.
- Direct communication between mobile device and base station (no direct communication between mobile devices).
- Each cellular base station is located in small geographic areas called cells, using a set of radio channels.
- Different frequency groups are used in neighboring cells.
- Ensuring the use of channel groups in different cells by limiting the coverage area to cells.
- Interference and noise levels are kept within tolerable limits.
- Frequency recovery and frequency planning.
- Wireless cellular network organizations.

6.4 1G Systems

A system with a public-switched telephone network (PSTN) interface, which was used in the 1940s, can be taken as the first cellular phone. The system in each city used a single and very powerful high-tower transmitter to enable wireless communication with a range of 50 km. The structure, which initially worked in one way (push and talk/half duplex) with 120 kHz FM modulation, turned into two way and automatic calls in the 60s. Over time, the FM band was reduced to 30 kHz with the developments in radio frequency (RF) filters and noise-sensitive amplifiers. This radio communication system used until the beginning of the 1980s was insufficient in terms of spectrum and geographically unusable everywhere. This noncellular mobile system can be considered a "0G" system. It can be said that communication generations started with 0G systems.

In line with research and development in the 50s and 60s, companies such as AT&T, Motorola, NTT in Japan, and Ericsson in Europe developed various cellular mobile systems in the 70s. All these systems were analog-operated FM systems. For the first time, the Advanced Mobile Phone System (AMPS) using 30 kHz channels on frequency division multiple access (FDMA) was designed and commissioned in North America in 1983. This infrastructure is known as the first analog cellular mobile system. These analog cellular mobile systems were called 1G (first-generation) wireless communication systems. In addition to the AMPS system, Total Access Communication System (TACS), Nordic Mobile Telephone (NMT), Narrowband Advanced Mobile Phone (NAMPS), Japanese Mobile Cellular System (MCS), CNET, and MATS-E systems are first-generation analog mobile communication systems. Figure 6.3 shows the evolution from the first generation (1G) to the present [4].

1G systems operating in the ISM 800 MHz band are designed to provide voice and circuit-switched low-speed (9.6 kbps) data services [5]. 1G systems are ineffective due to poor voice quality, massive mobile devices, low calling times, and low battery life (Figure 6.4).

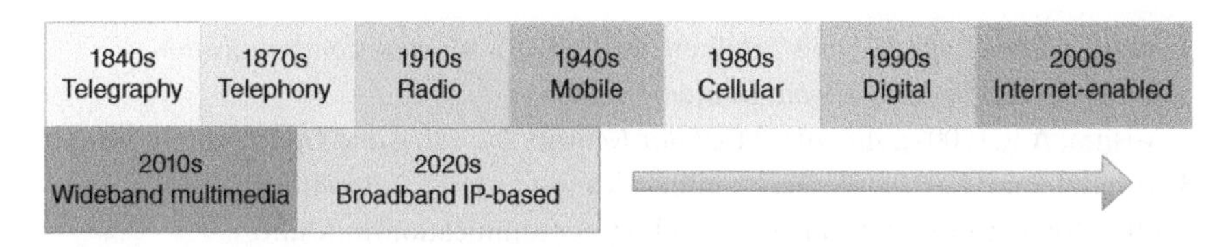

Figure 6.3 Development of communication systems.

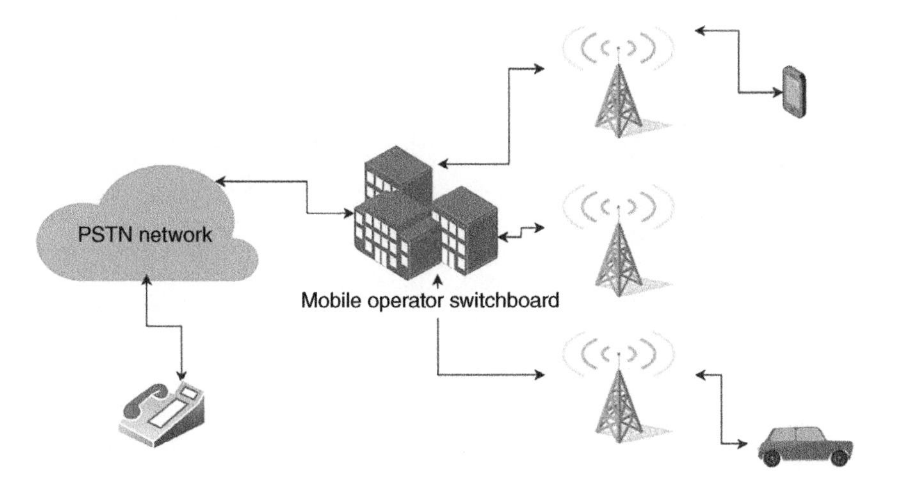

Figure 6.4 1G communication.

1G analog system features are summarized as follows:

- **AMPS**
  - Analog FM modulation, frequency division duplexing (FDD) communication, FDMA-based multiple access
  - Channel bandwidth up to 10 kHz
- **NMT**
  - Analog FM modulation, FDD communication, FDMA-based multiple access
  - 12.5 kHz or 25 kHz bands
  - Roaming in European countries
- **TACS**
  - Variant of the Japan development
  - Analog FM modulation, FDD communication, FDMA-based multiple access
  - 30 kHz channel bandwidth

References

1 Beard, C., Stallings, W., and Tahiliani, M.P. (2015). *Wireless Communication Networks and Systems, Global Edition*. Pearson.
2 Mishra, A.R. (2007). *Advanced Cellular Network Planning and Optimisation*. Wiley.
3 Tutorial Point (2023). Wireless communication tutorial. *Tutorial Point*. https://www.tutorialspoint.com/wireless_communication/index.htm (accessed 14 February 2023).
4 Grami, A. (2015). *Introduction to Digital Communications*. Academic Press.
5 Garg, V. (2010). *Wireless Communications & Networking*. Elsevier.

7

2G Systems

7.1 Introduction

We have mentioned that first-generation cellular networks (for example, Advanced Mobile Phone System [AMPS]) offered poor sound quality. Second-generation (2G) cellular communication networks provided higher signal quality, data rates, and digital service capacity than 1G. The second-generation wireless network technology, commonly known as global system for mobile communication (GSM) and called 2G, has been standardized by the third-generation partnership project (3GPP). In this section, besides explaining the 2G specifications, the differences with 1G will also be given.

7.2 1G and 2G Comparisons

The comparison between 1G and 2G is given as follows [1]:

- **Digital traffic channels:** We can say that the main difference between 1G and 2G is that 2G is built from digital systems, although 1G is entirely analog. First-generation systems are built to support audio channels using frequency modulation. In 1G, digital traffic is seen only as digital data conversion to analog form via modems. 2G works on digital traffic channels. In 2G, voice traffic is encoded into digital form before transmission. Of course, user traffic (data or digitized voice) must be converted to analog form for communication between the mobile terminal and the base station. This process is given in Figure 7.1.
- **Encryption:** Since user and control data are transmitted in digital form in second-generation systems, encryption is relatively easy to prevent eavesdropping. In 1G systems, user traffic is sent and received openly to eavesdropping,

Evolution of Wireless Communication Ecosystems, First Edition. Suat Seçgin.

Published 2023 by John Wiley & Sons, Inc.

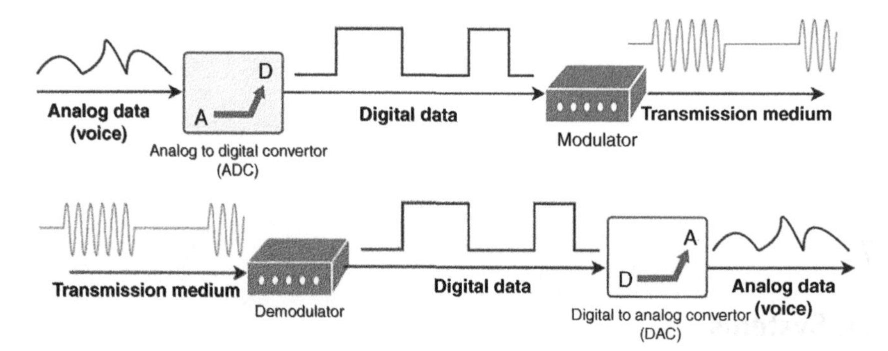

Figure 7.1 Digital communication.

while encryption is performed in 2G systems. This way, the security vulnerability in 1G systems was also eliminated.
- **Error detection and correction:** Since 2G systems are digital, error detection and correction mechanisms have also become applicable. Thus, higher quality audio transmission compared to 1G systems is achieved.
- **Channel access:** Each cell supported several users in the first-generation systems. Since multiplexing techniques were not used in these systems, a channel could only be allocated to one user at any time. With 2G systems, a multichannel structure per cell has been established using time division multiple access (TDMA) or code division multiple access (CDMA). So each channel can be dynamically allocated among users (Table 7.1).

While the IS-136 and IS-95 in the table became the widespread standard in the United States, GSM was accepted in Europe.

Table 7.1 Comparison of mMIMO, LOS MIMO, and OAM for fixed wireless link.

	GSM, DCS-1900	IS-54/IS-136, PDC	CDMA One, IS-95
Uplink frequencies (MHz)	*890–915 (EU)* *1850–1910 (US PCS)*	*800, 1500 (JP)* *1850–1910 (US PCS)*	*824–849 (US Cellular)* *1850–1910 (US PCS)*
Downlink frequencies (MHz)	*935–980 (EU)* *1930–1990 (US PCS)*	*824–849 (US Cellular)* *1930–1990 (US PCS)* *800, 1500 (JP)*	*869–894 (US Cellular)* *1930–1990 (US PCS)*
Duplexing	*FDD*	*FDD*	*FDD*
Multiple access	*TDMA*	*TDMA*	*CDMA*

Table 7.1 (Continued)

	GSM, DCS-1900	IS-54/IS-136, PDC	CDMA One, IS-95
Modulation	*GMSK*	*DQPSK*	*BPSK*
Carrier separation	*200 kHz*	*30 kHz (IS-136) (25 kHz for PDC)*	*1.25 MHz*
Channel data rate	*260.833 kbps*	*48.6 kbps (IS-136) (25 kHz for PDC)*	*1.2288 Mchips/s*
Voice channels per carrier	*8*	*3*	*64*
Speech coding	*RPE-LTP at 13 kbps*	*VSELP at 7.95 kbps*	*CELP at 13 kbps EVRC at 8 kbps*
Year	*1990*	*1991*	*1993*

7.3 2G Architecture

The 2G architecture and details will be explained using Figure 7.2. The main components of the system include the base station subsystem, the mobile switching center (MSC), and the data communication network.

The abbreviations in the figure are:

- **AuC:** Authentication center
- **EIR:** Equipment identity register
- **HLR:** Home location register
- **ME:** Mobile equipment
- **PSTN:** Public-switched telephone network
- **SIM:** Subscriber identity
- **VLR:** Visitor location register

If we explain the structure in Figure 7.2, the device called Mobile Equipment (ME) is the user terminal (mobile phone). This device consists of a radio transceiver, digital signal processor, and subscriber identity module (SIM card). The SIM card is a smart card that can be removed from the device and provided to the user by the operator. Subscriber number, operator network information to which the user will be connected, encryption keys, and other customer-specific details are kept on the SIM card [2].

Base Station Subsystem (BSS)

- **Base transceiver station (BTS):** BTS is the basic structure that defines a cell. It consists of a radio antenna, radio transceiver, and connection subcomponents

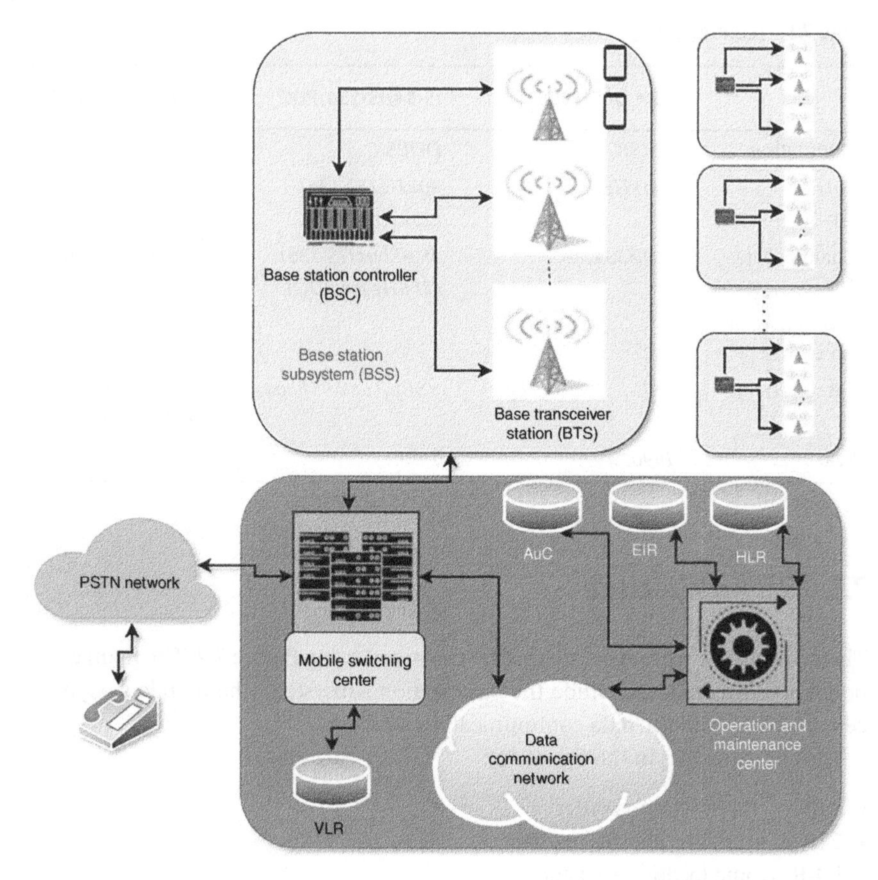

Figure 7.2 2G network architecture.

to the base station controller structure. In a GSM system, a cell diameter varies between 100 m and 35 km, depending on user density and environmental factors.

- **Base station controller (BSC):** BSC is used to control one or more BTS structures. The BSC is responsible for preserving radio frequencies, roaming the user between cells under the same BSS, and the "paging." If we briefly explain the "paging" process, it determines the location of the mobile device (subscriber) before the call is established. The process takes place one-to-one between the base station and the mobile device. Again with paging, the mobile device is notified of an incoming call.
- **Network subsystem (NSS):** NSS is responsible for wireless wide area network (WWAN) operations such as central switching, databases required for subscribers, and mobility management. As we mentioned, the mobile device roaming between BTSs is organized by BSSs, and NSS controls the roaming process

between BSSs. Besides roaming, NSS is responsible for user authentication, authorization, and activity monitoring (accounting) functions. At the heart of the network subsystem, there is the MSC. MSC (in a sense, the switchboard) uses four different databases to provide user connections, which we will give next. MSC's gateway will enable communication with conventional telephone systems on 2G wireless networks.

- **Home location register (HLR):** HLR is the database where the mobile device's location information is kept. When the mobile user changes a cell, the location information in the HLR is also updated. Also, HLR plays a vital role in sending short message service (SMS). Before the message is sent to the intended recipient, the HLR is scanned to find which MSC the recipient used last. If the target MSC reports that the mobile device is out of range or turned off, a flag is set to indicate a message waiting on the HLR for the relevant user. For example, let us take a user traveling to another city in a car. As the user's location changes, their location continues to be updated via the MSC to which they are currently connected. HLR works integrated with Gateway Mobile Switching Center (GMSC), visitor location register (VLR), and authentication center (AuC) structures.
- **Visitor location register (VLR):** Records are kept in the VLR similar to those found in the HLR. The VLR also maintains the international mobile subscriber identity (IMSI) number and mobile subscriber integrated services digital network number (MSISDN) information. The main difference between data held in HLR and VLR is that the data in the VLR frequently changes, while the data in the HLR is more persistent. When a subscriber enters a different MSC domain, the relevant record is updated in the VLR, and a notification is automatically sent to the HLR. VLR is also responsible for monitoring the subscriber's location in the VLR jurisdiction, whether it has access to a particular service, allocating roaming numbers against incoming calls, deleting inactive user information, and recording the information received from the HLR.

  Let us give an example of the interaction between HLR and VLR. Incoming calls toward the primary center are directed to the user according to the location information recorded in the VLR. The user's HLR record has information about the VLR where it was last recorded. Thus, for an incoming call, the VLR is used to initiate the call.
- **Authentication center (AuC):** Authentication and encryption key information of all subscribers registered in both HLR and VLR are kept in the AuC database. This center also controls access to user information while the subscriber performs the necessary verification processes while registering in a network. We have mentioned in the previous sections that GSM transmission can be encrypted because it is digital. The transmission between the base station and the user is encrypted with the A5 stream encryption algorithm.

- **Equipment identity register (EIR):** Each mobile phone has an identification number stored in the EIR. This number distinguishes one mobile device from the others. Pirated, stolen, or under mobile surveillance devices are registered in the EIR. Using the mobile device can be prevented using the International Mobile Equipment Identity (IMEI) number over the EIR. The EIR has lists labeled white, gray, and black. In the white list, the IMEI numbers of the devices that will enter the network and have no harm in using them are recorded. In the gray list, the information of the devices under observation is kept. Finally, the identity information of the mobile device that is blocked from entering the wireless network (for example, with a theft report) is also kept in a black list. It should be noted that even if it is on the black list, 112 emergency calls can be made over a device. EIR lists of registered operators are kept centrally on the central EIR (central EIR [CEIR]) established in Ireland. This way, the use of a stolen or illegal mobile device anywhere in the world, through any operator, is prevented.
- **Gateway mobile switching center (GMSC):** Calls from outside to mobile network users or from mobile network users to outside of the mobile network (for example, PSTN) are routed through this gateway.

7.4 Detailed Infrastructure and 2.5G

The detailed infrastructure of 2G is given in Figure 7.3. As can be seen, there are two main components: BSS and NSS. The core network operations are performed on the NSS, while wireless access systems are installed on the BSS.

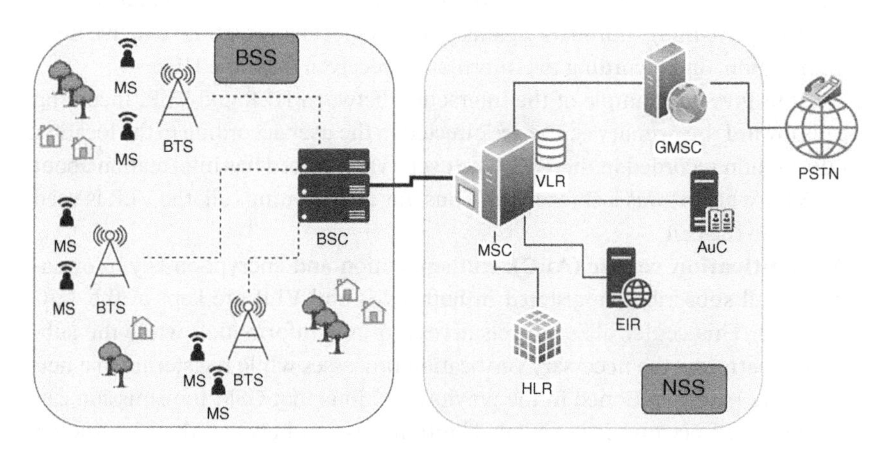

Figure 7.3 GSM infrastructure.

We can consider the wireless network technology known as 2.5G as a step between 2G and 3G. With the standard, also known as GPRS (General Packet Radio Service), data transmission rates, which were 28 kbps before, increased from 56 to 115 kbps. With the activation of the GPRS service, two new nodes have been added under NSS: Serving GPRS Support Node (SGSN) and Gateway GPRS Support Node (GGSN). With this part of NSS, the way for IP-based communication has been opened. With the creation of the package core, new elements such as IP routers, Domain Name System (DNS), and firewall were added to the system. With the introduction of Edge technology, the GSM radio access network has also become known as GERAN (GSM EDGE Radio Access Network) [3].

Concepts in architecture are explained as follows (Figure 7.4):

- **PCU:** The packet control unit directs the data traffic to the GPRS network.
- **PLMN:** Public land mobile network (in telecommunication, a public land mobile network is a combination of wireless communication services offered by a specific operator in a particular country).
- **GMSC:** Gateway Mobile Switching Center.
- **Um:** Air interface between GSM mobile station and GSM base station.
- **A-bis Interface:** Interface between BTS and BSS. A-bis interface allows control of the radio equipment and radio frequency allocation in BTS.
- **A interface:** Interface between Signaling System Num.7 (SS7) held by BSC and MSC.
- **Gn interface:** Interface between SGSN-GGSN.

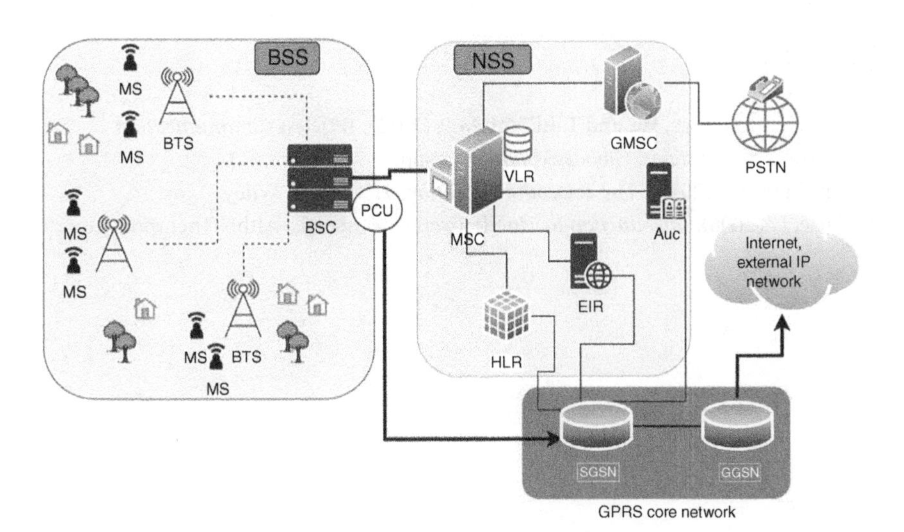

Figure 7.4 Evolution of GPRS network.

Subsequently, EDGE (enhanced data rates for GSM evolution) technology, defined as a GSM standard by 3GPP, was used. With EDGE technology, also known as 2.75G, GPRS service speeds have been increased to three times (384 kbps). Of course, with GPRS and EDGE technologies, the WAP (Wireless Application Protocol) application, which can be considered the ancestor of today's mobile web applications, has been developed. We can say that the packet-based communication brought by GPRS is the first step for mobile Internet.

Finally, the innovations brought by 2G compared to 1G are summarized as follows:

- Digital encryption of data and audio signals
- High voice quality
- Simultaneous services to more users
- A more efficient spectrum
- Smaller and cheaper phones
- Roaming support
- Messaging support
- Audio, video, and video transmission (GPRS and EDGE) with MMS (multimedia messaging service)
- Mobile devices become lighter and smaller in size with decreasing power consumption

Although 4G and beyond systems are being used today, some operators state that they will continue to support 2G until 2025.

References

1 Beard, C., Stallings, W., and Tahiliani, M.P. (2015). *Wireless Communication Networks and Systems, Global Edition*. Pearson.
2 Penttinen, J.T. (2015). *The Telecommunications Handbook*. Wiley.
3 Harte, L. (2006). *Introduction to Mobile Telephone Systems*. Althos Incorporated.

8

3G Systems

8.1 Introduction

Universal Mobile Telecommunications Standard (UMTS) systems were developed by International Mobile Telecommunication-2000 (IMT-2000). The specifications determined by International Telecommunication Union (ITU) are defined as third-generation mobile communication (3G). As it will be remembered, 2G systems are known as Global System for Mobile Communications (GSM), while UMTS has come into use as a 3G wireless communication standard. The IMT-2000 project implemented high-speed data transmission, IP-based services, global roaming, and multimedia services. Third-generation partnership project (3GPP) is responsible for developing GSM, UMTS, LTE, and 5G wireless networking standards. The 3GPP forum determines the specifications of these systems, the main framework of which is defined by the ITU through releases. For example, 3GPP has set all specifications of UMTS systems with Release 99.

8.2 2G and 3G Comparison

Although voice communication constitutes the most critical part of GSM systems, data services have emerged as an essential determinant in wireless networks and GPRS. The convergence of voice and high-speed data services into a single service was the driving force in the definition of UMTS. UMTS should be seen as a revolutionary transformation rather than an evolution on GSM. Although UMTS uses GSM components as core network components, the UMTS radio access network (UTRAN) incorporates new developments [1] (Table 8.1).

Compared to GSM, the most crucial innovation in UMTS has been redesigning the UMTS terrestrial radio access network (UTRAN). In addition, the time and

Evolution of Wireless Communication Ecosystems, First Edition. Suat Seçgin.

Published 2023 by John Wiley & Sons, Inc.

Table 8.1 Comparison of GSM and UMTS.

	2G (GSM)	3G (UMTS)
Access method	*TDMA/FDMA*	*WCDMA*
Maximum downlink	*10–156 kbps*	*384 kbps*
Maximum uplink	*10–150 kbps*	*128 kbps*
Multiple access	*TDMA*	*TDMA*
Bandwidth	*200 kHz*	*5 MHz*
Modulation	*200 kHz*	*30 kHz (IS-136)* *(25 kHz for PDC)*
Channel data rate	*OMSK*	*QPSK*
Core network	*Circuit switched*	*Circuit/packet switched*

frequency multiplexing and access (TDMA, FDMA) methods used in 2G have been replaced by the Wideband Code Division Multiple Access (WCDMA) method in 3G. This way, downlink speeds of 384 kbps per user have been achieved. These speed values change according to the circulation of the mobile device (for example, in a vehicle). A 384 kbps speed is given for 120 km/h speed. In picocells, the download speed can go up to 2 Mbps for 10 km/h. These speeds are provided in the UMTS 1900–2025 MHz band.

Until 3GPP release 4, GSM and UMTS circuit-switched voice calls were processed over E1 circuits with 64 kbps bandwidth. The concept of the bearer independent core network (BICN) was introduced in release 4. This version has replaced the circuit-switched 64 kbps time slots with Internet protocol (IP) and packet-switched communication methods. With the transition to the packet-switched system, mobile switching center (MSC) is divided into two components: MSC-server and media gateway (MGW) (Figure 8.1).

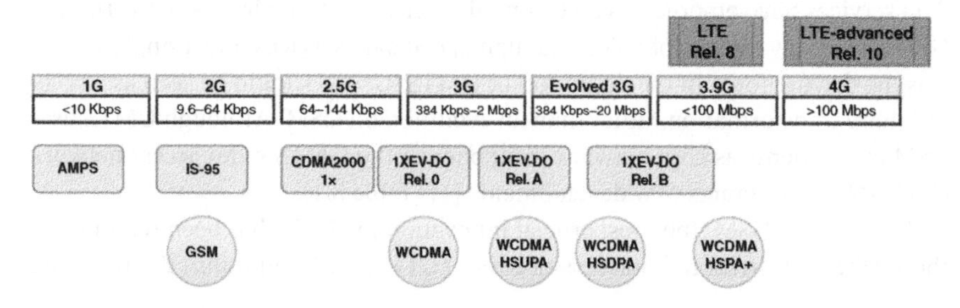

Figure 8.1 Evolution from 1G to 4G.

With 3GPP Release 4, UMTS radio access standards have been determined, and UMTS systems have reached the speeds stated before. Release 5 introduced the high-speed downlink packet access (HSDPA) standard. With HSDPA, the downlink capacity has been increased to the range of 1.8–14.4 Mbps. The mentioned speeds have been achieved by implementing adaptive modulation and coding, hybrid automatic repeat request (hybrid ARQ), and fast scheduling with HSDPA. This development is known as UMTS HSDPA, 3.5G, or evolved 3G.

With Release 6, high-speed packet access plus (HSPA+) technology was defined, increasing the uplink speed to 5.76 Mbps. Similar to the speed increase in the downlink channel in release 5, this development was also realized in the uplink channel (high-speed uplink packet access [HSUPA]). We can evaluate HSDPA and HSUPA technologies under a single heading high-speed packet access (HSPA).

Although speed improvements were achieved in the downlink and uplink channels with releases 5 and 6, power consumption was still the weakest point in these improvements. In standby situations with no data transfer, power consumption continued due to the signal sent to the base station (keep the link established) so the link would not be broken. The Continuous Packet Connectivity (CPC) recommended power consumption has been reduced with release 7. Again, with release 7, many antennas can be used simultaneously via multiple input multiple output (MIMO), which will gain an important place in 4G and beyond systems. Additionally, 64-QAM (6 bits/symbol) provides more efficient spectrum use.

8.3 3G Architecture

In the journey to 4G, to summarize, 3.5G HSDPA, 3.75G HSUPA, 3.8 HSPA+, 2.85G HSPA+/MIMO, and 3.9G LTE (long-term evolution) generation communication specifications have been defined. EV-DO (Evolution Data Optimized) is a development used on CDMA-based 3G systems for high-speed Internet access via mobile devices. EV-DO is a protocol that integrates with HSPA technology. It aimed to develop broadband Internet access over 3G rather than voice transmission (Figure 8.2).

While MSC-Server performs call control (CC) and mobility management (MM) functions, MGW performs the function of directly managing user traffic. MGW also transforms user data generated by different transmission methods (for example, conversion from circuit switching to packet switching-transcoding function). Thanks to this method, we can say that the circuit and packet switching systems have converged. Thus the operating and management costs of the operators are significantly reduced compared to the operational costs of the circuit-switched systems [2].

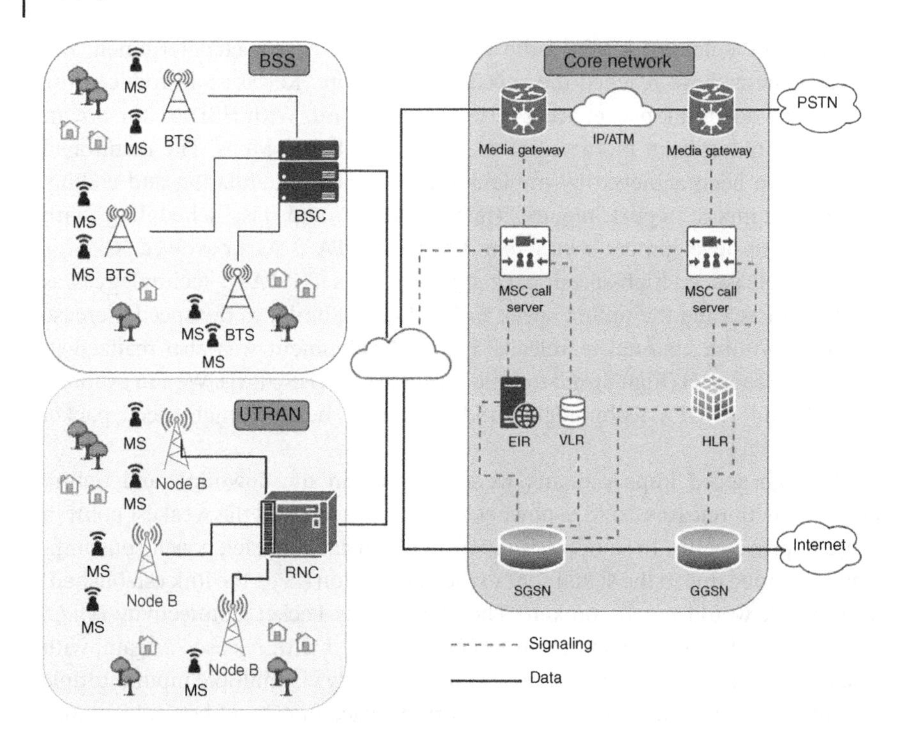

Figure 8.2 UMTS conceptual architecture.

Although the essential components were taken from GSM while UTRAN was being designed, some changes were made in the terminology. The structures called base transceiver station (BTS) and base station controller (BSC) in 2G entered the literature as Node-B and Radio Network Controller (RNC), respectively. In addition, mobile devices are called mobile stations (MSs), while these devices are called user equipment (UE).

The functions in the architecture developed with UMTS are summarized as follows:

- **Public Land Mobile Network (PLMN):** Wireless network communication services offered by other operators.
- **Charging Gateway (CG):** It determines user fees by using call detail record (CDR) data. Pricing (usage) information is pulled from SGSN and GGSN servers.
- **Border Gateway (BG):** The structure runs the comprehensive area network routing function called Border Gateway Protocol (BGP) and security protocols such as IPSec.

- **Service Control Points (SCP):** The intelligent network (IN) telephone system defines the geographical number to which the call will be made.
- **Service Creation Environment (SCE):** Function that defines all properties of all services running on IN.
- **Gateway Mobile Switching Center (GMSC):** It works as an "intermediary" between internal and external networks.
- **Service GPRS Support Node (SGSN):** Mobility management, session management, billing, and communication with other parts of the network.
- **Gateway GPRS Support Node (GGSN):** A complex router that manages all operations between the external packet-switched network and the system's internal packet switching network.

References

1 Sauter, M. (2017). *From GSM to LTE-Advanced Pro and 5G*. Wiley.
2 Penttinen, J.T. (2015). *The Telecommunications Handbook*. Wiley.

9

4G Systems

9.1 Introduction

International Mobile Telecommunications-Advanced (IMT-Advanced) Standard, developed in line with the requirements published by International Telecommunication Union Radiocommunication Sector (ITU-R), is known as 4G and 4.5G with its use in the field. Entirely packet-based network applications and 4G systems based on mobile broadband will be explained in a little more detail to understand 5G better and beyond systems due to the differentiation in both user and network planes. Figure 9.1 shows the evolution from Third-Generation Partnership Project (3GPP) Release 8 to Release 16.

4G technology has been built on the convergence of wired and wireless systems and has been implemented to provide mobile multimedia services everywhere. Working entirely on Internet protocol (IP), the system offers connection speeds of 100 Mbps in high mobility and 1 Gbps in low mobility or fixed connection. At all these speeds, it also provides end-to-end quality of service (QoS) requirements. Services such as IP telephony, ultra-broadband internet access, game services, and high-definition TV (HDTV) are the services provided to users [1] (Table 9.1).

With 4.5G, developed after 4G, higher download speeds have been achieved with effective spectrum usage (spectral efficiency). 7–12 Mbps download speed in 4G increased to 14–21 Mbps with 4.5G. In addition to this download speed, significant improvements have been made in latency times, which is a critical parameter, especially in real-time applications. The latency of 12 milliseconds (ms) in 4G has been improved by more than 50% and reduced to 5 ms with 4.5G. With the 4.9G standard, which was subsequently developed, the delay time reached 2 ms and speeds over 1 Gbps. With these delay times and rates, machine-to-machine (M2M) connection, Internet of Things (IoT), and virtual and augmented reality applications have been developed and used [2]. Figure 9.2 shows licensed 4G spectrums of selected sample operators from Asia, Europe, and North America.

Evolution of Wireless Communication Ecosystems, First Edition. Suat Seçgin.

Published 2023 by John Wiley & Sons, Inc.

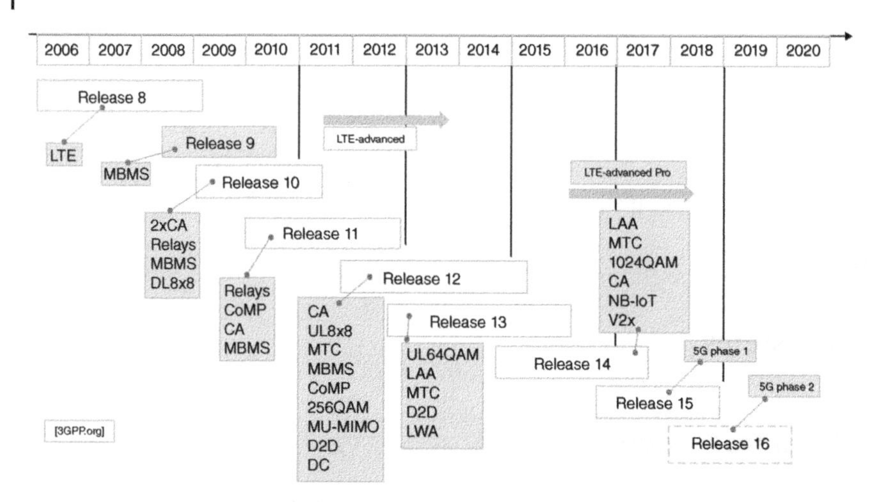

Figure 9.1 3GPP version road map.

9.2 Toward 4G

We have mentioned in the previous sections that the first-generation (1G/NMT, AMPS, TACS) and second-generation (2G/GSM, IS-136, PDC, IS-95) communication systems are utterly voice centered. With the third-generation (3G/WCDMA, HSPA, cdma2000, TD-SCDMA) communication systems, which we can consider as a transition between voice and mobile broadband, users are now acquainted with data-based communication (Table 9.2).

After the broadband Internet came into our lives with 3G systems, the increasing need for data-based systems accelerated the transition to packet switching (IP-based) systems. In fact, with high-speed packet access (HSPA) versions, also known as 3.5G, the movement toward forming a networked society has begun, with speeds at the level of megabits. As the following sections explain, this transition has reached maturity with 5G and beyond systems (Figure 9.3). Following are the releases and features provided by the 3GPP organization [3]:

- **Release 8 (2008):** It is the first version in which long-term evolution (LTE) was defined, and improvements have been made to form the basis for later versions.
- **Release 9 (2009):** Femtocell concept, self-organizing network (SON), evolved multimedia broadcast and multicast service (eMBMS), and new spectrum bands (800 MHz, 1500 MHz).
- **Release 10 (2011):** LTE Advanced, Carrier Aggregation (CA), enhanced multi-antenna downlink communications, improvements in MBMS services, enhancement on SON, relay for LTE, new bandwidth intervals.

Table 9.1 4G specifications.

Mobile-IoT family			**LTE-M**	**NB-IoT**
Technology generation	***4G*** *LTE 3GPP Rel.8*	***4G*** *LTE 3GPP Rel.8 Power Saving Mode (PSM)* *LTE-Advanced 3GPP Rel. 12*	***4.5G*** *LTE Advanced Pro* *3GPP Rel. 13*	***4.5G*** *LTE Advanced Pro* *3GPP Rel. 13*
Device category	*Cat. A*	*Cot 1*	*Cat M1*	*Cat. NB1*
Typical use	*Mainstream smartphone*	*Mid-data volume IoT*	*Low-mid-data volume IoT*	*Low-data volume IoT*
Max. coupling loss (dB)	*144*	*144*	*156*	*164*
Peak data rate downlink	*150 Mbps*	*10 Mbps*	*1 Mbps*	*170 kbps*
Peak data rate uplink	*50 Mbps*	*5 Mbps*	*1 Mbps*	*250 kbps*
Voice support	√	√	√	—
Used frequency bandwidth	*Carrier bandwidth 1.4–20 MHz*	*Carrier bandwidth 1.4–20 MHz*	*1.08 Mbps (1.4 MHz carrier bandwidth)*	*180 kHz (200 kHz carrier bandwidth)*
Usable frequency spectrum	*LTE carrier*	*LTE carrier*	*In-band in LTE carrier*	*In-band, quard band, standalone*
Modem complexity	*100%*	*80%*	*20%*	*15%*
Modem battery life	*<1 year*	*>10 years (with PSM)*	*>10 years*	*>10 years*

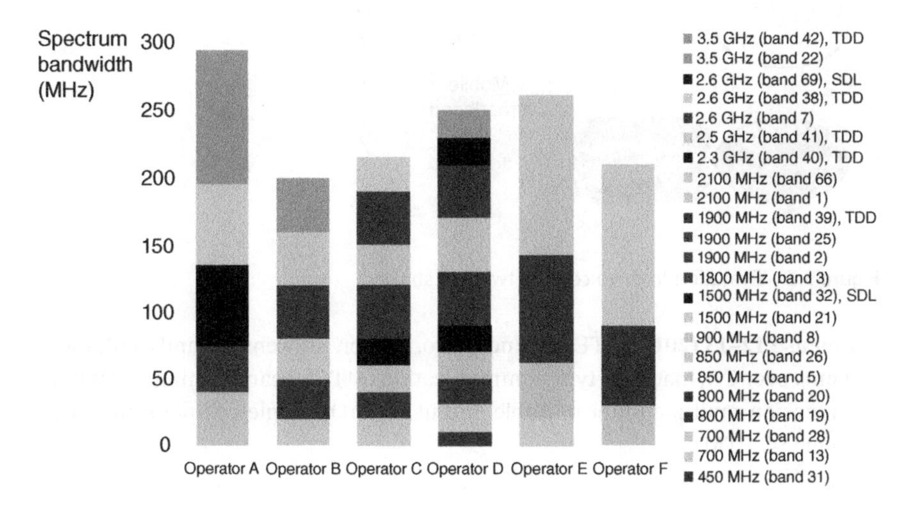

Figure 9.2 4G spectrum.

Table 9.2 4G evolution from 1G to 4G.

	1G	2G	2.5G	3G	4G
Year	*1984*	*1991*	*1999*	*2002*	*2012*
Services given	*Analog voice*	*Digital voice*	*High capacity data packets*	*High capacity, broadband*	*Fully IP based*
Data rate	*1.9 kbps*	*14.4 kbps*	*384 kbps*	*2 Mbps*	*200 Mbps*
Multiplexing method	*FDMA*	*TDMA, CDMA*	*TDMA, CDMA*	*CDMA*	*OFDMA, SC-FDMA*
Core network	*PSTN*	*PSTN*	*PSTN, packet network*	*Packet network*	*IP backbone*

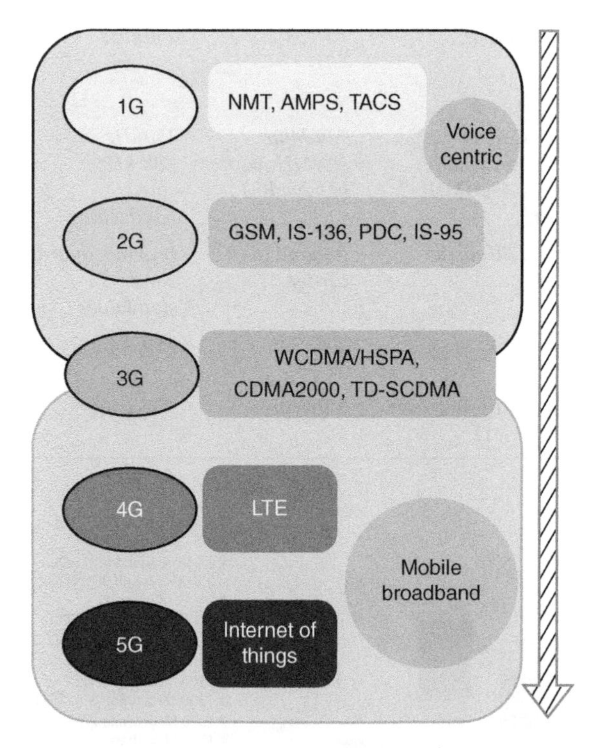

Figure 9.3 Evolution from voice to networked society.

- **Release 12–13 (2015):** LTE Advanced Pro, LTE on unlicensed bands, enhancements over CA, machine-type communication (MTC), beamforming and whole dimension multiple input-multiple output (MIMO), single-cell point to multipoint (SC-PTM).

Despite the additions such as HSPA, the 3GPP organization has announced the LTE structure with the release of eight instead of Universal Mobile Telecommunication Standard (UMTS) systems that have difficulties meeting the increasing user needs. With this new structure, the core network architecture used in 3G has also been changed to a large extent. This new structure, enhanced packet core (EPC), provides data rates, capacity, spectrum usage, and flexible installation opportunities [4].

With this system, which was used commercially in 2009, speeds of up to 300 Mbps were achieved. Concepts such as developments in antenna technologies, multi-site coordination, and machine communication started to become widespread rapidly.

There are two initiatives for 4G standardization: LTE developed by 3GPP and WiMAX (worldwide interoperability for microwave access) recommended by IEEE 802.16 committee. Both developments have a lot in common regarding performance and technology use. From 2008 to 2009, WiMAX has been used for fixed broadband wireless access, while LTE has become a universal standard for 4G wireless systems.

We said that LTE had become a standard in 4G wireless communication. For this reason, the following sections will focus on LTE architecture and technologies. Mastery of these technologies and architecture will also play an essential role in understanding the 5G and 6G architectures, which will be explained in the following sections.

The main change in the transition from 3G to 4G is roughly given in Figure 9.4. As can be seen, a new radio access architecture has emerged with the advanced packet core (evolved packet core), IP Multimedia Subsystem (IMS), home subscriber server (HSS), and evolved Node B (eNB) structure due to being full IP based. Additionally, 3G UTRAN (UMTS terrestrial radio access network) have evolved into E-UTRAN (evolved-UMTS terrestrial radio access network) with 4G.

9.3 Services and Servers

With 4G/LTE, significant changes have been made in the evolved packet core (EPC) layer, where the core components are located, and in the radio access layer (E-UTRAN). If we briefly explain the structures in the EPC layer [5]:

- **Mobility Management Entity (MME):** We can define the MME as a user radio access network control point. It is responsible for setting up/closing the mobile device carrier line and controlling user mobility. It organizes user authentication (with HSS), S-GW selection for the user, application of roaming restrictions, and mobility between LTE and GSM/UMTS access networks.

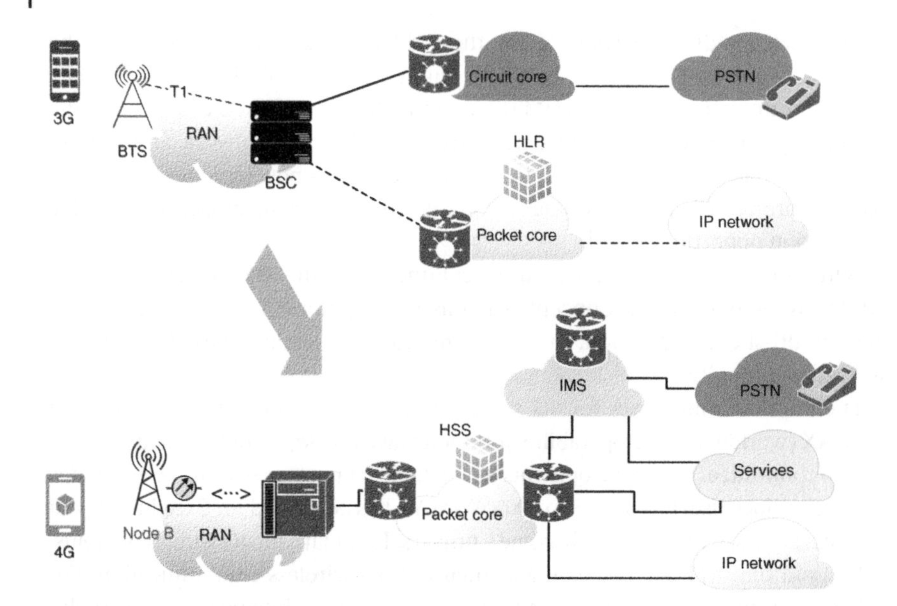

Figure 9.4 3G core and 4G core.

When the user moves from one cell to another (handover), the MME performs the necessary monitoring and organization (via paging) (Figure 9.5).

- **Serving Gateway (S-GW):** The primary function of the serving gateway is to forward user data packets and send them to the next service. In addition, it manages user roaming between GSM/UMTS and 4G networks. In other words, the S-GW acts as an interconnection point between the radio side and the EPC. The forwarding of a data packet from one eNodeB2 to another eNodeB is also handled by the S-GW. Finally, S-GW is responsible for packet buffering and transmission level marking functions in uplink and downlink channels.
- **PDN-Public Data Network-Gateway (PGW):** PDN gateway acts as a connection point between mobile devices and external IP networks (e.g. the Internet). To access multiple PDNs, the UE connects to multiple PGWs simultaneously. It also performs policy enforcement, charging support, packet filtering, legal eavesdropping, and packet monitoring.
- **Policy and Charging Rules Functions (PCRF):** Provides QoS information to PGW. This information includes pricing rules, flow control, and traffic prioritization details.
- **Home Subscriber Server (HSS):** Stores customer profile information and forwards user authorization information to MME. This server also keeps the

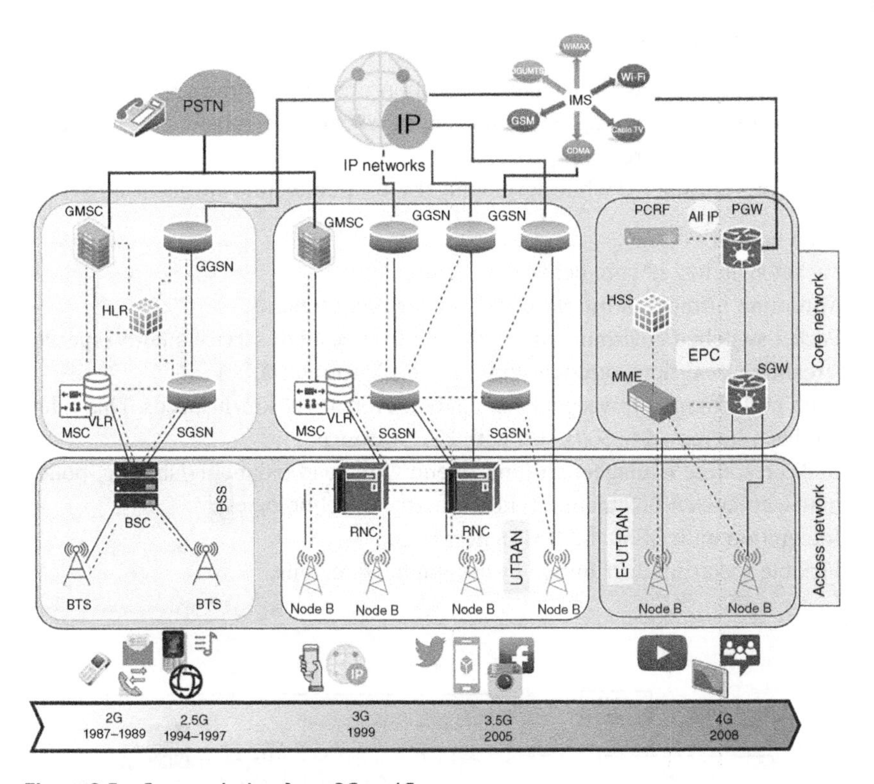

Figure 9.5 Core evolution from 2G to 4G.

PDN information that the user can connect to. HSS also maintains the MME information to which the user is added and registered. HSS works with the authentication center (AuC) that holds the user authentication and security key information. HSS also implements support functions for mobility management, call, login/setup, user authentication, and access authorization.

9.4 Architectural Structure and Novel Concepts

Since 4G has completely IP-based packet-switched communication, some changes and innovations have been made both in the core network and in the servers where the services running on this network are provided. These applications had to be developed both for communication between backward systems (2G/3G) and for continuing communication with circuit-switched systems that are still operating. This section will describe these improvements.

9.4.1 Architectural Structure

The 4G architecture with detailed connection drawings is given next. Protocols operating on each line (e.g. SIP) and interfaces between components are again expressed in the Figure 9.6 [6].

The topics considered while developing the 4G architecture are given as follows:

- Lean network design.
- Packet-switched IP core network.
- Minimum number of interface and network components.
- Packet-switched environment suitable for each level of service quality (speech, streaming, real-time, non-real-time, and background traffic).
- High performance, low signal degradation even at 120 km/h speeds. This value has been increased to 500 km/h with LTE Advanced.
- Radio resource management (end-to-end QoS, load balancing/sharing, policy management/enforcement for different access technologies).
- Integration with GSM and UMTS networks.
- Flexible spectrum distribution by geographical regions.

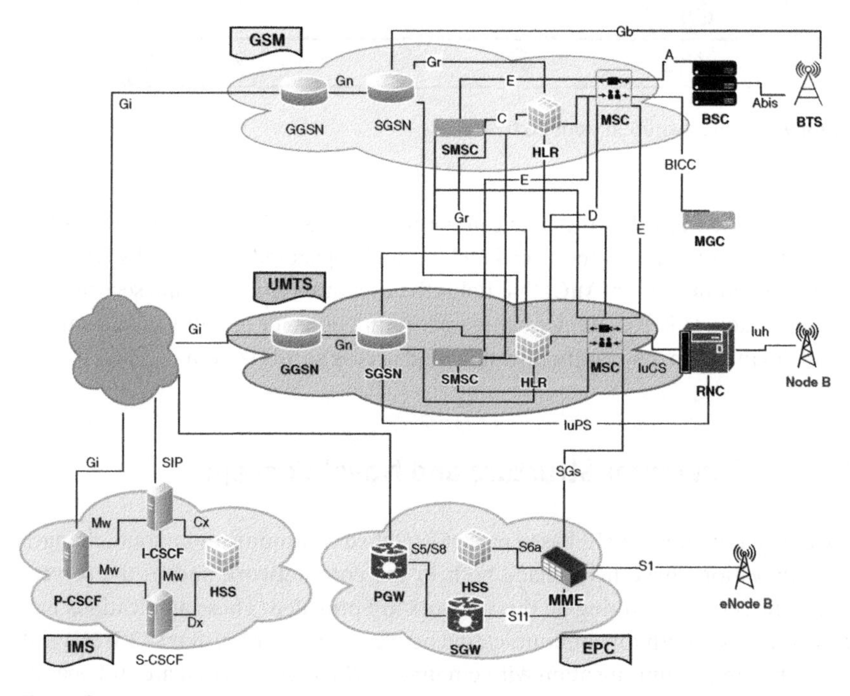

Figure 9.6 4G core network.

- Broadcast and multicast services for emergencies.
- Scalable bandwidth from 1.4 to 20 MHz.
- Aggregation of all bands up to 100 MHz.
- Using both FDD and TDD.
- Low cost, high spectrum efficiency, reuse of existing reserved spectrum ranges, flat network architecture for fewer network components, base stations with lower power consumption and smaller size, self-configuration, and self-optimization.

9.4.2 IP Multimedia Subsystem (IMS)

Especially in the past 30 years, there have been revolutionary developments in wired and wireless communication systems. Wireless cellular communication systems, which started with voice-based 1G systems, also offered some data services with 2G and continued to develop with high-speed data services and multimedia applications with 3G and 4G. Until the third generation, wireless communication, voice, and Integrated Services Digital Network (ISDN) networks have taken place with their dominant usage areas in practice. With the widespread use of Asymmetric Digital Subscriber Line (ADSL) running on fixed telephone lines in Internet access, users started to use chat, messaging, online gaming, and voice over IP (VoIP) real-time applications intensively. Users now have always-on and always-connected devices in their hands and homes [7].

Developments in wired and wireless communication devices and changes in communication structures have brought many application and communication structure variations. An orchestration mechanism was needed so that different applications such as multiple network types, multimedia applications, and fixed and wireless device communication could work together (we can call it convergence in a way). IMS is a system that has reached its maturity with 4G even though it started with 3G as a structure to control and organize this convergence.

The IMS architecture is designed according to the next-generation networks (NGN) principles based on the session initiation protocol (SIP) for session control. The IMS specifications define functions to manage all multimedia session controls on top of the UMTS architecture. In this sense, IMS is a structure for converging data and voice in fixed and mobile networks. IMS is a structure that runs on many packet-based technologies (GPRS, ADSL, WLAN, cable, WiMAX, and EPS) for real-time operation of all these together [7]. It is not limited to IMS voice transmission but also manages group management, push-to-talk service, messaging, conference, and IMS multimedia telephony applications (Figure 9.7).

In summary, we can say that IMS is the critical structure for the convergence of device, network, and service perspectives in fixed and mobile communications.

IMS has been designed in a layered structure for simplicity in communication between users or between users and content providers and to gather the standard

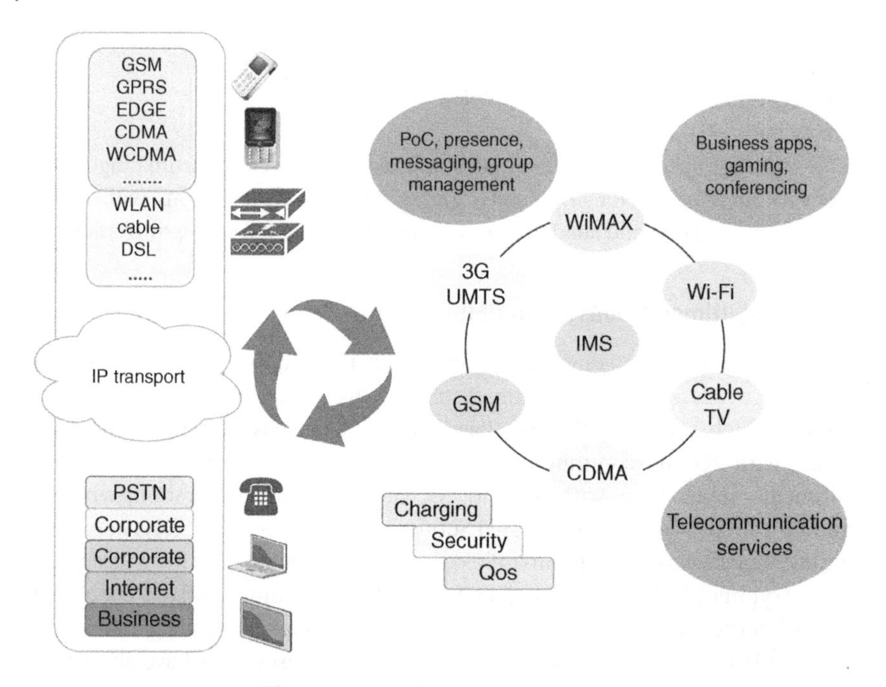

Figure 9.7 IMS for converging networks.

functions of multiple applications at a certain point. From the bottom up, these layers are the transport, session, and control and the application layer at the top. Another layer, which we will call the access and user endpoint layer (endpoint and access layer), is considered at the same level as the transmission layer [8] (Figure 9.8).

- **Transport and endpoint layer:** In this layer, incoming data in different formats (analog, digital, and broadband) are combined into real-time transport protocol (RTP) and SIP formats over media gateways and signaling gateways. Media processing servers such as notification/announcement, in-band signaling, and conferencing are also located in this layer. In terms of its functions, we can consider it as a network access layer. This layer also abstracts the access layer and network resources such as fixed lines and packet-switched radio. All systems running on this layer are IP based.
- **Session and control layer:** As the name implies, this layer is responsible for authentication, routing, and orchestration of distribution traffic between the transmission and application layers. We can say that most of the traffic here consists of SIP protocol traffic related to VoIP. In addition to forwarding SIP messages to the appropriate services, this layer acts as an interface between

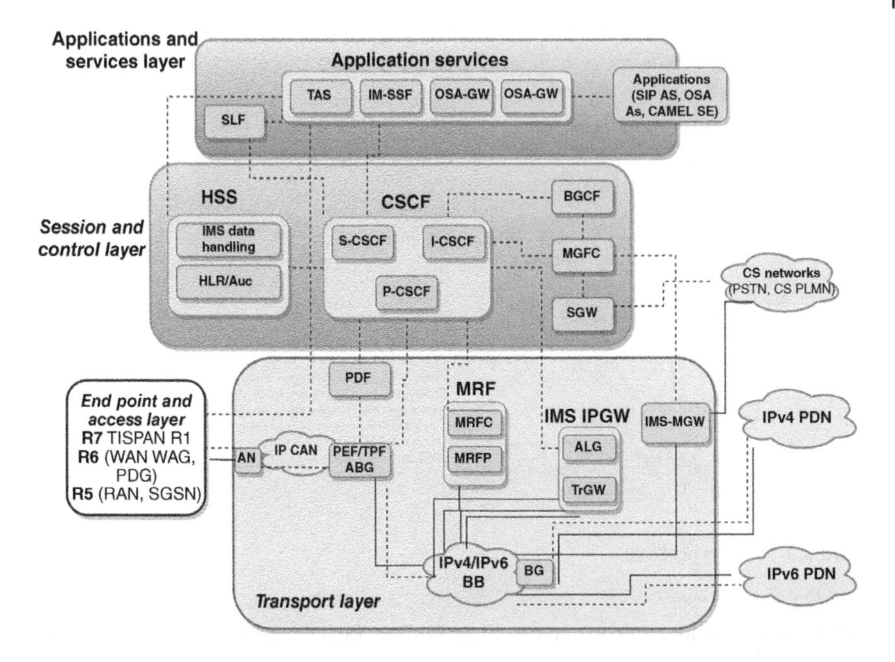

Figure 9.8 IMS layered structure.

the application layer and other layers. For example, in the pay-per-download service that provides video purchasing services to users, functions such as billing services, communication, and the requested QoS level also pass through this layer.

The main component of this layer is the call session control function (CSCF), which enables interaction between application servers, media servers, and HSS. Media resource function (MRF), which provides features such as announcement, tone, and conference call for voice over LTE (VoLTE); WiFi calls and fixed VoIP solutions; IP address assignment; activity tracking; QoS functions; content transfer; and policy can be counted as the functions of this layer.

- **Application service layer:** The layer where the actual services are run. Primary services such as multi-application services (Application Servers), telephone application services (Telephony Application Server [TAS]), IP Multimedia Services Switching Function (IM-SSF), and Open Service Access Gateway (OSA-GW) are located in this layer. To put it briefly, it functions as a bridge between IM-SSF 800 services and local number portability, TAS IP Centrex service, and OSA-GW phone services and back office applications (messaging, calling, etc.) (Figure 9.9).

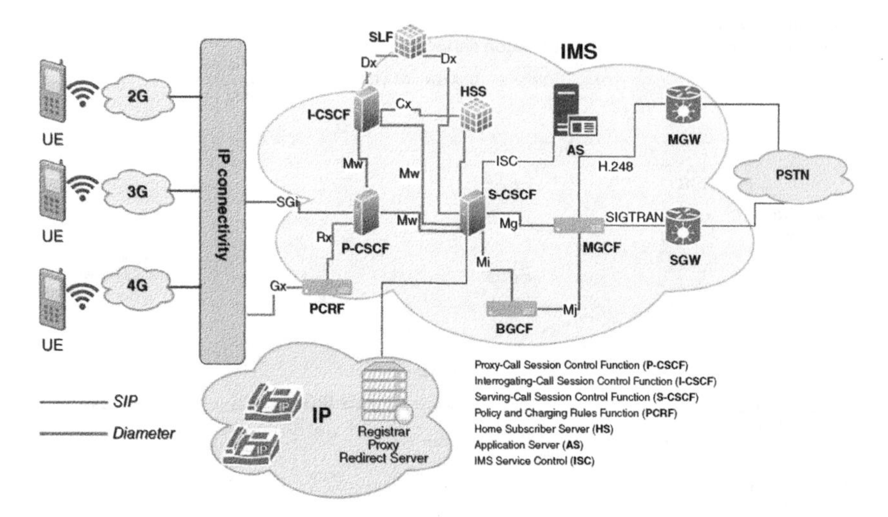

Figure 9.9 IMS in practice.

Thanks to the layered structure described so far, the convergence has been achieved in devices, networks, services, and applications. With the convergence of mobile or fixed devices, smartphones, tablets, computers, or television sets have become manageable by the network.

With network convergence, wireless access networks, public-switched telephone network (PSTN), and broadband fixed networks have become services that can be managed from the perspective of IMS. Convergence of network services includes all necessary network functions to enable subscriber-level applications. Accessing user profiles, authentication and billing, location services, and creating media control services through open and standards-based Application Programming Interfaces (APIs) are also considered within the scope of network convergence.

Finally, converged IMS applications can be hosted anywhere on the service provider network or domain. Thus, users in different networks and other devices can be accessed transparently using the advantages of standard network services.

9.5 Voice over LTE (VoLTE)

In the IMS section, we described the evolution of the radio access network from GERAN (2G)/UTRAN (3G) to LTE. With this change, an essential milestone in the transition from voice-based to data-based communication, voice has also switched to packet-based communication. We have explained that IMS systems have been developed for both the management and orchestration of these IP packet-based systems and the transparent communication environment between hybrid systems (e.g. Plain Old Telephone System [POTS] and LTE). Thus, the voice core network was transferred to IP multimedia systems, and with IMS, both voice and data traffic began to be provided in the packet-switched core network. IMS-based VoLTE solutions have been adopted as the most popular VoIP solutions. VoLTE has found a usage area with LTE Advanced Pro, also known as 4.5G.

IP multimedia sessions on the IMS support IP multimedia applications. Therefore, VoLTE solutions will replace systems using legacy circuit switching techniques. With this gradual transition, the IP core network has received high-definition video and voice services and high-speed Internet service [9] (Figure 9.10).

While connecting to LTE, the VoLTE client (user device) gets an IP address from the P-CSCF (Proxy Call Session Control Function) server. The P-CSCF server is the entry point for IMS signaling and connects to the VoLTE device (user equipment [UE]) via SIP. The P-CSCF is also responsible for handling security protocols between it and the UE.

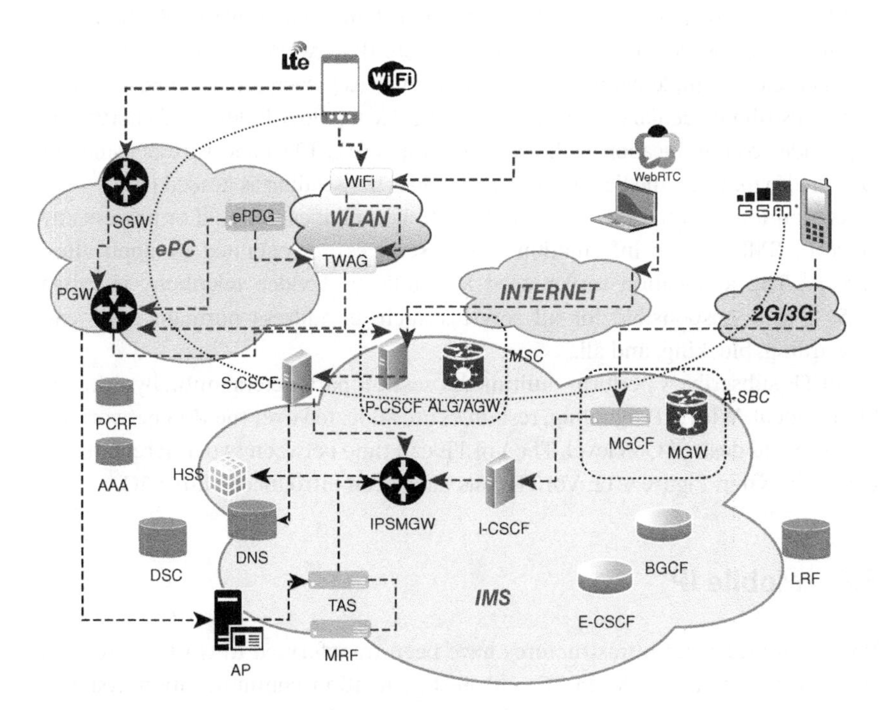

Figure 9.10 VoLTE architecture.

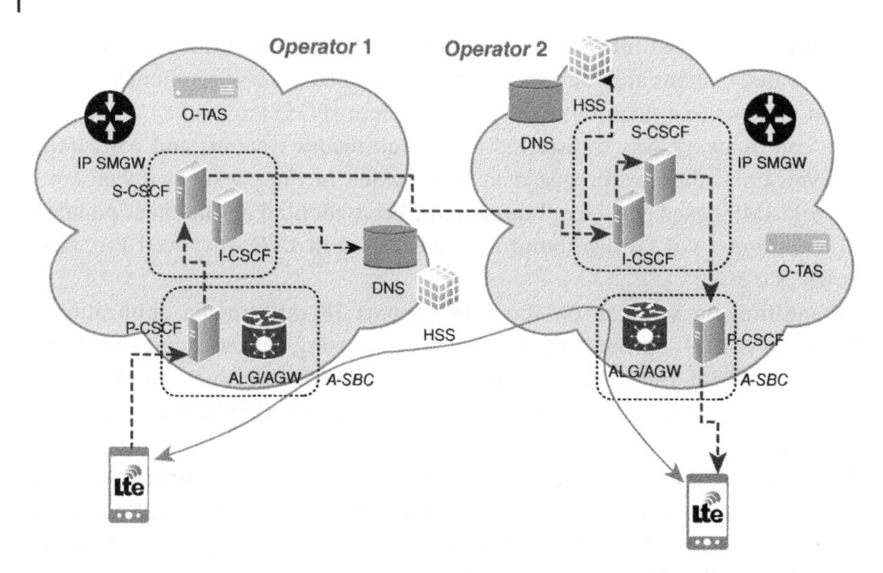

Figure 9.11 VoLTE call.

P-CSCF is usually part of the Access Session Border Controller (A-SBC) system. A-SBC is responsible for connecting two or more IP networks (IPv4 and IPv6, NAT traversal, etc.). It implements security features such as prevention of denial of service (DoS), distributed denial of service (DDoS) attacks, topology hiding, and encryption. It provides communication with access networks (e.g. LTE) and is responsible for QoS. It also manages media services and provides transcoding as needed [10].

As mentioned in the IMS section, the IMS kernel does not hold or process any voice or SMS service information. These services are operated by application servers. The application server used for audio and video telephony is called TAS. TAS is responsible for all services such as address normalization, call forwarding, blocking, and all.

VoLTE subscribers perform multimedia session parameters mutually over the SIP protocol. With SIP signaling, resources are allocated over the IMS network for a secure and desired QoS level. The VoLTE call time between two different operators is shown in Figure 9.11. VoLTE calls have been introduced with 4.5G.

9.6 Mobile IP

Voice and video call infrastructures have been transformed to work entirely on IP-based packet networks in 4G and next-generation communication systems, starting from 3G. In classical usage, devices connected to any network are included in the communication domain by giving a fixed IP. With this IP address and subnet mask, the device's identity and the network where the device is located are determined. In this way, packet (datagram) communication is carried out.

Suppose a mobile device is assigned an IP. When the mobile device switches from its cell to another network, its current IP address will not be recognized in the network where it is logged into. In other words, its data will no longer be routable in the network it passes through, and the network connection will be broken. In this case, the mobile device will have to be reconfigured by the network IP address pool in which it is located, which means an extra configuration load.

To overcome these situations, a mobile IP configuration has been developed. In the mobile IP configuration, two different IP addresses (given by the network) are defined for the mobile device: a fixed home address (home address) and an IP address called the "care-of address." The so-called care-of address is renewed (changed) with each gateway. Mobile IP keeps the same home address of a device unchanged and works without losing connection to the Internet or related network. The network changes the care-of address at each new mobile device location. Figure 9.12 shows the mobile IP topology [11].

Mobile IP provides forwarding of IP datagrams to mobile nodes. The mobile node's home address always identifies the mobile node, regardless of the current port to the Internet or an organization's network. While away from home, the temporary address associates the mobile node with the home address, providing information about the mobile node's current port to the Internet or an organization's network. Mobile IP uses a registration mechanism to register the temporary address with a home agent.

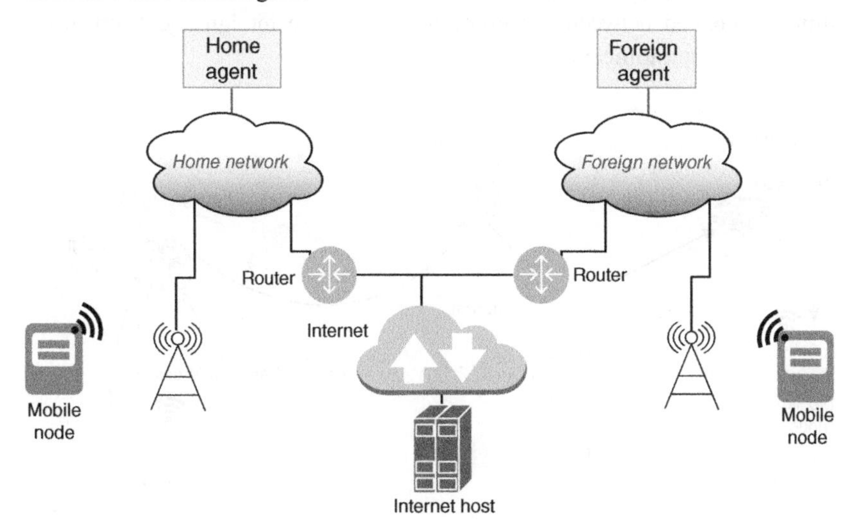

Figure 9.12 Mobile IP topology.

The home agent creates a new IP header containing the temporary address of the mobile node as the destination IP address and forwards the datagrams from the home network to the temporary address. This new header then encapsulates the original IP datagram, causing the mobile node's home address not to affect the encapsulated datagram's routing until it reaches the temporary address. This type of encapsulation is also called tunneling. After getting the temporary address, each datagram is de-encapsulated and delivered to the mobile node.

Three different functional innovations have come here with mobile IP.

- **Mobile Node (MN):** Router that changes port from one network to another
- **Home Agent (HA):** Router that captures packets destined for the mobile device and forwards them via the care-of address. The home agent also keeps the current location information of the mobile device.
- **Foreign Agent (FA):** Router on the other network to which the mobile device is logged in. When the mobile device is registered to the new network, it provides the necessary routing services.

Let us explain the issue with an example. The mobile IP home address identifies the mobile device regardless of the Internet or company network point to which the device is currently connected. When the device leaves the home network, the care-of address is associated with the home address of the mobile device. The home agent creates a new IP header with the care-of address as the destination address of the mobile device and forwards the packets from the home network to the care-of address.

As can be seen in Figure 9.13, when the mobile device enters another network, a tunnel is created between the home agent and the foreign agent, ensuring uninterrupted communication.

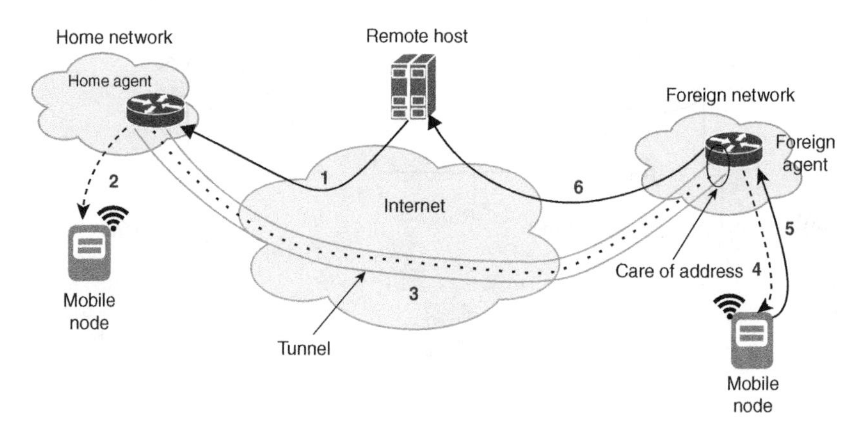

Figure 9.13 How mobile IP works.

9.7 Multiple Access Techniques

This section will explain the access interactions and techniques between the base station (eNodeB) and the mobile user device (UE). Frequency division duplexing-time division duplexing (FDD-TDD) distributed orthogonal frequency division multiple access (OFDMA)-based access is provided in the downlink channel, while single carrier-frequency division multiple access (SC-FDMA)-based access is provided in the uplink channel. The reasons for the development of both methods are given in the following section.

9.7.1 OFDM Access

LTE uses orthogonal frequency division multiplexing (OFDM) method in multiple access. OFDM is also known as multicarrier modulation. It is similar in principle to frequency division multiplexing, but in OFDM, many subcarriers are assigned to a single data source.

We will refer to the OFDM multiplexing method, which we explained in the previous sections, especially in orthogonality, from a conceptual point of view. We will explain how the properties data stream is assigned to subcarriers and its advantages. Let us say we have a data stream at R bps. Let our bandwidth be Nf_b and our center frequency f_0. Let us assume that all the bandwidth is used in this data stream, with each bit time being $1/R$. Alternatively, let us separate the data stream from serial to parallel converter into N substreams (these subchannels are created with 15 kHz gaps). In this case, each subdata stream will be at an R/N bps rate, and let us transmit each one with a different subcarrier (f_b adjacent subcarriers). In this case, the bit duration will be N/R (significantly longer), and mechanisms will be obtained to overcome multipath attenuation [12].

As can be seen in Figure 9.14, the OFDM scheme distributes data over multiple carriers using advanced digital signal processing techniques. There is an orthogonal relationship between the subcarriers. Although the carrier frequencies seem to overlap, discrimination can be made at the frequency at which a subcarrier peaks since the other carriers are at zero value.

Since information is carried by different subcarriers (QPSK, 16QAM, and 64QAM can be used as modulation here), losses due to signal fading affect only the information held on that subcarrier, not the entire information signal. If the data stream is protected with forwarding error-correcting code, losses due to fading can be easily tolerated. Moreover, OFDM also overcomes intersymbol interference (ISI), which has excellent effects, especially at high data rates, in an environment where the signal is scattered in multiple ways (reflection, scattering, etc.). While practicing OFDM, this is achieved by inverse fast Fourier transform (IFFT) and cyclic prefix (CP) operations.

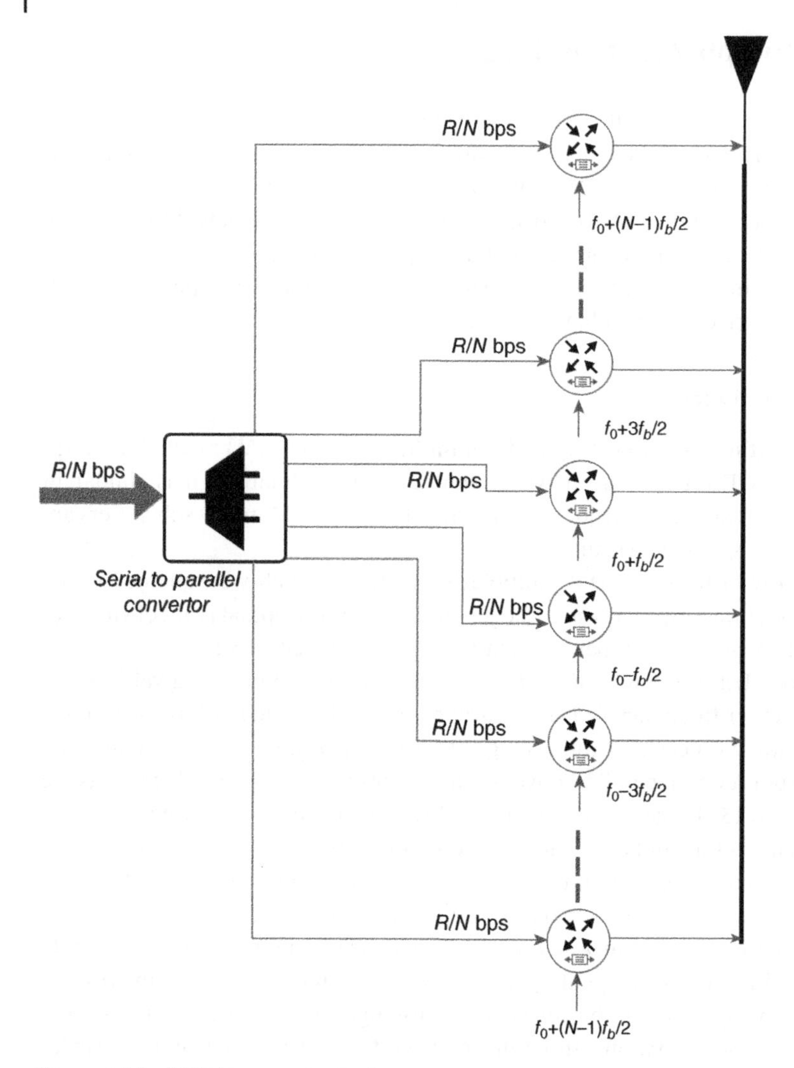

Figure 9.14 OFDMA conceptual schema.

Mobile radio communication is vulnerable to losses due to frequency fading, distance-related attenuation, interaction with other cells, and interference from user devices. A channel-dependent scheduling mechanism has been developed to prevent all these negativities in LTE technology. In this method, time and frequency resources are dynamically shared among users. This way, resource needs that can change quickly in packet-based communication are met dynamically.

Depending on the channel state, the scheduler allocates time-frequency resources to users. Timing decisions are made every 1 ms at the 180 kHz frequency

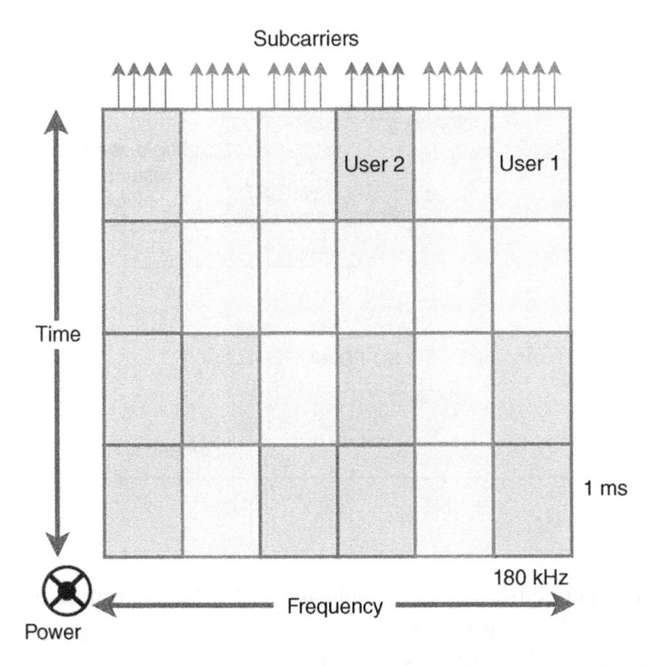

Figure 9.15 Channel-dependent timing in time and frequency domains.

domain granularity. The scheduler performs resource allocation by looking at the Channel State Information (CSI) received from the user device (Figure 9.15).

OFDM multiple access methods can be categorized as OFDM-FDMA, OFDM-TDMA, OFDM-CDMA, and OFDM-SDMA. In practice, mixtures of these techniques are used as an access mechanism [13]. While WiMAX uses OFDM on both channels, LTE uses OFDM on the downlink channel and SC-FDMA on the uplink. In the following section, SC-FDMA technology will be discussed.

9.7.2 Single Carrier-FDMA

The peak-to-average power ratio (PAPR) of a signal's waveform is an important metric. The smaller this value, the more efficiently the signal works the power amplifier. This means that the battery life is extended on the user's side. Although the PAPR value on the base station side is not a critical parameter (due to lack of stations in number, being fed by fixed power lines, etc.), it is an essential parameter for mobile devices in terms of complexity and power consumption.

Due to the high PAPR values, one of the critical problems of the OFDM system, analog-to-digital converter (ADC) and digital-to-analog converter (DAC) structures, becomes more complicated. An OFDM signal has a high PAPR value

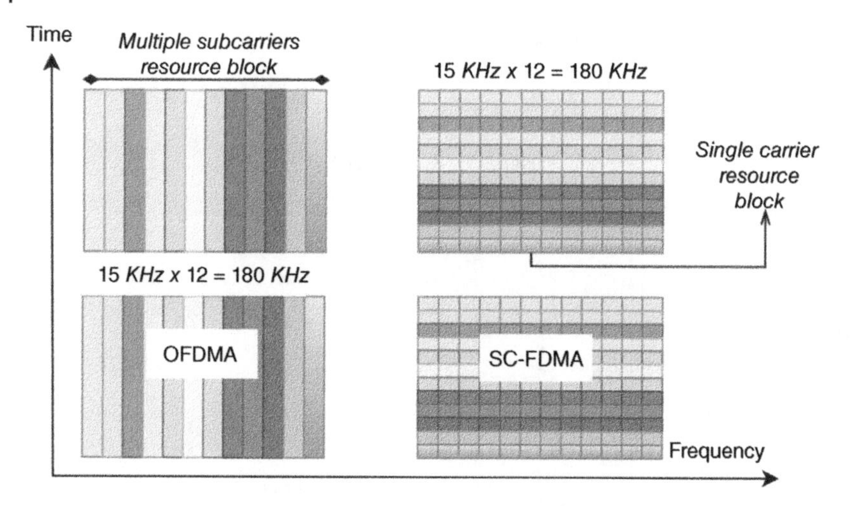

Figure 9.16 OFDMA and SC-FDMA.

because it consists of independently modulated subcarriers. SC-FDMA is used in the LTE uplink (the channel from the device to the eNodeB) to avoid the negative aspects of device complexity and power consumption on the user device.

Although the term single carrier is used here, multi-carrier is used. The specific point here is that all subcarriers in the uplink channel are modulated with the same data (Figure 9.16). Thus, a reliable and flexible connection is established under all conditions.

9.8 Multiple Input-Multiple Output (MIMO) Antenna Systems and SDM Access

Multiple antenna systems, which can be applied at the transmitter and the receiver, are used for spatial multiplexing of the radio channel in wireless communication. The space division multiplexing (SDM) process mentioned in the previous sections is carried out with such advanced antenna systems. With this technology, SDM is also used with multiplexing in time and frequency domains. This way, it aims to increase coverage, capacity, and end user throughput [14].

Directed sector antennas have been used for a long time since the first analog wireless communication systems were installed. These antennas were installed as antenna columns in the early days to expand the coverage area. In addition to these antenna systems, which are still used today, especially with the introduction of LTE technology into our lives with 4G, dynamic bandwidth, and high-efficiency radio communication with directional beamforming features have become available.

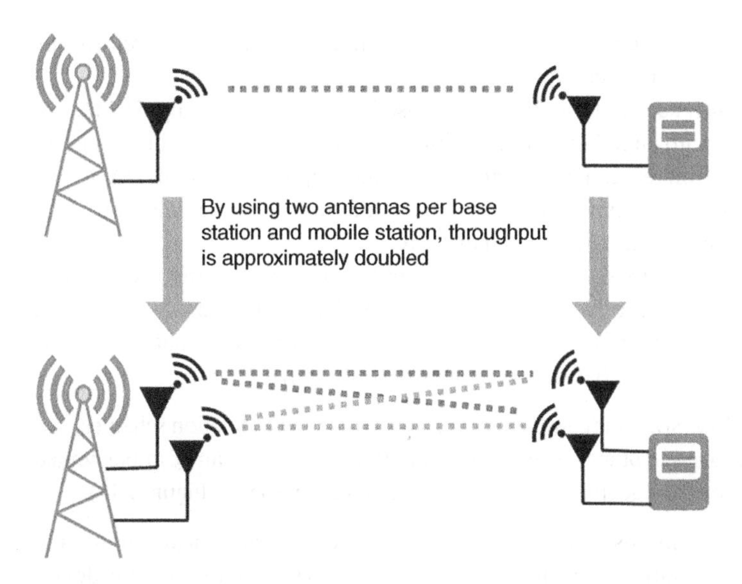

By using two antennas per base
station and mobile station, throughput
is approximately doubled

Figure 9.17 MIMO performance.

Since parallel data transmission is possible with the antenna arrays used in both the receiver and the transmitter, the data transmission rate is increased roughly by the number of antennas (Figure 9.17).

The quality of a wireless channel is expressed by three parameters: transmission rate, transmission distance (coverage), and reliability. The reliability parameter should be considered the lossless transmission of the transmitted information (or packet) to the other party. While the transmission distance can be increased by reducing the transmission rate, the transmission reliability can be increased by reducing both the speed and the distance. In addition, these three parameters can be increased simultaneously with the introduction of MIMO systems.

With SDMA access used in MIMO, the same bandwidth can be given to multiple users in different geographical locations. The advantages achieved with SDMA are given as follows [15]:

- **Distance expansion:** With the help of antenna arrays, signals can be transmitted over longer distances by beaming on individual antenna systems. Cell numbers covering a geographic area can be significantly reduced in an SDMA system. For example, a 10-element array provides a gain of 10, doubling the cell radius and thus quadrupling the coverage.
- **Multipath mitigation:** During transmission, electromagnetic waves can reach the target receivers in many ways due to physical reasons, reflection, scattering, etc., along the way. This causes losses, and therefore speed drops to tolerate the

failures. SDMA systems utilizing MIMO architecture significantly reduce the adverse effects of multipath propagation.

- **Capacity increase:** With a limited increase in system complexity, SDMA can be integrated into already running multiple access systems, providing a significant capacity increase. For example, with the application of SDMA to the traditional TDMA system, two or more users can use the same time slot. Thus, the system capacity can be doubled or more.
- **Interference suppression:** Interference with other systems and user interactions can be significantly reduced with SDMA. This is done using the unique and user-specific channel impulse responses (CIR – the value that defines the period properties of a channel) of the desired user [16].
- **Compatibility:** SDMA is compatible with most existing modulation schemes, carrier frequencies, and other specifications. Additionally, SDMA can be implemented using different array geometries and antenna types than MIMO (Figure 9.18).

As seen in the figure, users with a single antenna can communicate simultaneously with a base station with an array of antennas. As seen here, a mobile device with an antenna array will thus have a much higher bandwidth transmission medium. The MIMO channel representation is given in Figure 9.19 [17].

9.9 Voice over WiFi (VoWiFi)

With the LTE technology starting to use the 5 GHz bands, LTE/WiFi aggregation has also been implemented. We can say that the structure where smartphones can exchange data from both the cellular network and the WiFi network is

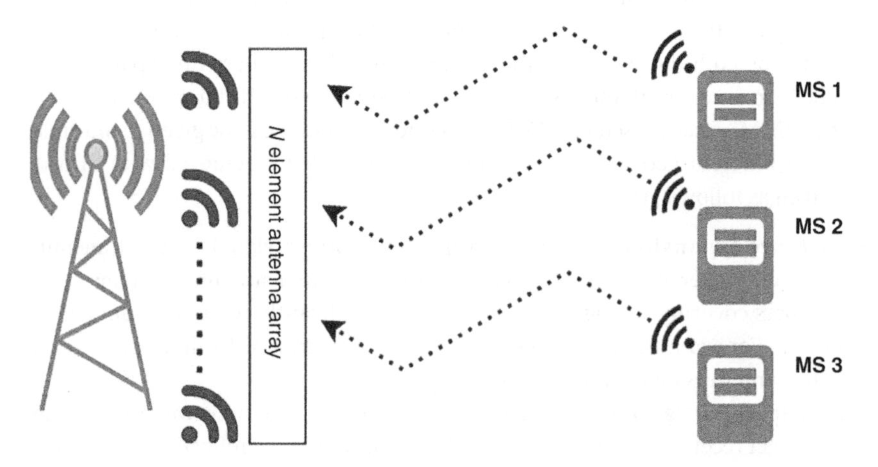

Figure 9.18 Communication of single antenna devices with the MIMO base station.

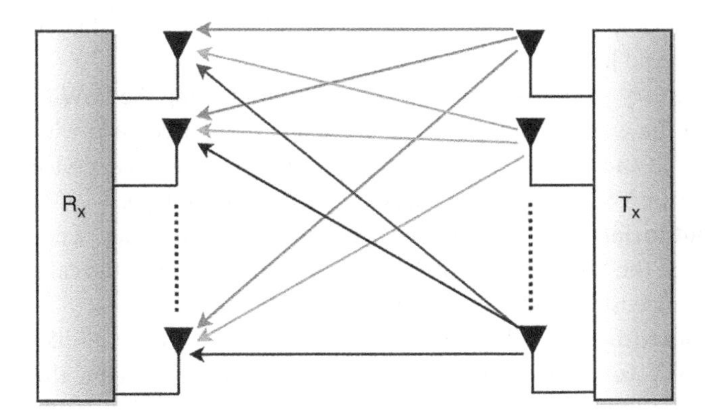

Figure 9.19 MIMO structure.

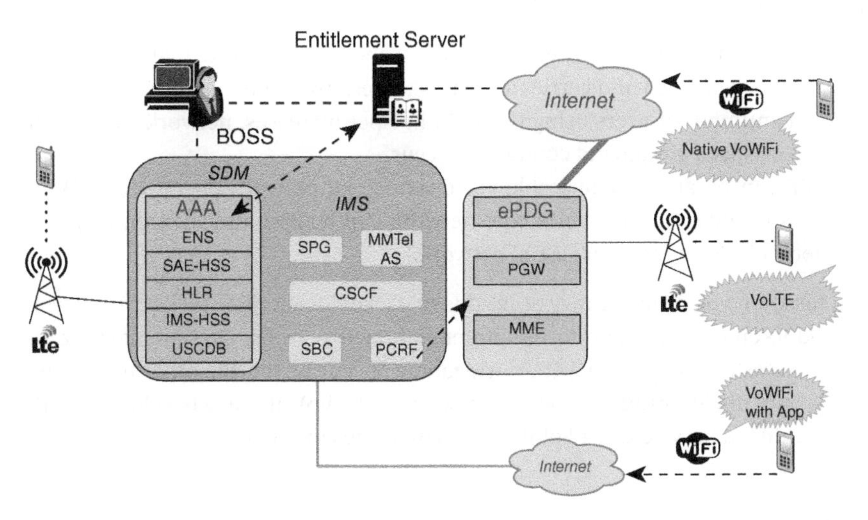

Figure 9.20 VoWiFi infrastructure.

simultaneously suitable for installing small cells. With VoWiFi, apart from the operator network, speech and other multimedia services can be accessed via this interface when there is a WiFi connection. From this point of view, we can say that VoWiFi and IMS services have been expanded to new access using existing infrastructure (Figure 9.20).

Today, VoWiFi is being developed as an integral part of VoIP development. VoWiFi advantages are given as follows [18]:

- **WiFi is a long-lasting ecosystem:** Today, smartphones in the hands of nearly all users consume almost all of their data over WiFi connections. Considering the ease of installation of WiFi hotspots, we can say that the use of VoWiFi will become widespread and long lasting.

- **It is a complementary structure for LTE services:** Integrating the VoWiFi infrastructure into the existing system has also become more accessible with installing IMS. VoLTE and VoWiFi can work together on EPC. The voice service availability outside the VoLTE coverage area is a significant advantage.
- **New business potentials:** It also solves the access problems that may occur indoors by extending the VoWiFi, LTE, or 2G/3G coverage area. Since the calls will be local, roaming costs will also decrease.
- **New service opportunities:** Mobile operators require WiFi services, unlike cable-connected OTT (over-the-top) devices (e.g. IPTV). With VoWiFi, the services of OTT clients are now given to smartphones.

In addition to the advantages mentioned earlier, VoWiFi technology may have problems in providing QoS levels:

- WiFi operates on unlicensed frequencies; therefore, a particular bandwidth cannot be allocated and carries a high risk of electromagnetic interference.
- As the number of users using the WiFi hotspot increases, network congestion and low-speed broadband connection occur.
- 3GPP can be an unmanageable voice service by network operators, as WiFi access to the operator's voice core network can be accomplished through the Internet Service Provider (ISP's) network.

Decreases in voice quality may occur due to packet losses, jitter (delays in delays), or end-to-end latency due to the aforementioned reasons. QoS architecture in WiFi access is based on packet prioritization. In VoLTE, on the other hand, QoS levels are made through resource reservations. VoWiFi needs additional IP Differentiated Service Code Point (DSCP) markings in the transmission layer.

References

1 Sauter, M. (2017). *From GSM to LTE-Advanced Pro and 5G*. Wiley.
2 Curwen, P. and Whalley, J. (2021). *Understanding 5G Mobile Networks: A Multidisciplinary Primer*. Emerald Group Publishing.
3 3GPP (2022). Portal > Home. *3GPP*. https://portal.3gpp.org/ (accessed 15 February 2023).
4 Dahlman, E., Parkvall, S., and Skold, J. (2016). *4G, LTE-Advanced Pro and the Road to 5G*. Academic Press.
5 Condoluci, M. and Mahmoodi, T. (2018). Softwarization and virtualization in 5G mobile networks: benefits, trends and challenges. *Computer Networks* 146: 65–64.

6 GL (2022). Enhances end-to-end wireless network LAB solutions. *GL Communications*. https://www.gl.com/newsletter/enhanced-end-to-end-wireless-network-lab-solutions.html (accessed 15 February 2023).
7 Poikselkä, M. and Mayer, G. (2013). *The IMS*. Wiley.
8 Akram R. (2022). IP Multimedia Subsystem (IMS) overview. *Rauf's Knowledge Portal*. https://raufakram.wordpress.com/2013/12/10/ip-multimedia-subsystem-ims-overview/ (accessed 15 February 2023).
9 Elnashar, A. and El-saidny, M.A. (2018). *Practical Guide to LTE-A, VoLTE and IoT*. Wiley.
10 Real Time Communication (2022). VoLTE in IMS. *Real Time Communication*. https://realtimecommunication.wordpress.com/2015/03/06/volte-in-ims/ (accessed 15 February 2023).
11 DOCS (2022). Overview of mobile IP. Mobile IP Administration Guide. Oracle Inc. https://docs.oracle.com/cd/E19455-01/806-7600/6jgfbep0v/index.html (accessed 15 February 2023).
12 Beard, C., Stallings, W., and Tahiliani, M.P. (2015). *Wireless Communication Networks and Systems, Global Edition*. Pearson.
13 Rohling, H. (2011). *OFDM*. Springer Science & Business Media.
14 Asplund, H., Astely, D., von Butovitsch, P. et al. (ed.) (2020). Advanced antenna system in network deployments. In: *Advanced Antenna Systems for 5G Network Deployments*, 639–676. Academic.
15 Cooper, M. and Goldburg, M. (1996). Intelligent Antennas: Spatial Division Multiple Access. Annual Review of Communications, 1996.
16 Coll, F.J. (2014). Channel characterization and wireless communication performance in industrial environments. Doctoral dissertation. KTH Royal Institute of Technology.
17 Sibille, A., Oestges, C., and Zanella, A. (2010). *MIMO*. Academic Press.
18 Elnashar, A. and El-saidny, M.A. (2018). *Practical Guide to LTE-A, VoLTE and IoT*. Wiley.

10

5G Systems

10.1 Introduction

Significantly, the rapid increase in mobile data traffic based on video streams, the increasing number of connections due to multiple devices of each user, the connection need of billions of devices due to IoT applications, and the need for efficient energy management due to this intense connection and traffic, the reduction of mobile operators' operating costs and headlines such as new revenue opportunities for operators due to developing applications have been the main driving force in the transition to 5G.

We are witnessing that the amount of traffic in mobile networks progresses exponentially. Between 2010 and 2020, this traffic increased 1000 times [1]. We can say that this data "hunger" is mainly due to the widespread use of smartphones and the increase in multimedia applications running on these phones. In addition to ultra-high-definition mobile TV, the transition to 5G communication systems has become a growing need over time due to the increasing use of many 3D applications (Figure 10.1).

Let us explain the "5G Flower" 5G key business indicators put forward by Chinese Mobile and accepted by the ITU [2] (Figure 10.2):

Petals (six performance indicators of 5G):

- 100× (0.1–1 Gbps) user speed compared to 4G
- Much higher connection density (1 million/km^2)
- Reduced interface latency (one-fifth of 4G-1 ms)
- 4× accelerated mobility of 4G (500+ km/h)
- About 20 times the peak rate of 4G (10–20 Gbps)
- High flow density (10–100 Tbps/km^2)

Evolution of Wireless Communication Ecosystems, First Edition. Suat Seçgin.

Published 2023 by John Wiley & Sons, Inc.

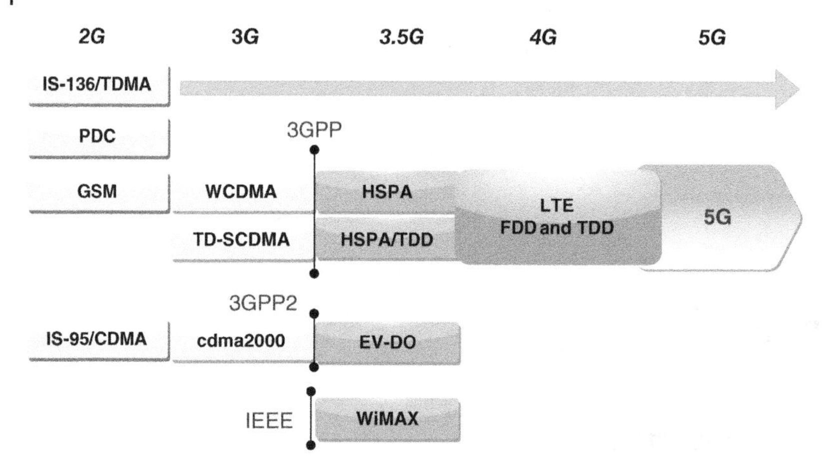

Figure 10.1 Evolution to 5G.

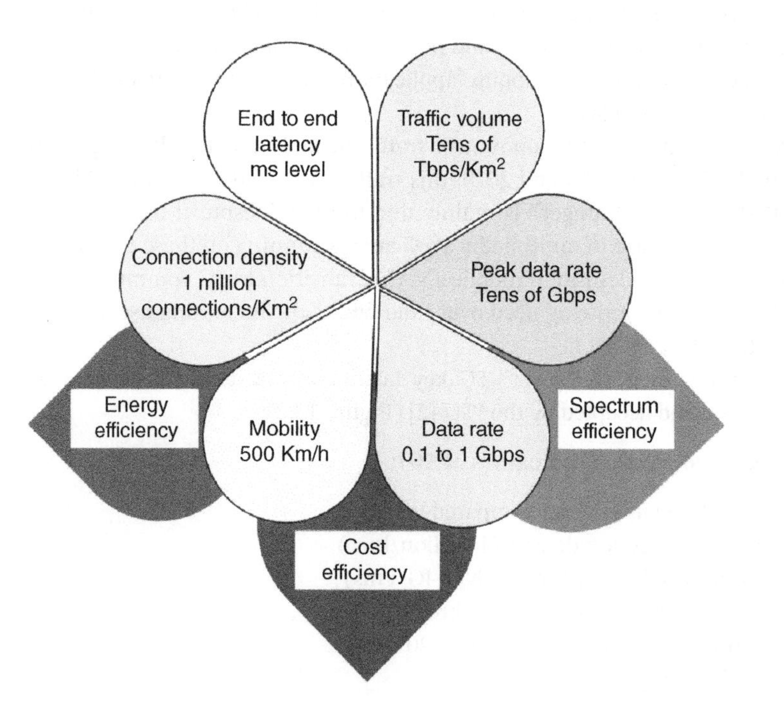

Figure 10.2 5G flower.

Green leaves (three separate productivity indicators):

- Energy efficiency
- Cost efficiency
- Spectrum efficiency

International Telecommunication Union (ITU) has defined International Mobile Telecommunication 2000 (IMT-2000) for 3G requirements, IMT-Advanced standard for 4G, and IMT 2020 for 5G requirements. ITU Radio Communication Sector (ITU-R) has foreseen three basic usage scenarios in the ITU-R M.2083 document published for IMT-2020 and beyond systems: enhanced mobile broadband (eMBB), massive machine-type communication (mMTC), and ultrareliable and low-latency communications (uRLLC). These scenarios and corresponding applications are given in Figure 10.3 [3].

- **Enhanced Mobile Broadband (eMBB):** Dealing with significantly increasing data rates, supporting hotspot scenarios that will provide high user density and very high traffic capacity, and providing high mobility opportunities and uninterrupted coverage.
- **Massive Machine-Type Communication (mMTC):** Connecting large numbers of devices to IoT systems that require low power and low data rates.
- **Ultrareliable and Low-Latency Communication (uRLLC):** Provides security and mission-critical applications infrastructure (Figure 10.4).

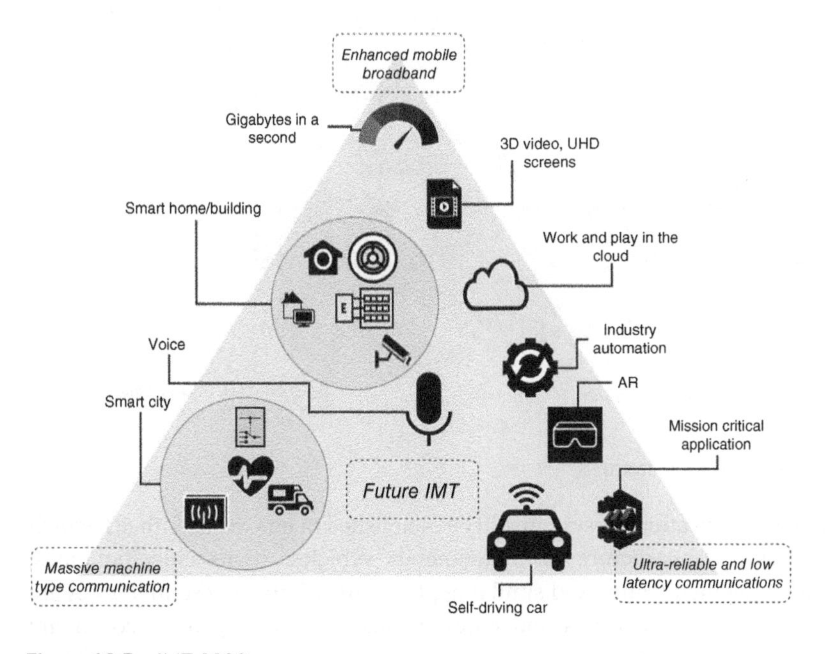

Figure 10.3 IMT-2020 use cases.

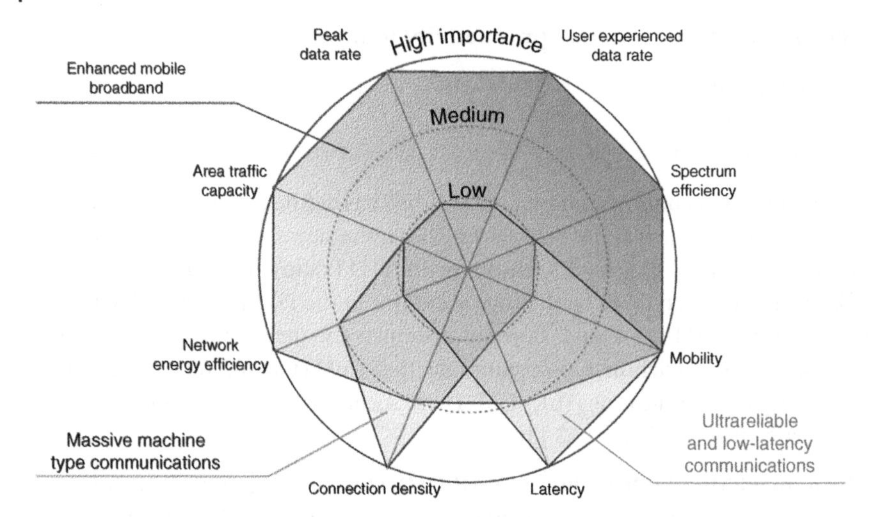

Figure 10.4 5G network capacity [4].

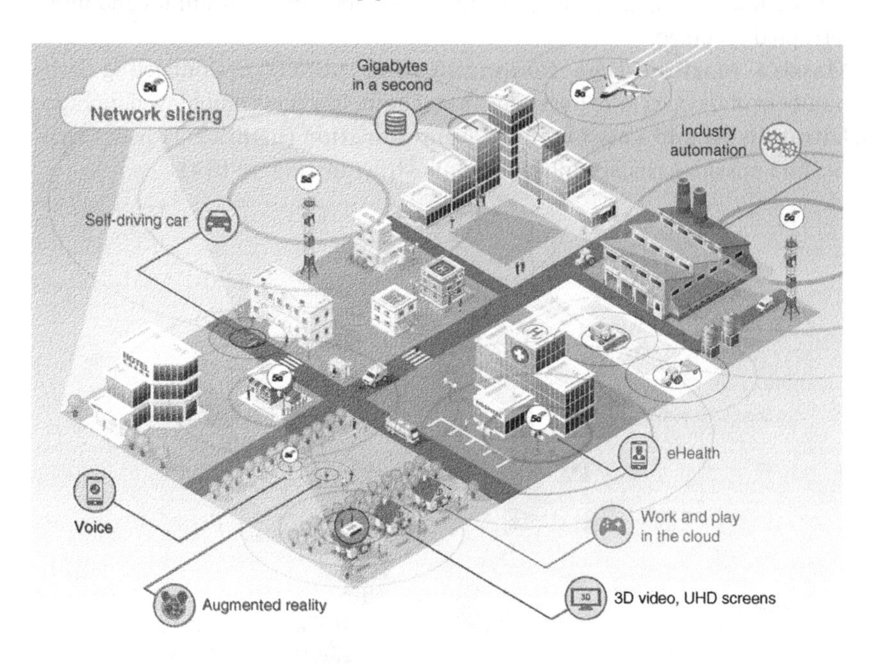

Figure 10.5 5G ecosystem.

A vast communication ecosystem will be established in our world with 5G, which will provide infrastructure for autonomous vehicles, smart cities/buildings, virtual/augmented reality, and similar applications, where almost all devices are designed to meet the needs and requirements we have explained so far are connected (Figure 10.5) [5].

10.2 5G Cell Structure

Due to the increasing number of users and the increasing data traffic of these users, a "small cell" structure has been developed for high-bandwidth usage and error-free handoff in densely populated places such as shopping centers, sports centers, airports, train stations, with fifth-generation wireless communication systems. Small cells are low-power, short-range wireless transmission systems that cover small geographic areas. A small cell is a small base station that divides a cell site/area into much smaller segments [6]. The cell structure, starting from the globe to the femtocell structure, is given in Figure 10.6.

The concept of small cell is used as a collective term encompassing picocell, microcell, and femtocell structures. These structures are briefly described as follows:

- **Femtocell:** They are small base stations that extend the mobile network for home users. Femtocell is a good solution when the signals from the operator are weak. We can use the term home base station for these structures. Femtocell network connections are made via Digital Subscriber Line (DSL), fiber, or cable Internet (Figure 10.7).
- **Picocell:** Another category of small cells, picocells, is a solution developed for enterprise applications with extended network coverage and high data transfer capacity (throughput). The connection scheme is similar to the femtocell.
- **Microcell:** It is a cell structure designed to serve more users than femtocell and picocell. Thanks to their high transmission power, they have wider cell diameters, and with these aspects, they provide a suitable infrastructure for smart cities, intelligent transportation, etc., applications (Table 10.1).

As for the answer to the question of why 5G has such a complex cell infrastructure, we can say that this infrastructure should be established for three scenarios for 5G, which we mentioned in the previous section. The need for connection

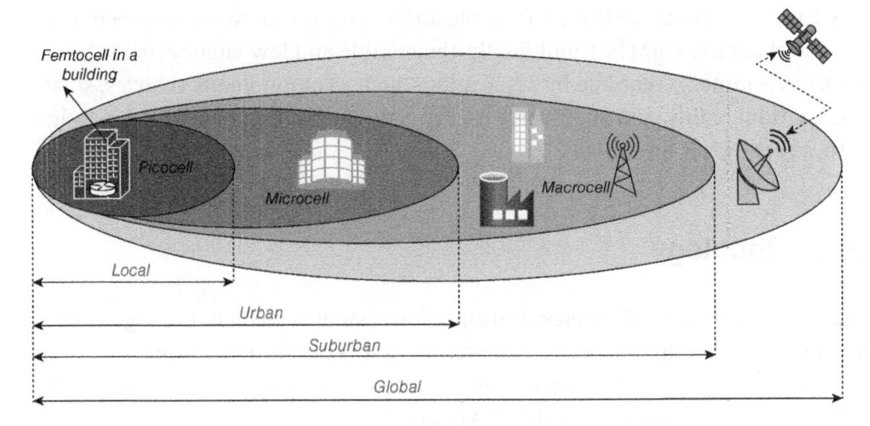

Figure 10.6 5G cell structure.

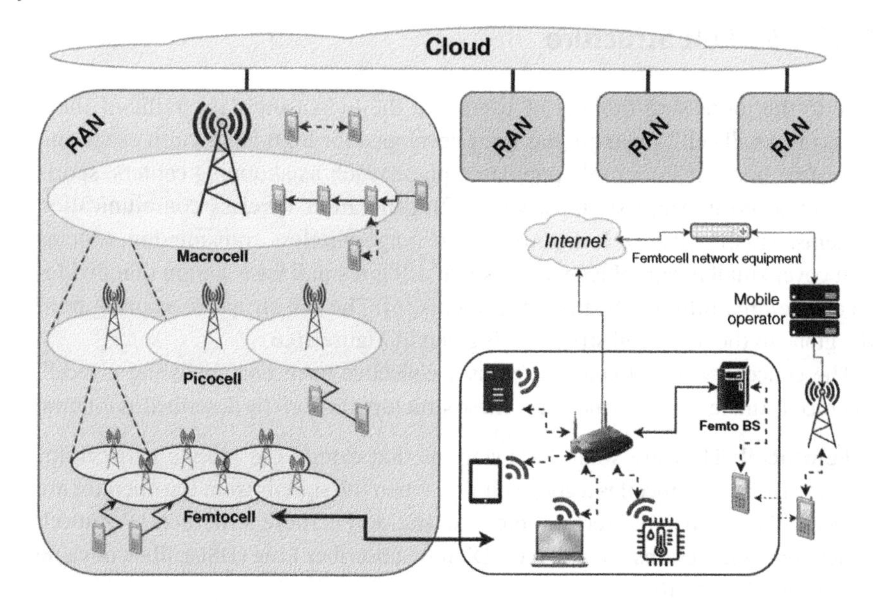

Figure 10.7 5G cell structure (different view).

Table 10.1 4G metrics of cells.

Cell type	Output power (W)	Cell radius (km)	Number of users	Location
Femtocell	*0.001–0.25*	*0.10–0.1*	*1–30*	*Indoor*
Picocell	*0.25–1*	*0.1–0.2*	*30–100*	*Indoor/Outdoor*
Microcell	*1–10*	*0.2–2*	*1000–2000*	*Indoor/Outdoor*
Macrocell	*10–50*	*8–30*	*>2000*	*Outdoor*

speed up to gigabits (eMBB), IoT applications and the network connection of billions of devices (mMTC), and finally, the reliable and low-latency (urLLC) network environments required by applications such as autonomous vehicles, smart cities, virtual reality, and augmented reality is the motivation points in the design of these cells [7] (Figure 10.8).

10.3 Topology

The access layer of a 5G wireless transmission system is given in the Figure 10.9. As can be seen, all hybrid access systems are first taken from the front haul, then collected in the middle layer, and delivered to the core 5G network via the Multiprotocol Label Switching (MPLS) network.

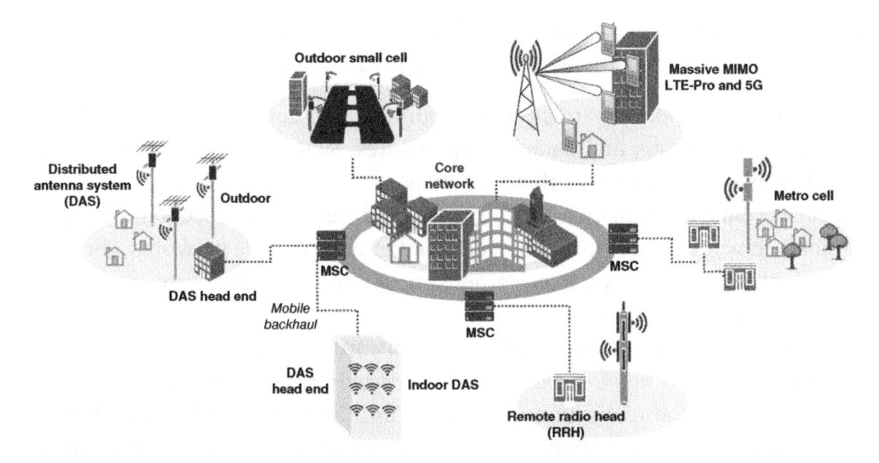

Figure 10.8 5G hybrid network.

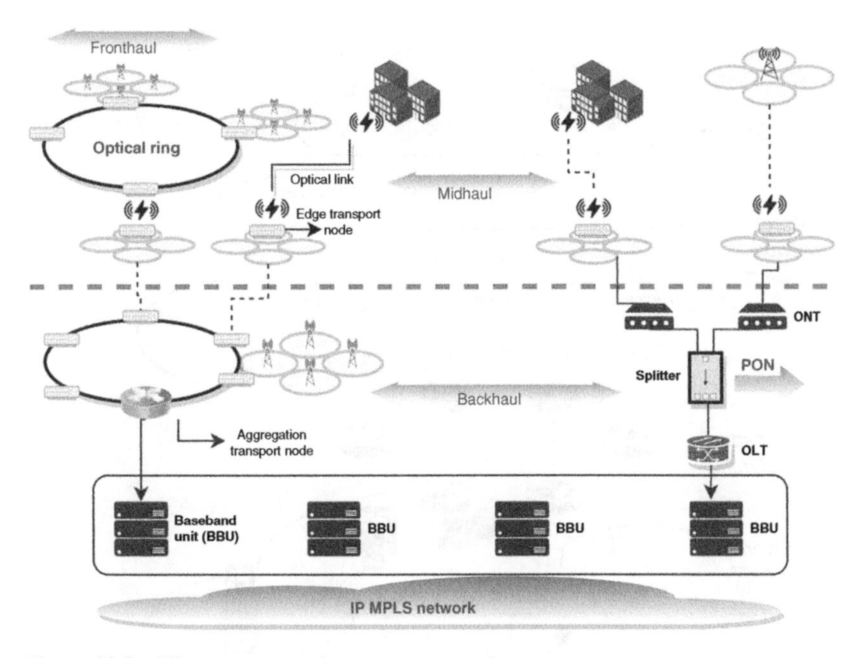

Figure 10.9 5G access network.

The most crucial change in 5G access networks has been the removal of the Radio Network Controller (RNC) layer to minimize latency and the connection of base stations directly to the core network via baseband units.

Compared to 4G, there have been revolutionary changes in architecture, platform, function, protocol, and other areas in the 5G core network to meet the network customization, service, high capacity, high performance, and low-cost headings that came with 5G (Figure 10.10). We can examine these changes from four aspects [8]:

- **IT based:** Evolution toward cloud computing-based developments, functional software, computing, and data parsing features. Switching from previous network devices to cloud networking function.
- **Internet based:** Transition from a fixed rigid network connection structure among network devices to dynamically configurable flexible networks. Service-based structure and new core network protocol architecture based on http/2.0 Internet protocol.
- **Minimal based:** Minimal design for data forwarding/routing and user access.
- **Service based:** 5G network design is for vertical industries. At the center of its technology lies the realization of Network as a Service (NaaS). Through network slicing, edge computing, low-latency connection, etc., the transformation of the network from universal service to personalized and customized service has been realized.

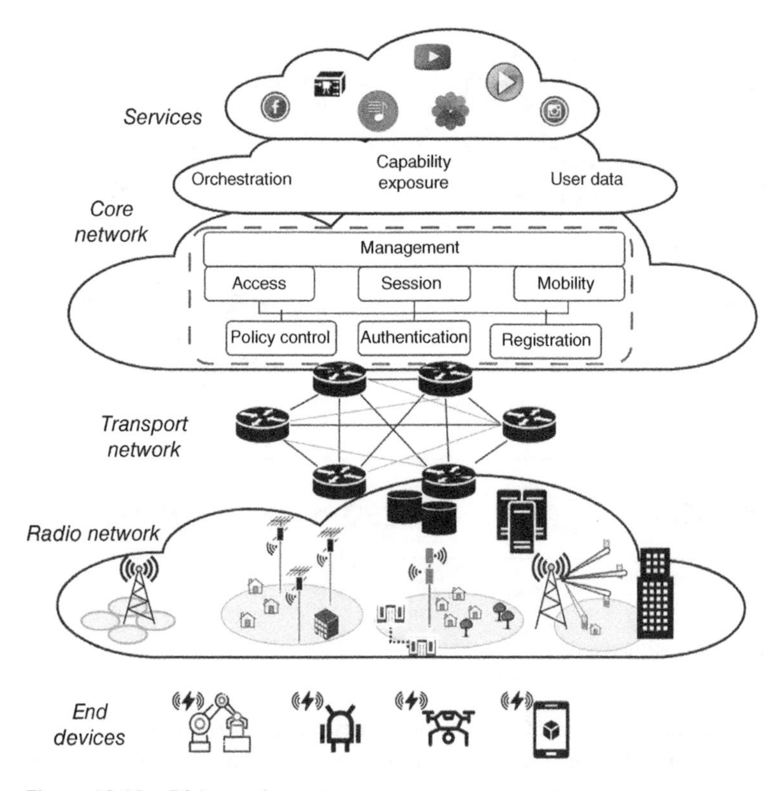

Figure 10.10 5G layered structure.

The architecture will be examined from two different perspectives:

- **Service-based architecture:** The 5G core network has evolved from a closed box to a service-based network and opened to the edges. On the other hand, service-based architecture provides flexible orchestration, parsing, openness, etc. It has unique advantages over traditional network architecture, such that it is an essential tool to rapidly meet the needs of vertical industries in the 5G era. The service-based architecture breaks up complex "single network elements" into modular services to improve network agility. This means that each network function consists of many services. On the other hand, such an architecture provides an agile network structure with small service modules shaped according to needs.
- **Software-based architecture:** What we can call a software-based architecture, or software-defined network (SDN), means the separation of software running on the network and algorithms running on the network control plane. In other words, the control plane where the network management is provided and the user plane where the user data flows are separated. With SDN, it is ensured that the network devices are managed by software-based control or application programming interfaces (APIs), regardless of the manufacturer. SDN can create and control a virtual network or traditional hardware via software. Network virtualization allows organizations to segment different virtual networks within a single physical network or connects devices in various physical networks to create a single virtual network. Thus, the functions of SDN, routing, and control of data packets through a central server are fulfilled. With this approach, high-speed and flexible network device management, customizable network infrastructure, and a strong level of security are achieved. This way, hardware-based traditional network management has evolved into software-based network management.
- **Network function virtualization (NFV):** We can define network function virtualization (NFV) as separating software and hardware layers of communication networks. We can transparently manage functions such as routing, firewall, and load balancing through virtual servers created by dividing a network device's hardware and software functions over an available server and virtualization technology.

The 5G network has the capability of integrated/converged management of the entire network and the whole region (Figure 10.11). The complete lifecycle management of the network is carried out flexibly. In addition, network connections and routing operations between and within data centers are flexibly configured. Thanks to the innovations described, internally open and developed architecture and customer-oriented services can be established quickly.

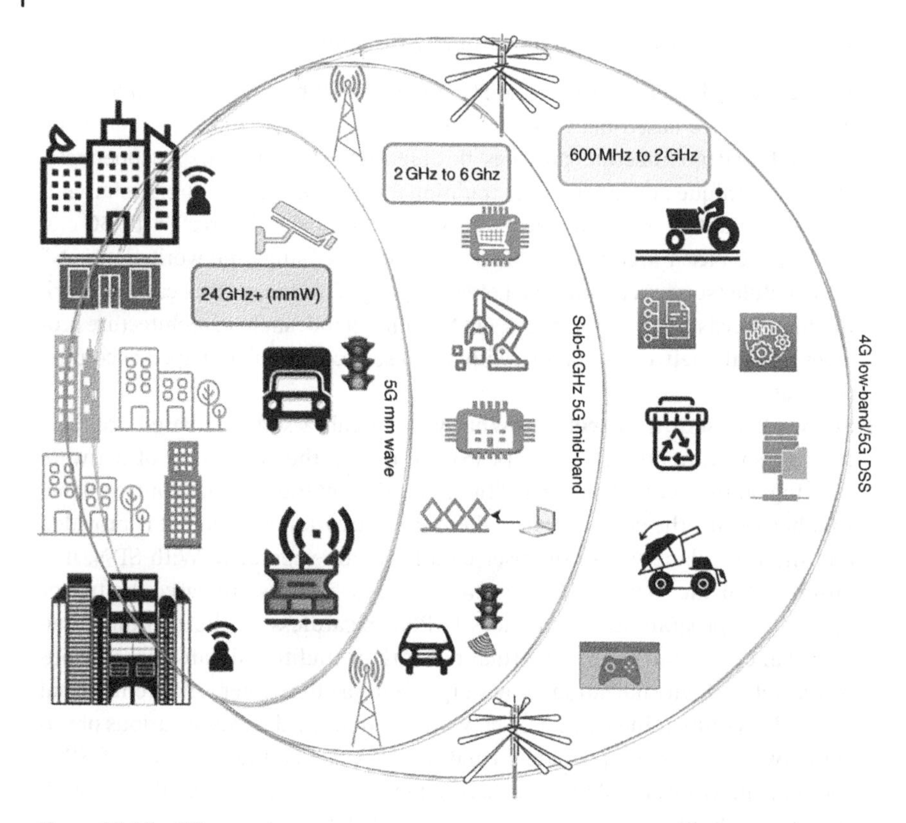

Figure 10.11 5G ecosystem.

10.4 Millimeter Wave

Signals in the 30 and 3000 GHz frequency range are called millimeter Wave (mmWave). mmWave requires more transmit power and is weak to attenuation caused by the atmosphere and physical barriers. It uses a massive MIMO technique to deal with this limitation (Figure 10.12).

Because mmWave bands are at high frequencies, they provide huge bandwidths (hence high speeds). As can be seen from the Figure 10.13, many technologies (GPS, WiFi, 3G, 4G, WiMax, L-band satellite, S and C bands, etc.) operate in the 1–6 GHz band. However, the 30–300 GHz frequency band is used much less. In this sense, it can be considered a new field. For this reason, 24–100 GHz is suggested as the frequency range for 5G.

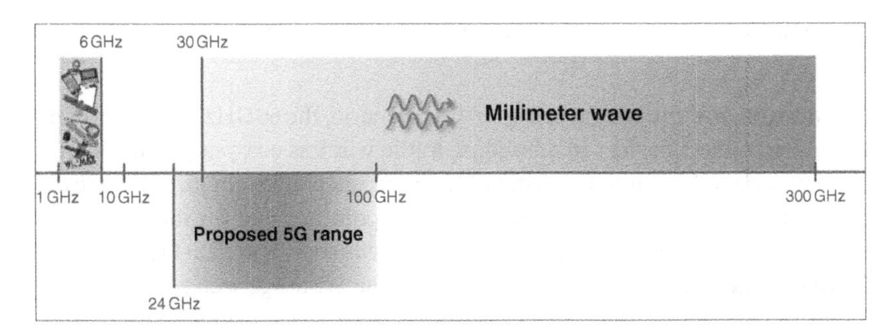

Figure 10.12 mmWave spectrum.

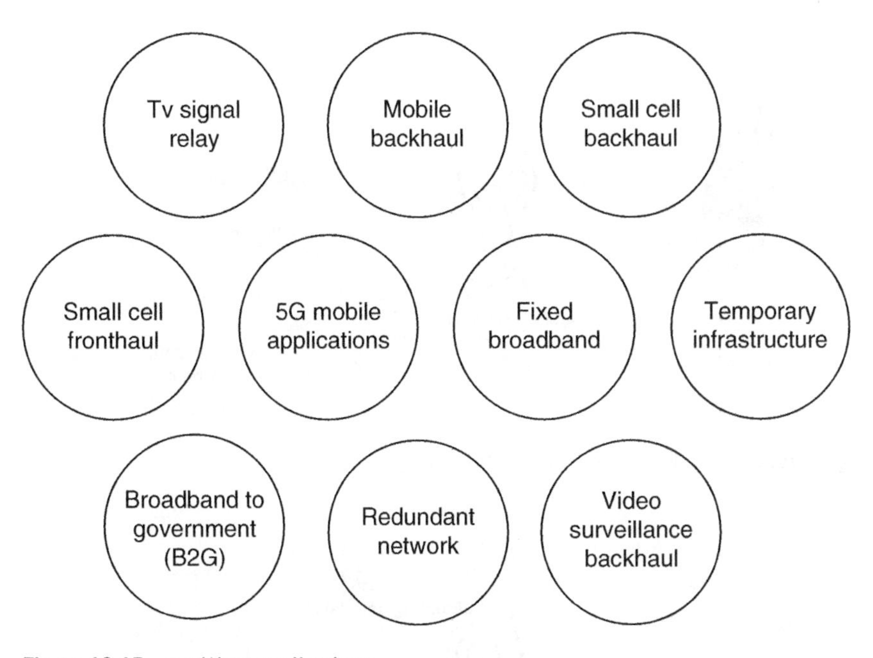

Figure 10.13 mmWave applications.

Although it requires attenuation and high power, mmWave has advantages such as being in a new and less used band, high-frequency waves having much more data carrying capacity than low-frequency waves, and supporting massive MIMO antenna structures. In the Figure 10.13, mmWave application and usage areas visualized by ETSI are given.

The mmWave communication scenarios introduced with 5G are given as follows:

- **Residential and indoor layering:** In this scenario, the 60 GHz frequency band is used to connect devices in a building to the wireless ecosystem. The 60 GHz frequency is used because the distance to the access point is small, and there are few obstacles. The connection of the access point located on the top of the building to the base station is provided with a frequency of 4 GHz (Figure 10.14).
- **Device-to-device communication:** One of the 5G usage scenarios is device-to-device communication. The devices directly exchange data with each other with the orchestration of the microcell base station. 4 GHz links provide the connection to the core backhaul (Figure 10.15).

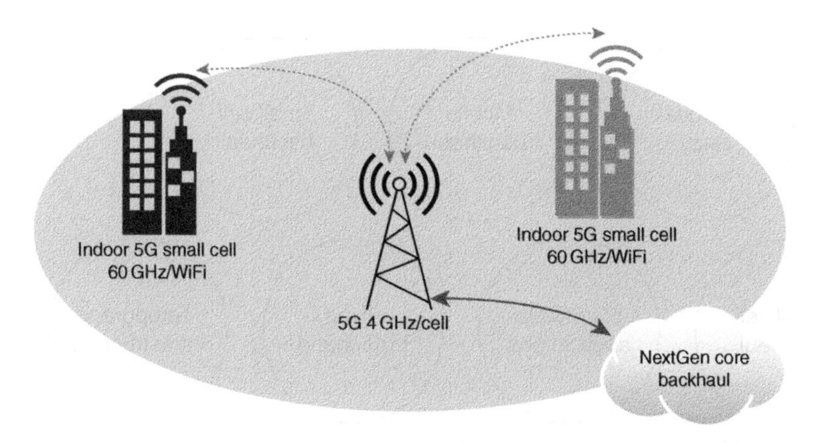

Figure 10.14 Residential and indoor 5G layering.

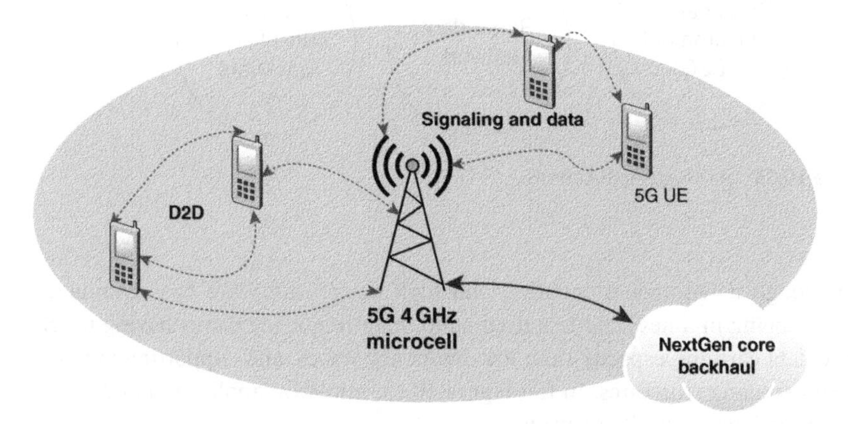

Figure 10.15 Forms of the device-to-device communication.

- **Dense small cells:** Cell structures operating at 30 GHz frequency are managed with a standard base station structure in a hierarchical control order. Starting from the femtocell, there is a hierarchical structure reaching the global cell structure at the macro level (Figure 10.16).
- **Device density:** Millions of devices can be connected per square kilometer, as the frequency in a wide range is flexible. While 60 GHz is used in small indoor cells, the range from 4 to 30 GHz is used in microcells with intensive IoT-based connections. Depending on the type of access and the need, the frequency bands used also change (Figure 10.17).
- **Dual connectivity:** While the mobile device connected to the system it receives control commands over the macro cell, and can perform data communication over the small cell base station. Synchronization establishes an inter-site transport link between the anchor and the small cell eNodeB (Figure 10.18).
- **Massive MIMO base station:** To realize the dense connection scenarios we have described, base stations with tens of antenna series are installed in macro cell centers. This structure will be further developed with 6G, and holographic MIMO applications will be used in spatial division multiplexing techniques (Figure 10.19).

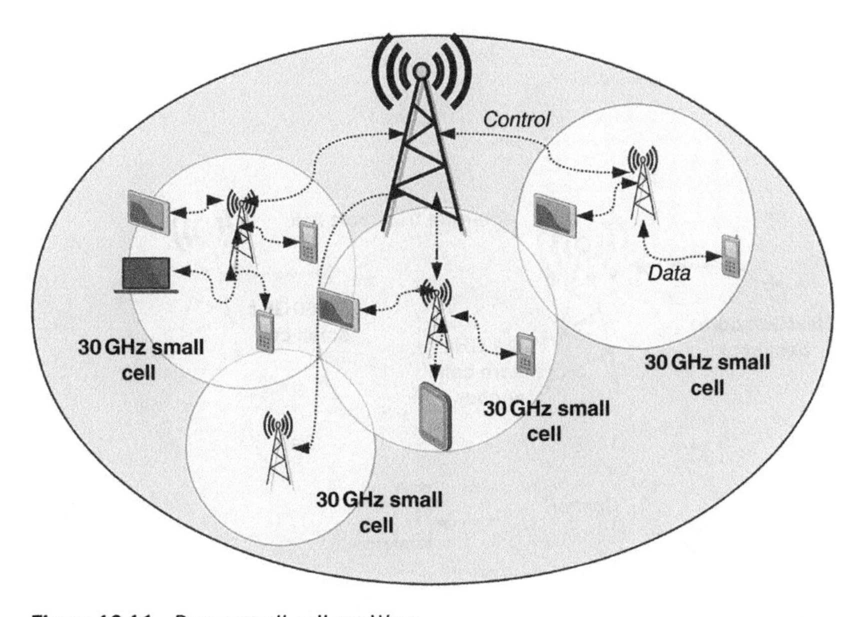

Figure 10.16 Dense small-cell mmWave.

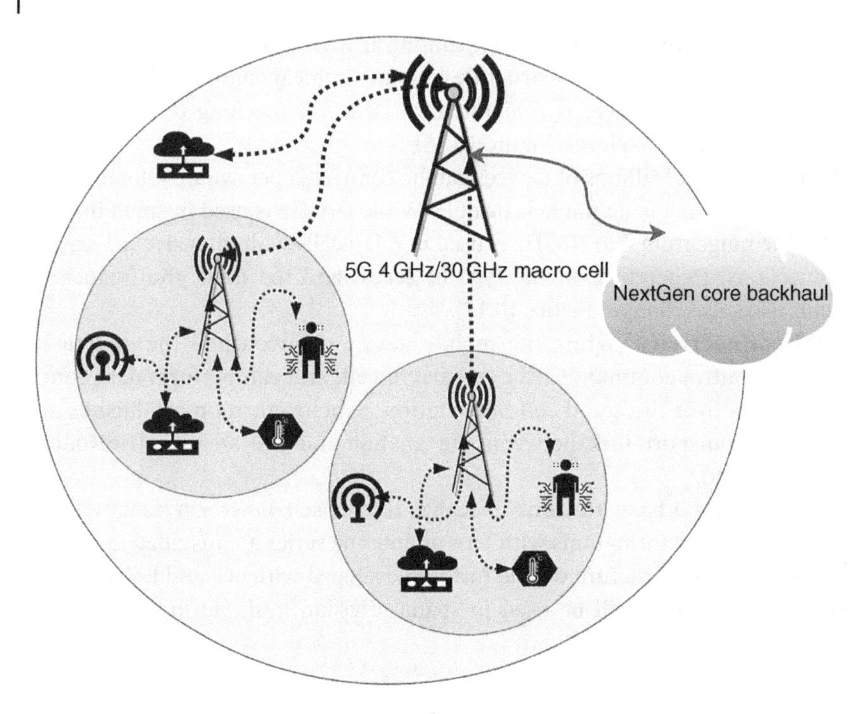

Figure 10.17 One million IoT devices/km^2.

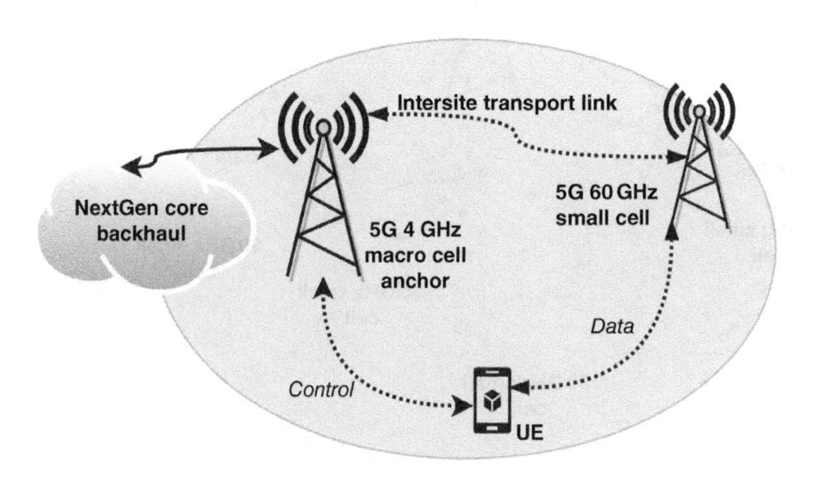

Figure 10.18 Dual connectivity.

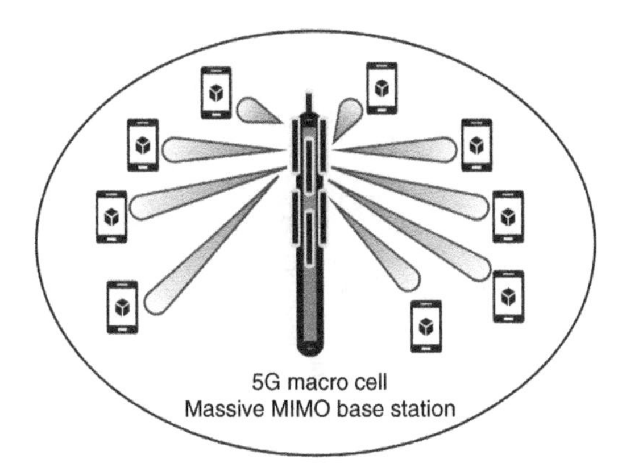

Figure 10.19 Massive MIMO with beamforming.

The holistic view of the 5G ecosystem, which consists of the subcomponents described so far, is given in Figure 10.20. It is hierarchically positioned in mmWave frequency bands with all related systems. The structures described so far are connected to the next generation core backhauls and form the 5G communication ecosystem covering the world.

10.5 Network Slicing

We have mentioned in the previous sections that 5G wireless communication networks are built on the provision of three services: massive machine-type communication (mMTC), ultrareliable low-latency communication (uRLLC), and eMBB (Figure 10.21).

Wireless communication infrastructures are evolving toward an ecosystem where humans and machines are always connected and interact with multiple devices simultaneously. To provide this densely connected infrastructure, next-generation wireless networks must be flexible, scalable, and capable of combining various architectures and standards. In addition, these network infrastructures are expected to provide suitable solutions for different traffic types and heterogeneous network types with basic requirements such as efficiency, several connected devices, latency, and reliability [9].

While using high-density devices that communicate with each other with mMTC, features such as high battery life and low cost must be considered. URLLC should provide ultrareliability as well as low latency. eMBB, on the other hand, should support high data rates and coverage. There is a need to redesign the wireless communication architecture (different from 4G) to offer these three services effectively to the users [10] (Figure 10.22).

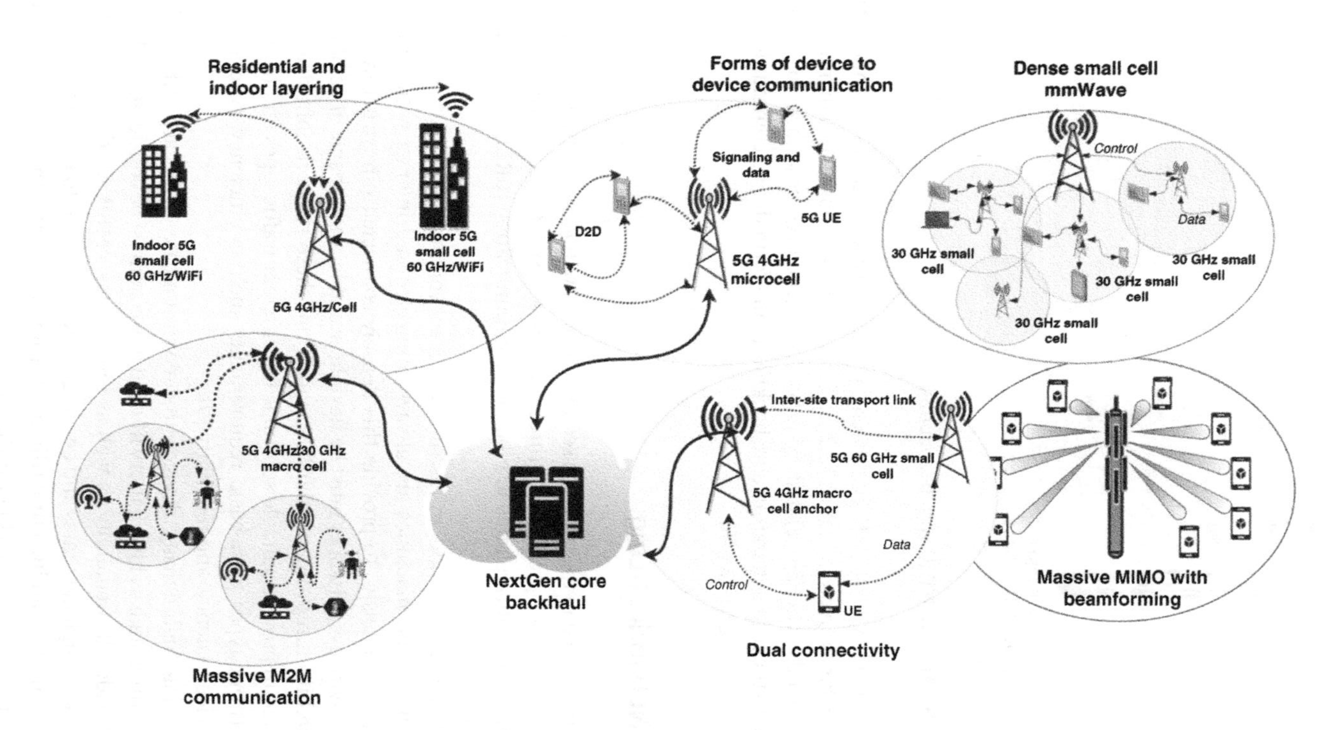

Figure 10.20 A holistic view of 5G subcomponents.

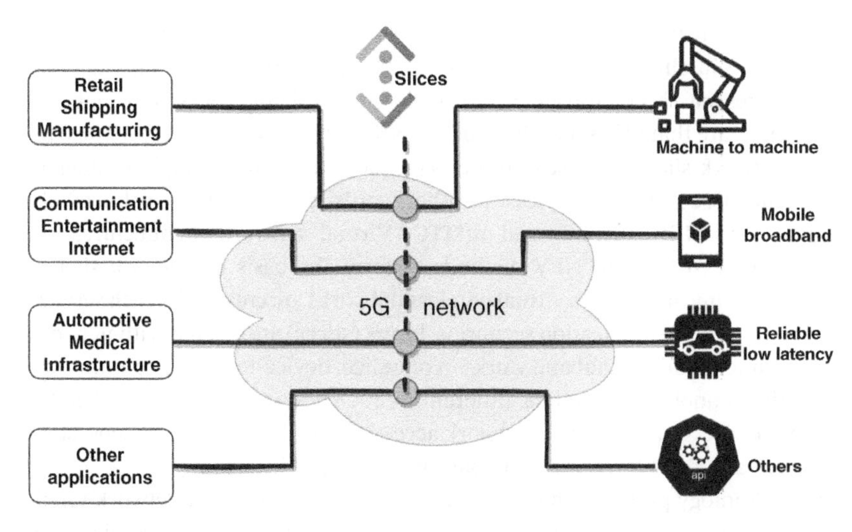

Figure 10.21 Network slicing.

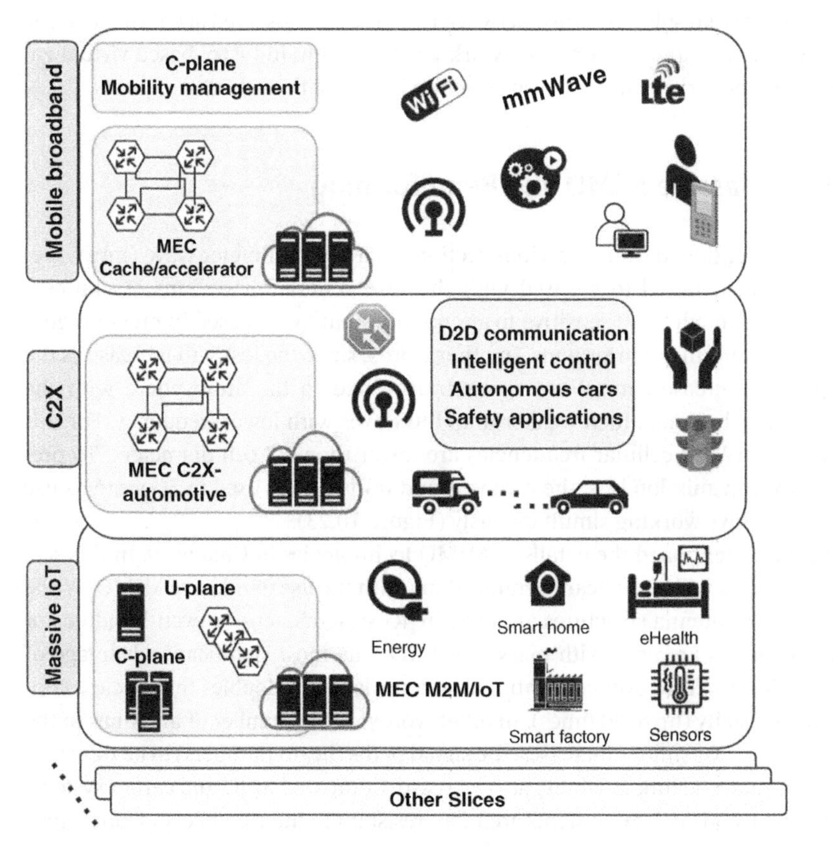

Figure 10.22 Mobile edge computing (MEC) layered network slices.

With eMBB, uRLLC, and mMTC, services such as eHealth, augmented reality, virtual reality, intelligent transportation systems, smart cities, smart buildings, and mobile games have become available to users. Network slicing refers to an approach and architecture that offers an infrastructure solution for all these needs. The concept of network slicing is based on the principle of creating multiple subnets to implement these different services. In other words, different slices (subnets) are designed for eMBB, uRLLC, and mMTC. Virtual network architectures are designed with SDNs, and NFV methods to form the basis of network slicing. Separate dynamic bandwidth virtual wireless networks are created for each service.

Depending on the application scenarios, layers (slices) are created with network slicing. While mobile broadband works in one slice, device-to-device communication works in another. Of course, different frequencies are used in each slice for both end device access and core network access. Thus, the 5G wireless ecosystem is designed as a flexible, robust, scalable, and manageable system [9].

This technology provides an optimized resource allocation and network topology for each situation, providing a wireless infrastructure at a certain service level agreement (SLA). In summary, network slicing converts some network infrastructures into virtual subnets. Thus, network resources are reserved according to basic customer needs. In addition to network virtualization, industry-based virtualization with network slicing can also be set up in line with customer requests.

10.6 Massive MIMO and Beamforming

We have mentioned in the previous sections that the millimeter wave (mmWave) structure has started to be used with the fifth-generation systems. As known, mmWave is much more sensitive to attenuation than waves used in previous generation communication systems. In other words, since the loss will increase as the frequency increases, mmWave waves transmitted to the atmosphere with the same power become much weaker than the waves with lower frequency. For this reason, mmWave cellular frequencies are lower than 4G cell diameters. To prevent this transmission loss, the receivers and transmitters used in 5G systems use antenna arrays working simultaneously (Figure 10.23).

We have mentioned the details of MIMO technologies in Chapter 9. In this section, the focus will be on beamforming along with the use of massive MIMO. While two or four antenna structures are used in 4G systems, tens or even hundreds of antenna arrays are used with massive MIMO. The most significant advantage of the MIMO structure over conventional systems is that it doubles the wireless connection capacity (up to 50 times). In other words, as the number of antennas in the receiver and transmitter increases, the signal paths (beam numbers) to be transmitted, data rate, and link reliability also increase. Compared to 4G and earlier systems with massive MIMO, 5G systems are more resistant to interference and jamming.

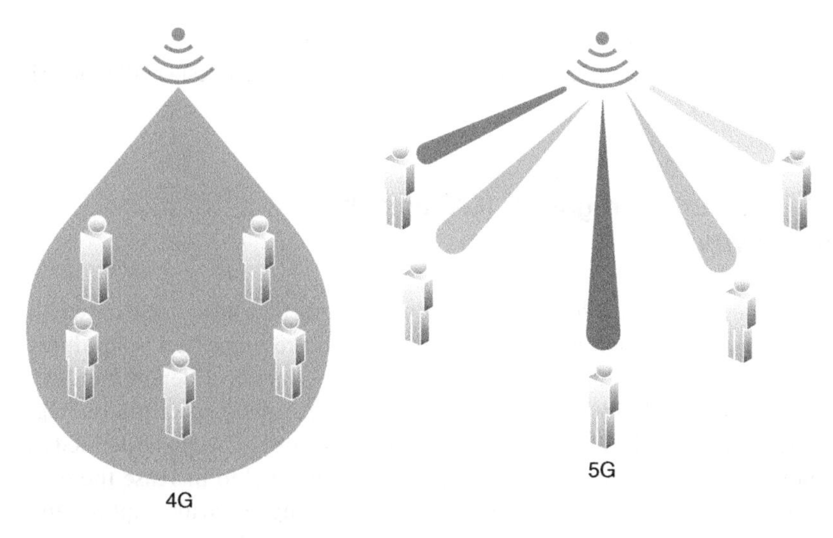

Figure 10.23 Massive MIMO and beamforming.

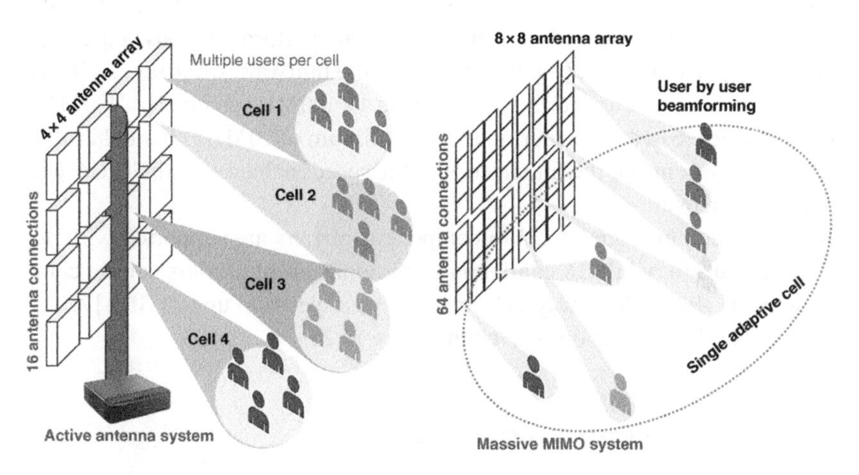

Figure 10.24 MIMO vs. massive MIMO.

Spectral efficiency is also increased by using frequency backup with beamforming created with MIMO. Here we also see a terminal-based space division multiplexing application. Since existing mobile networks (4G and earlier) use a single "beam" for all users in the cell, performance bottlenecks may occur in dense user areas. With massive MIMO and beamforming, communication processes are executed more intelligently and effectively, providing stable speed and latency values [11] (Figure 10.24).

Massive MIMO is seen as an essential milestone for 5G technologies. With massive MIMO serving multiuser and multidevice environments simultaneously, high data rates and stable performance metrics are provided in dense user environments. With this feature, 5G offers an excellent ecosystem, especially for IoT applications [12].

10.7 Carrier Aggregation (CA) and Dual Connectivity (DC)

Carrier aggregation (CA) allows for scalable expansion of the adequate bandwidth available to users through simultaneous use of radio resources across multiple carriers. These carriers can be aggregated from the same or different bands to maximize interoperability and utilization of the scarce radio spectrum or fragmented spectrum available to operators [13]. In other words, CA can be defined as combining two or more carriers into a single data channel to increase the data capacity of the wireless network. By using the existing spectrum, uplink and downlink capabilities can be improved with the CA application.

With CA used in 5G new radio (NR) systems, multiconnection functions with asymmetric upload and download infrastructure are performed. Thus, a user can be given a high bandwidth of up to 700 MHz in the mmWave spectrum. At lower (for example, 7 GHz) frequencies, by combining 4 × 100 MHz channels, a 400 MHz channel can be created. Considering that a maximum of 20 MHz channel is allocated to each user in 4G, the importance of the speeds achieved with 5G CA will be better understood.

As can be seen in Figure 10.25, 5 component carriers are supported with 4G LTE-Advanced Pro, while 16 contiguous and nonadjacent carriers can be combined with 5G NR CA. This way, 5G can combine a spectrum up to 1 GHz [7].

An example CA application is given in Figure 10.26.

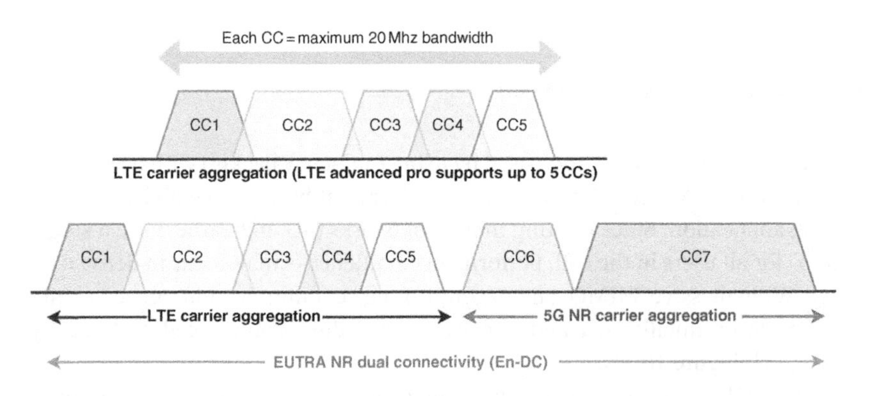

Figure 10.25 LTE and 5G NR carrier aggregation.

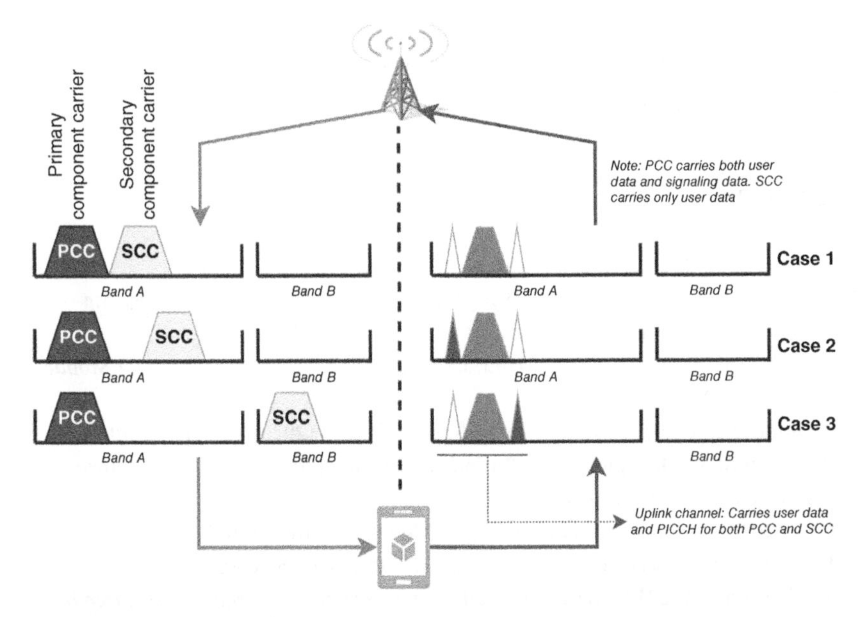

Figure 10.26 CA example.

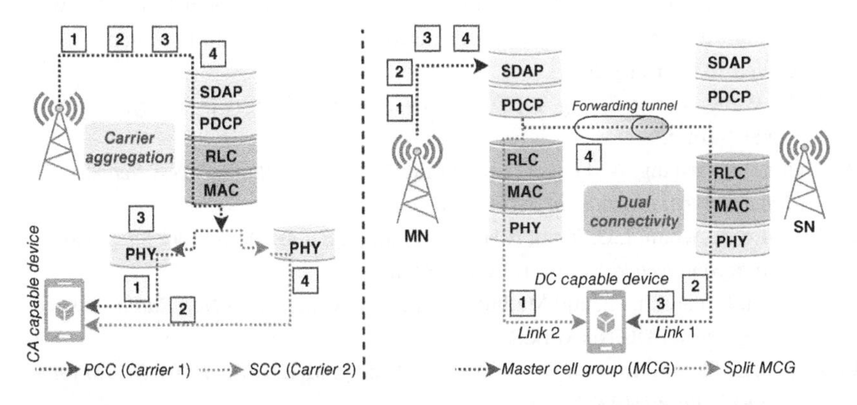

Figure 10.27 Carrier aggregation and dual connectivity.

This is achieved by combining (CA) two or more carriers. With 5G NR DC, it is a method applied especially in nonstandalone (NSA) 5G networks. While 5G supported mobile devices using mmWave frequencies for increased data capacity, it uses 4G infrastructure for voice transmission. It is a structure developed to access 5G advantages in networks where pure 5G infrastructure is unavailable. Standalone (SA) 5G NR uses end-to-end mmWave and sub-GHz frequencies. In SA 5G mode, the existing 4G/LTE infrastructure is not used, and thus, the infrastructure for eMBB, uRLLC, and mMTC services, which are the cornerstones of 5G, is established (Figure 10.27).

DC allows a mobile device to transmit and receive data on two cell groups and multiple component carriers via the main eNodeB (Master Node) and secondary eNodeB (Secondary Node).

References

1 Rodriguez, J. (2015). *Fundamentals of 5G Mobile Networks*. Wiley.
2 Zhengmao, L., Xiaoyun, W., and Tongxu, Z. (2020). *5G+: How 5G Change the Society*. Springer Nature.
3 Dahlman, E., Parkvall, S., and Skold, J. (2013). *4G: LTE/LTE-Advanced for Mobile Broadband*. Academic Press.
4 Asplund, H., Astely, D., von Butovitsch, P. et al. (ed.) (2020). Advanced antenna system in network deployments. In: *Advanced Antenna Systems for 5G Network Deployments*, 639–676. Academic Press.
5 ETSI (2022). Mobile technologies – 5g, 5g Specs | Future technology. *ETSI*. https://www.etsi.org/technologies/5G (accessed 15 February 2023).
6 Sambanthan, P. (2017). Why femtocell networks? *Global Journals of Research in Engineering* 17 (4): 1–8.
7 Qorvo (2022). Small cell networks and the evolution of 5G. *Qorvo*. https://www.qorvo.com/design-hub/blog/small-cell-networks-and-the-evolution-of-5g (accessed 15 February 2023).
8 Li, Z., Wang, X., and Zhang, T. (2020). *5G+: How 5G Change the Society*. Springer Nature.
9 Zhang, L., Farhang, A., Feng, G., and Onireti, O. (2020). *Radio Access Network Slicing and Virtualization for 5G Vertical Industries*. Wiley.
10 Kazmi, S.A., Khan, L.U., Tran, N.H., and Hong, C.S. (2019). *Network Slicing for 5G and Beyond Networks*, vol. 1. Berlin: Springer.
11 Newson, P., Parekh, H., and Matharu, H. (2018). *Realizing 5G New Radio Massive MIMO Systems*. EDN Network.
12 Hanzo, L., (Jos) Akhtman, Y., Wang, L., and Jiang, M. (2010). *MIMO-OFDM for LTE, Wi-Fi and WiMAX*. Wiley.
13 Grami, A. (2015). *Introduction to Digital Communications*. Academic Press.

11

6G Systems

11.1 Introduction

With the completion of 5G standards with version 15 of the 3GPP organization in 2018, the focus in this area started to shift to the sixth-generation (6G) wireless communication. In July 2018, the ITU Telecommunication (ITU-T) standardization sector established the ITU-T Focus Group Technologies for Network 2030 (FG NET-2030) working group. FG NET-2030 group started working on network requirements for 2030 and beyond [1]. As a result, the vision of the 6G wireless communication began to be drawn. In the first step, it was envisaged that 6G (unlike conventional wireless networks) would be a human-centric structure rather than a machine, application, or data center structure [2].

In Chapter 10, we mentioned that 5G systems fit into three pillars: ultrareliable low-latency communication (uRLLC), enhanced mobile broadband (eMBB), and massive machine-type communication (mMTC). Among these specific areas, subareas such as eMBB and end user device energy consumption, MIMO improvements, Quality of Experience (QoE), multicast and broadcast services, coverage and high mobility, uRLLC and IoT, location services, RAN slicing and cross reality (XR), and finally with mMTC, subareas such as low-dimensional data transfer, high-volume devices, and M2M communication are addressed [3].

The journey from 1G to 6G in Figure 11.1 and 5G–6G technical specifications in Table 11.1 are given comparatively [4].

In 4G and 5G systems, bands between 6 and 300 GHz were used in satellite communication, radio astronomy, remote sensing, radar, and similar applications. The developments in antenna technologies in recent years have put forward the idea that the spectrum containing these high frequencies can also be used for mobile communication. The availability of this frequency band, called the millimeter

Evolution of Wireless Communication Ecosystems, First Edition. Suat Seçgin.

Published 2023 by John Wiley & Sons, Inc.

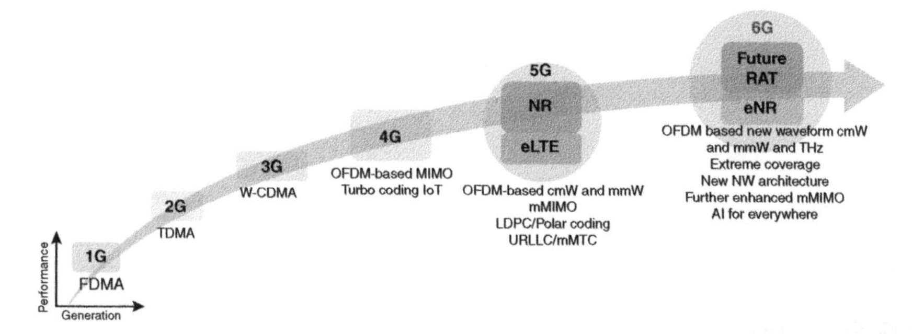

Figure 11.1 Evolution to 6G.

Table 11.1 Comparison of 5G and 6G.

Parameters	5G	6G
Downlink	*20 Gbps*	*>1 Tbps*
Uplink	*10 Gbps*	*1 Tbps*
Traffic capacity	*10 Mbps/m²*	*1–10 Gbps/m²*
Latency	*1 ms*	*10–100 μs*
Reliability	*Up to 99.999%*	*Up to 99.99999%*
Mobility	*Up to 500 km/h*	*Up to 1000 km/h*
Connectivity density	*10^6 devices/km²*	*10^7 devices/km²*
Security and privacy	*Medium*	*Very high*

wave (mmWave) with 1–10 mm (30–300 GHz), has provided hundreds of times the bandwidth compared to 6 GHz [5].

As it is known, mobile data traffic has been increasing exponentially for more than 10 years. This trend seems to continue as IoT devices enter supply chains, healthcare applications, transportation, and vehicle communication [6] (Figure 11.2).

According to ITU's estimations, 38.6 billion devices by 2025 and 50 billion devices by 2030 will be connected to the network with IoT. Three different scenarios can be put forward to meet such intense data traffic [4]:

- Developing better signal processing techniques for higher spectral efficiency of the channel
- Overcondensation of cellular networks
- Using additional spectrum

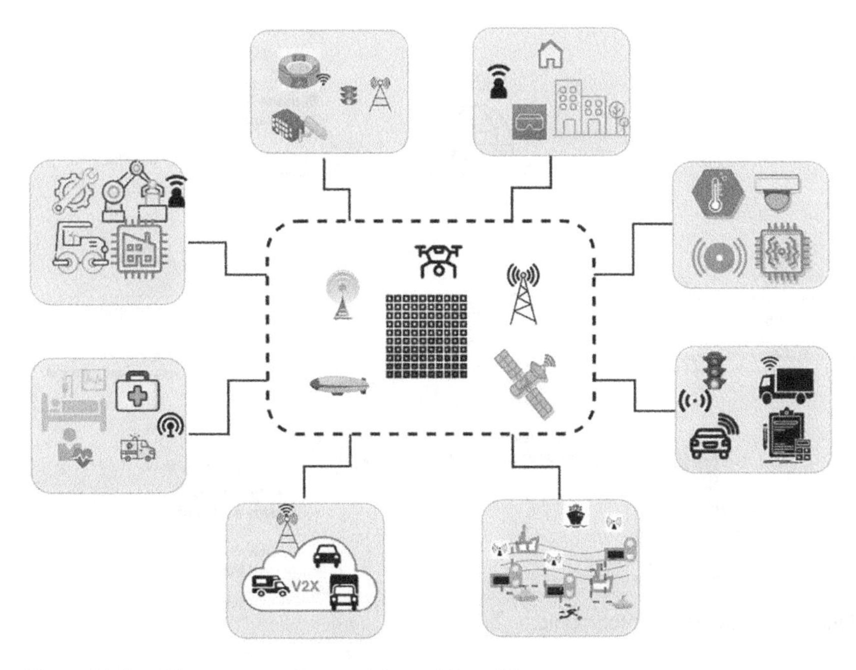

Figure 11.2 6G ecosystem. *Source:* Adapted from [6].

Applications for the first item have already been developed for 5G. In the second item, the intensification of cellular networks brings the problem of interference. mmWave frequencies provide nearly 100 times the spectrum compared to 6 GHz and below frequencies. As delay-sensitive applications (eHealth, space surgery, autonomous vehicles, augmented reality, etc.) come into our lives, the required bandwidth is also increasing. For example, extended reality needs Tbps speed. As such, 6G and terahertz bands have been on the agenda.

Conceptually, the 6G network is designed to expand the human experience in the physical, biological, and digital worlds while enabling a new-generation industrial operation environment beyond Industry 4.0 in performance dimensions such as positioning, sensing, ultrareliability, and energy [7]. Essential topics and comparative transitions from 5G to 6G are given in Figure 11.3.

This technical transformation journey, starting from 1G to 6G, will increase the integration of the physical world (all connected devices, wearable devices, etc.) with the cyber world in 6G. All kinds of information (Big Data) that will be produced by everything connected to the network (Internet of Everything [IoE]) will be compiled and collected using artificial intelligence (AI) techniques and transferred to the servers in the cyber world. Future estimation and knowledge discovery/knowledge discovery methods will be applied when transforming the

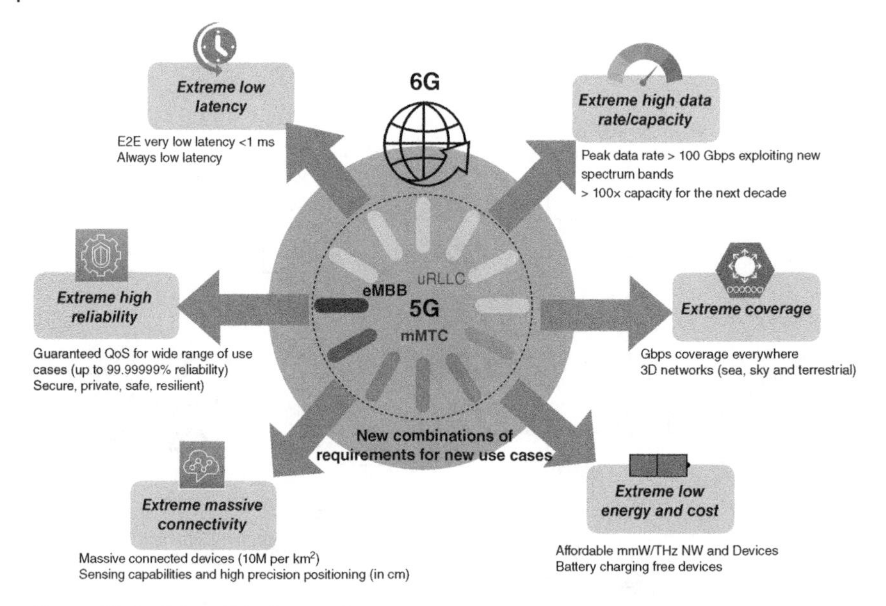

Figure 11.3 5G vs. 6G infographic.

data collected on these servers into value. The results will be practical guidance tools in the actions that will take place in the physical world. The applications, trends, and technologies driving 6G, which we can express as the 6G vision, are summarized as follows [8] (Figure 11.4).

The 6G wireless ecosystem will be a system that will contain many topics, from quantum communication to blockchain technologies. It is essential to manage this complex structure hierarchically to arrange the subheadings. These titles that will create the 6G taxonomy are key enablers, use cases, machine learning (ML) techniques, networking methods, and supplementary technologies. This taxonomy is given in Figure 11.5 [9].

This section will explain the infrastructure, applications, and services included in the 6G vision, although there is no clear standardization yet.

11.2 Network

6G provides an infrastructure that allows access to all geographic locations, from terahertz communication bands to visible light communication, from 3D network support to satellite communication infrastructure. Although 6G is a system built

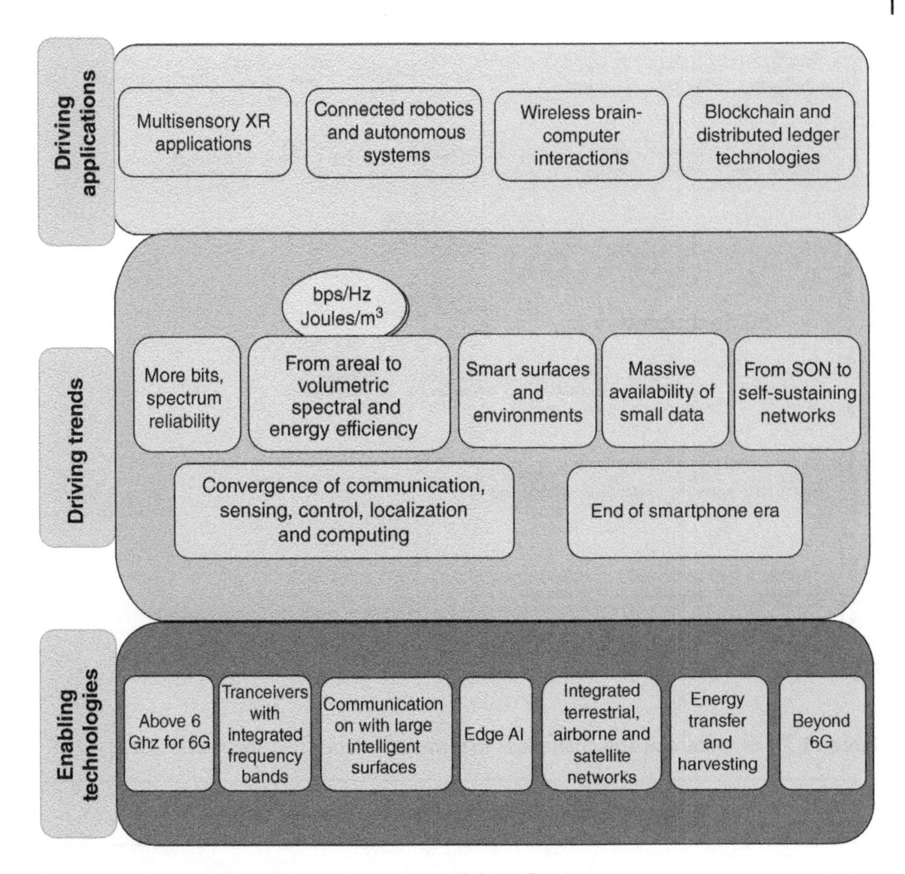

Figure 11.4 6G vision. *Source:* Macrovector/Adobe Stock.

on 5G, it offers new visions regarding communication and application. In summary, the essential requirements for the 6G architecture include network programmability, deployment flexibility, simplicity, efficiency, reliability, robustness, and automation. 6G potential and applications are summarized in Tables 11.2 and 11.3 [4].

The high data rate, small antenna size, and focused beam scenarios above 100 GHz frequencies are being investigated. It can be a transmission medium for visible light communication backhaul connections. More dynamic links will be provided with multispectrum flexibility.

By the way, it should be stated that 6G offers a four-layer network structure. 6G and submarine network structures have been added to the space, atmosphere, and

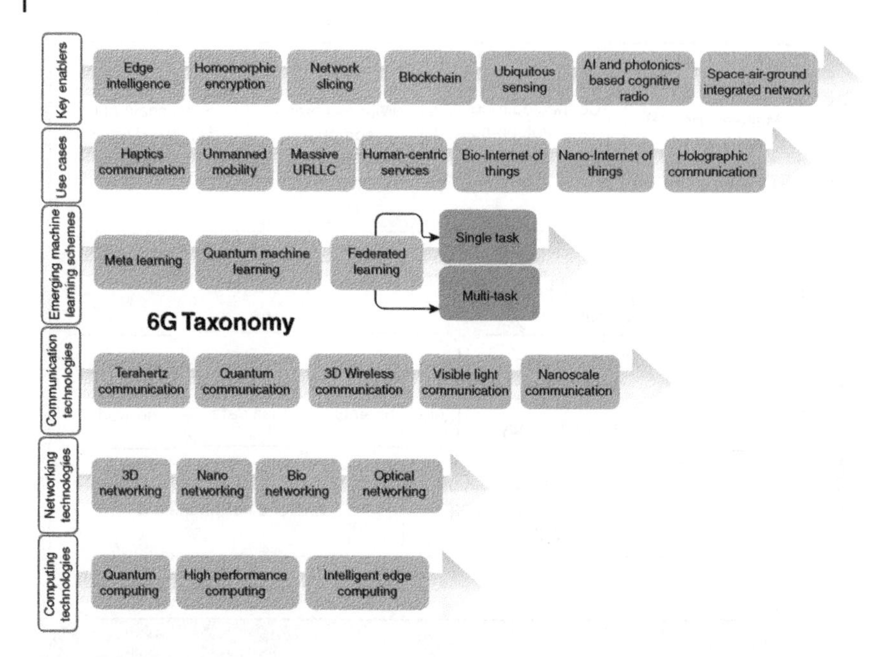

Figure 11.5 6G taxonomy.

Table 11.2 6G potential and challenges in terms of frequencies above 100 GHz.

Novel wireless paradigms and frequencies above 100 GHz			
Enabling technology	**Potential**	**Challenges**	**Use cases**
Terahertz	*High data rate, small antenna size, focused beams*	*Circuit design, propagation loss*	*Pervasive connectivity, Industry 4.0, teleportation*
VLC	*Low-cost hardware, limited interference, unlicensed spectrum*	*Limited coverage, need for RF uplink*	*Pervasive connectivity, eHealth*
Full duplex	*Relaying and simultaneous TX/RX*	*Interference management and scheduling*	*Pervasive connectivity, Industry 4.0*
Out-of-band channel estimation	*Flexible multispectrum communications*	*Need for reliable frequency mapping*	*Pervasive connectivity, teleporting*
Sensing and localization	*Novel services and context-based control*	*Efficient multiplexing of communication and localization*	*eHealth, unmanned mobility, Industry 4.0*

Table 11.3 6G potential and challenges in terms of network architecture.

Multidimensional network architectures			
Enabling technology	**Potential**	**Challenges**	**Use cases**
Multiconnectivity and cell-less architecture	*Seamless mobility and integration of different kinds of links*	*Scheduling, need for network design*	*Pervasive connectivity, unmanned mobility, teleporting, eHealth*
3D Network architecture	*Ubiquitous 3D coverage, seamless service*	*Modeling, topology optimization, and energy efficiency*	*Pervasive connectivity, eHealth, unmanned mobility*
Disaggregation and virtualization	*Lower costs for operators for massively dense deployments*	*High performance for PHY and MAC processing*	*Pervasive connectivity, teleporting, Industry 4.0, unmanned mobility*
Advanced access-backhaul integration	*Flexible deployment options, outdoor-to-indoor relaying*	*Scalability, scheduling, and interference*	*Pervasive connectivity, eHealth*
Energy harvesting and low-power operations	*Energy-efficient network operations, resiliency*	*Need to integrate energy source characteristics in protocols*	*Pervasive connectivity, eHealth*

surface network layers supported in the 5G infrastructure. These concepts are explained as follows [1]:

1) **Space-network tier:** This network layer supports applications for space travel and wireless coverage via satellites, as well as orbital or space Internet services. Laser communication is being considered for long-range intersatellite connectivity in space. For integration into the terrestrial communication network, using mmWave for high-capacity satellite-to-satellite connectivity is another potential solution. Three main topics need to be resolved in this integration study: high propagation delays, Doppler effects from high-speed moving satellites, and path losses in mmWave transmission [10].
2) **Air-network tier:** This layer operates in the low-frequency, microwave, and mmWave bands, providing a flexible and reliable connection. Thus, applications such as supporting dense user areas with flying base stations (BSs), maintaining communication with drone BSs during a disaster, or providing a communication environment to geographically tricky points can be developed.
3) **Terrestrial-network tier:** Similar to 5G, this network layer provides the central infrastructure and wireless coverage for most human activities.

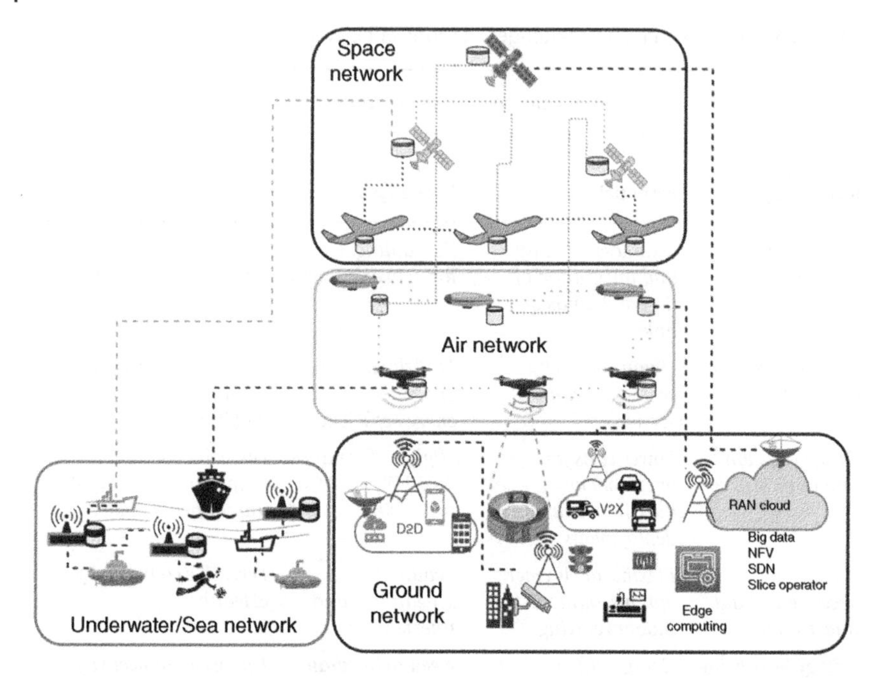

Figure 11.6 Four-layer 6G architecture.

Ultrahigh capacity backhaul (wireless backbone) infrastructures are offered by supporting low-frequency, microwave, mmWave, and THz bands. Although wireless backbone solutions remain attractive, fiber-optic remains essential for 6G [11].

4) **Underwater network tier:** This layer will provide coverage and Internet services for offshore and deep-sea activities for military or commercial applications. Since water has different propagation properties, it is planned to use acoustic and laser communication to provide high-speed data transmission for bidirectional underwater communication [12].

Figure 11.6 illustrates the four-layer 6G network and the interaction between these networks.

11.3 Terahertz Communication

The frequency range from 100 GHz to 3 THz is the potential bandwidth for next-generation wireless network communication. This frequency interval is called terahertz region. This spectrum's frequency range of 100–300 GHz is called sub-THz or sub-mmWave. Short-wavelength frequencies in the mmWave and THz range are used in dense spatial multiplexing and wireless backbone communications. Since the range from 100 GHz to 3 THz enables high gain and shallow dimensional antenna design, a secure communication environment is provided.

Local wireless networks and cellular structures obtained with mmWave and THz will be essential in developing applications such as computer communication, autonomous vehicles, robotic control, holographic gaming, and high-speed wireless data distribution for data centers, with superfast download speeds. These applications can be grouped under the main headings of wireless cognition, sensing, imaging, wireless communication, and positioning/THz navigation [13].

Especially in critical applications (virtual and augmented reality, ultra-HD video conferencing, 3D games, brain-machine interfaces, etc.), constraints such as throughput, reliability, and latency are also significant in addition to speed. Speed and bandwidth have become one of the most critical topics in wireless transmission infrastructures due to the vast increases in data requirement and usage required for these and similar applications. For this reason, it is planned to overcome these limitations by using the terahertz band for 6G systems. Existing and unlicensed bands at these frequencies are the solution for the high data rates that will come with 6G (Figure 11.7).

As for the usage classes, for example, mmWave frequencies can be used in backhaul connections of outdoor BSs instead of fiber-optic (FO) connections, especially in ultradense user areas. This way, more advantageous installation and maintenance can be made and save on building and rental operations. Wired data centers can be made entirely wireless by beamforming in pen widths. As for the THz band, in addition to providing tens of GHz bands, it allows for a low-interference transmission environment and thousands of submillimeter antenna integration due to its high-transmission frequencies [14].

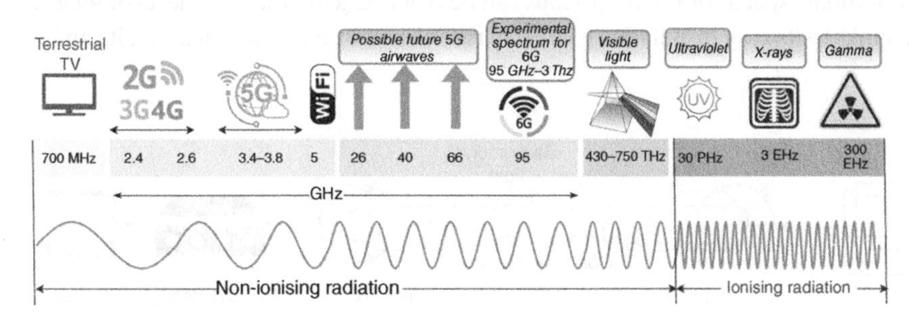

Figure 11.7 Frequency spectrum.

11.4 Visible Light Communication

Usually, when it comes to wireless communication, many people think of technologies and systems that use radio frequencies. This spectrum communicated with electromagnetic waveforms goes up to 3THz. From a usage perspective, speed limits, capacitance, and serious interference problems arise at low frequencies. At high frequencies, however, we encounter space propagation losses, the need for Line of Sight, and hardware prices. In this case, other parts of the electromagnetic spectrum (e.g. optical frequency ranges) are considered as a solution (Figure 11.8).

Optical frequency ranges can be divided into ultraviolet, visible, and infrared light. Working in such high-frequency ranges provides almost unlimited bandwidths. Therefore, frequencies in this range will be used for high-speed applications and services shortly. The advantages of optical wireless communication (OWC) compared to traditional RF (radio frequency) wireless communication is as follows [15]:

- Extremely high bandwidths.
- The advantage of not paying license fees for operators of unregulated bandwidths.
- Very high degree of secure communication: The optical beams used for transmission are primarily narrow and confined to a particular area. Therefore, it is difficult to disconnect communication and difficult to intervene.
- It is resistant to electromagnetic interference/distortion.
- OWC systems do not experience multipath fading.
- Since no electromagnetic radiation is experienced in OWC systems, systems do not harm health.
- OWC can be easily used in restricted RF usages, such as airplanes and hospitals.

Figure 11.9 shows the structure showing the optical wireless connection between the BSs and the core network. With this structure, despite the adverse geographical conditions, speeds of up to 200 Gbps can be reached with a security level of 99.9%. Thanks to its easy installation (only one receiver and transmitter) feature, it can be

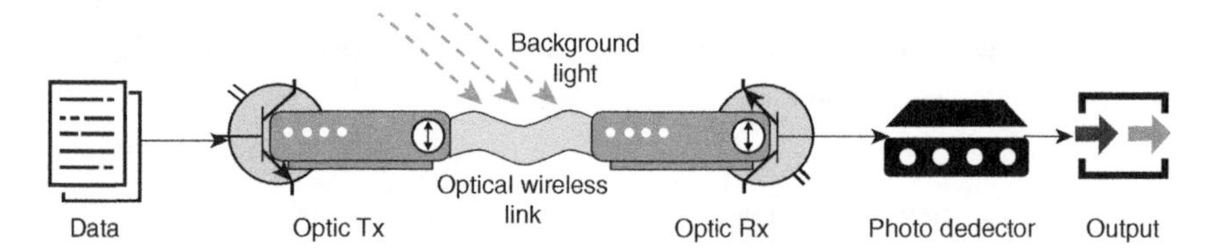

Figure 11.8 Optical communication diagram.

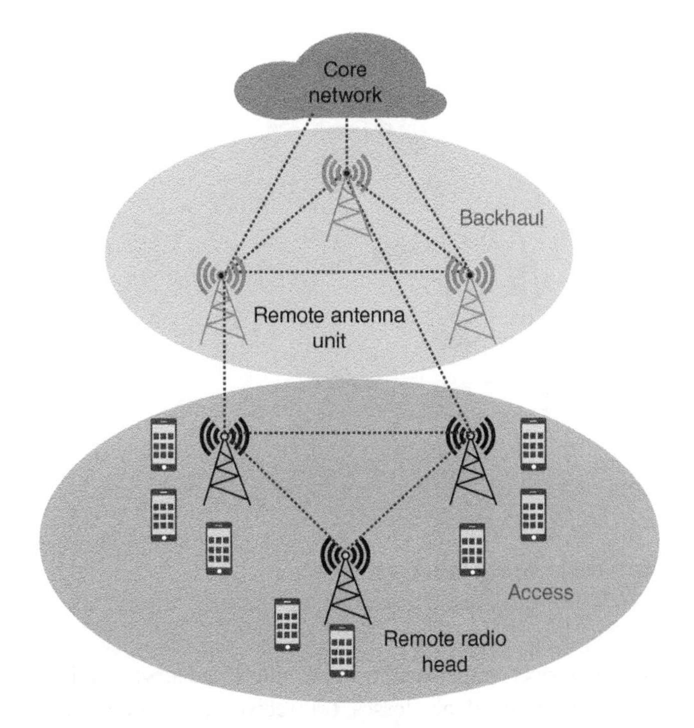

Figure 11.9 OWC for mobile backhaul.

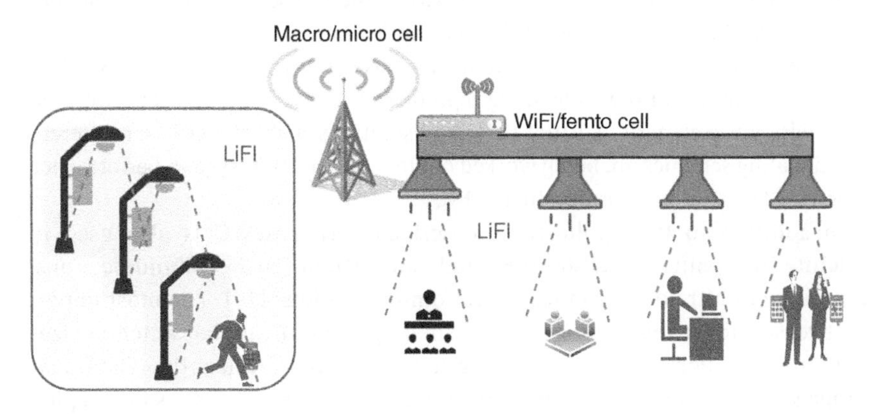

Figure 11.10 LiFi.

used as a quick solution to replace operator infrastructures that have become unusable due to disasters. The application, which appears as Light Fidelity (LiFi) in local use, is seen as an alternative to WiFi communication in environments such as offices. A typical LiFi application is given in Figure 11.10 [16].

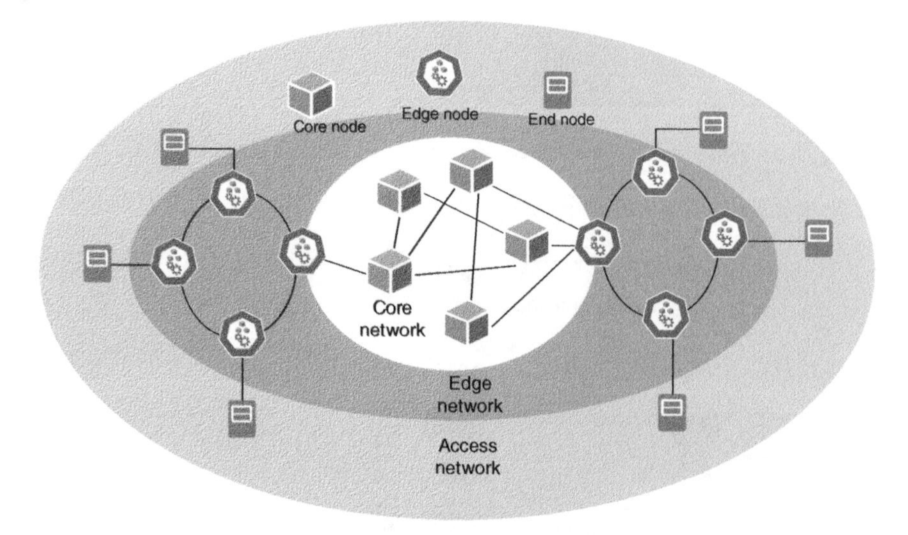

Figure 11.11 The architecture of a typical optical network.

Despite all these positive aspects, transmission losses due to adverse weather conditions in OWC, atmospheric turbulence losses, transceiver adjustment difficulties due to long distance and narrow beams, attenuation caused by ambient light, and any material that will enter between the transceiver could be encountered that will cause interruptions in communication.

To overcome these negative situations, various studies are carried out by scientists. For example, for the ambient light problem, the researchers used far-infrared light with a wavelength of 10 μm [17]. And solutions such as dual-laser differential signaling schemes are being worked out to improve OWC system performance when background noise is dominant [18].

In addition to the applications described earlier, OWC is also used in satellite-to-satellite, satellite-to-ground, and Moon-Earth communications, especially in ultralong distance space communications [19]. The most important reason for using light (laser) for such applications is that antenna sizes are much smaller than RF systems (due to very low wavelength). The linear propagation property of light is one of the other important reasons for keeping OWC. As an example application in space-space communication, we can show the Moon-Earth communication carried out by NASA in 2013. With this application, data transmission between the Earth and the Moon was realized at a speed of 622 Mbps at a distance of 384 600 km. Typical optical network architecture is given in Figure 11.11 [20].

11.5 Satellite Integration

With 5G wireless communication networks, the connection and speed requirements for very high-density IoT devices are increasing daily. Therefore, access support to terrestrial and satellite networks is a critical issue. Even though a ubiquitous communication network infrastructure will be established with 5G, there are significant problems in terms of investment cost in including mountains, seas, and uninhabited areas, which we can describe as terminal units, into the coverage area with the existing BS logic. At this point, satellite networks are an essential solution point with their ability to provide a global coverage area. Especially with the integration of terrestrial networks with satellite networks, next-generation communication will reach every point globally.

When it comes to satellite communications, there are many parameters to consider. First, a satellite link covers much longer distances than a terrestrial link. To tolerate the path loss caused by this distance, ground terminals must be equipped with high-power transmitters and sensitive receivers. Second, overlapping signals are highly prone to intersatellite signal interference due to the beam spacing of neighboring satellites. In addition, the cost of broadband service via satellites is very high. Therefore, satellite communication within 6G systems is one of the critical areas of interest [21].

The satellite communication infrastructure, which will be integrated with 5G systems until 2025, is also an essential component for 6G in establishing 3D networks. In satellite communication, band capacities will increase from 100 Gbps to Tbps speeds with hundreds of beam spots and back-use of very high frequencies. In addition, improvements in satellite payload technology through optimized designs and new materials will increase the payload power from 20 to 30 kW. Techniques such as adaptive beam hoping and shaping and interference management will be used to improve connectivity and flexibility to varying traffic demands and patterns. With the optical links between satellite-to-satellite and satellite-to-ground, essential steps will be taken to integrate satellites in different orbits into the 3D wireless network [22]. As a result of the integration of these air-based networks with terrestrial networks, the data speed and coverage required by unmanned aerial vehicles (UAVs) can be established [23, 24].

Example architecture of this integrated satellite, ground, and maritime wireless communication network is given in Figure 11.12. As can be seen from the figure, the satellite network consists of satellites in various orbits. The network to be established in the air will be formed by aircraft, UAVs, balloons and similar vehicles, terrestrial network ground cellular mobile networks, satellite ground stations, and mobile satellite terminals. Finally, the submarine network will be built from underwater BSs, sensors, and other components [25].

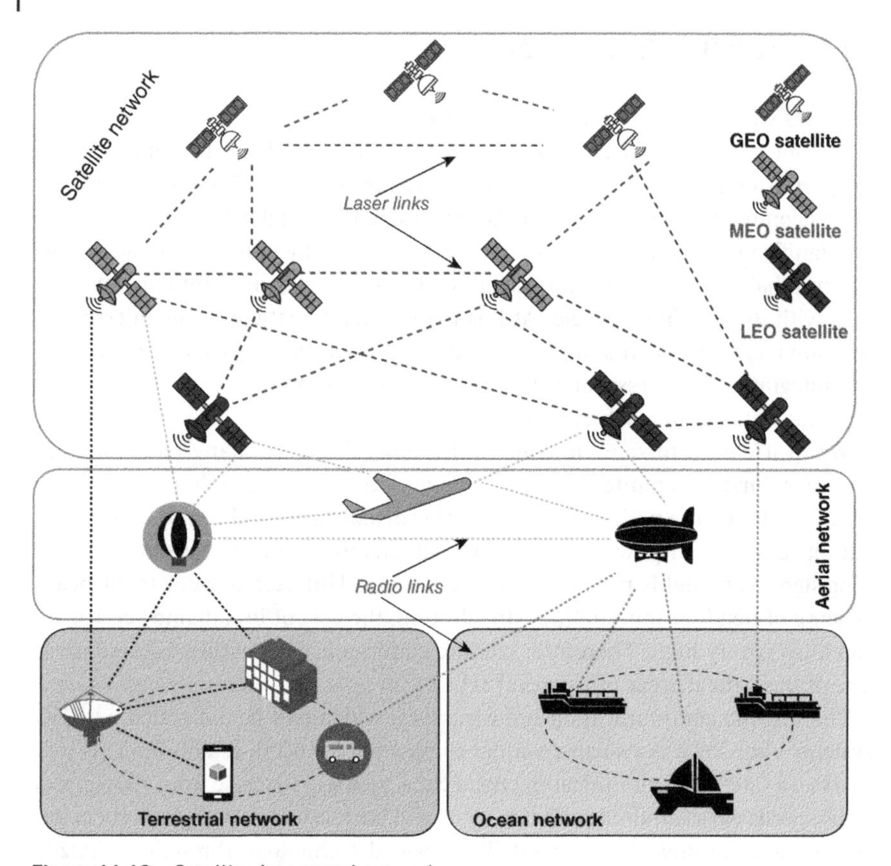

Figure 11.12 Satellite-integrated network.

11.6 Cloud Radio Access Network

A centralized or cloud radio access network (C-RAN) has been proposed as a new cloud computing architecture to support 5G and B5G (beyond 5G) systems. While C-RAN supports 2G, 3G, and 4G infrastructures, it also meets 5G requirements. C-RAN retrospectively has subcomponents such as central processing, collaborative radio, and a real-time radio network. This structure will also be available for 6G systems [26].

In traditional systems, the radio access network consisted of structures that offer wireless access to users called BSs or NodeB (NB). BSs are designed as structures that provide communication between users' mobile phones and the core network. After a mobile device is connected to the nearest BS, the BS transmits voice, data, and video traffic to the core network of the respective operator. The BS

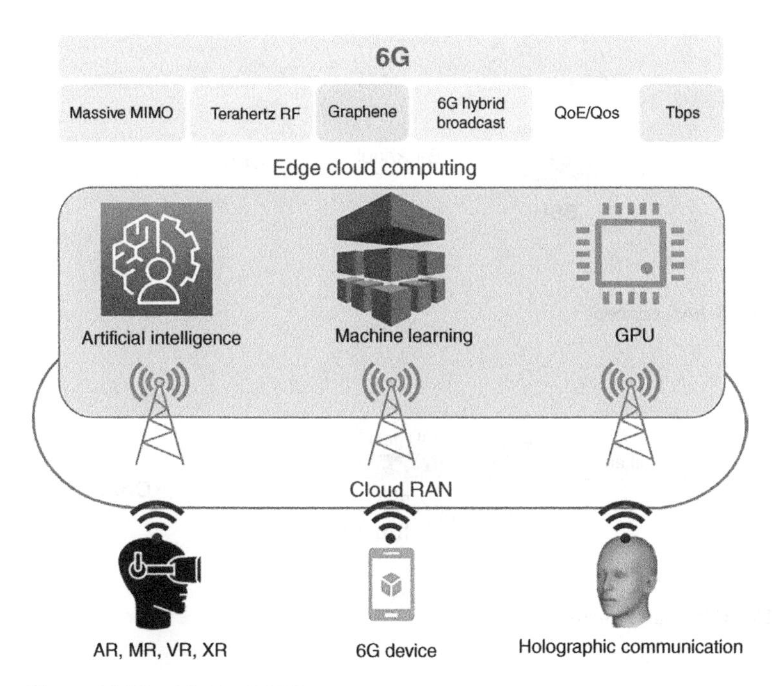

Figure 11.13 6G cloud RAN concept.

consists of two subcomponents: the baseband unit (BBU) and the remote radio unit/remote radio head (RRU/RRH). When the BS receives a signal from the mobile device, the RRU receives these signals from the antenna, converts them to analog radio frequency, and transmits them to the BBU. BBU handles digital signal processing, while RRU handles analog signal processing. The combination of these structures on the cell side is called the distributed radio access network (D-RAN) architecture (Figure 11.13).

In D-RAN architecture, all BS structures are located in cell towers. The RRU is situated at the top of the tower at the base of the antennas, while the BBUs are installed in equipment rooms located at a different location from the tower (Figure 11.14).

The BS is subdivided into the C-RAN architecture. BBUs are combined in a central office. So digital signal processing can be centralized while the RRU remains adjacent to the cell antennas (Figure 11.15).

With this structure, the one-to-one connection structure between RRU and BBU has been eliminated. Thus, baseband calculations with high processing loads have been virtualized, and a more manageable network infrastructure has been created for the operator. Another benefit of concentrating BBUs in the same place is the effective use of information resources and the reduced maintenance costs for operators (Figure 11.16).

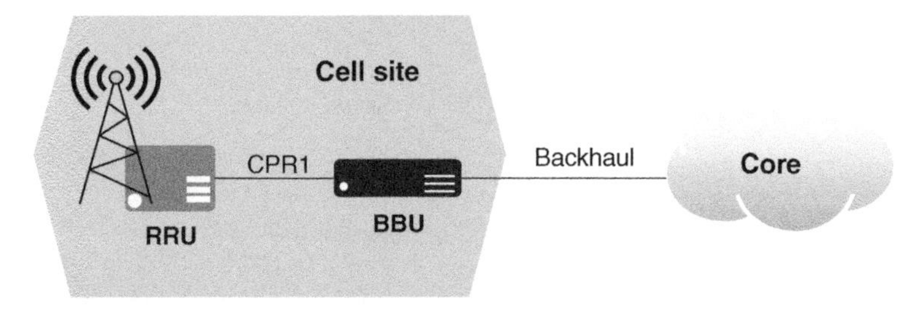

Figure 11.14 D-RAN concept.

Figure 11.15 C-RAN architecture.

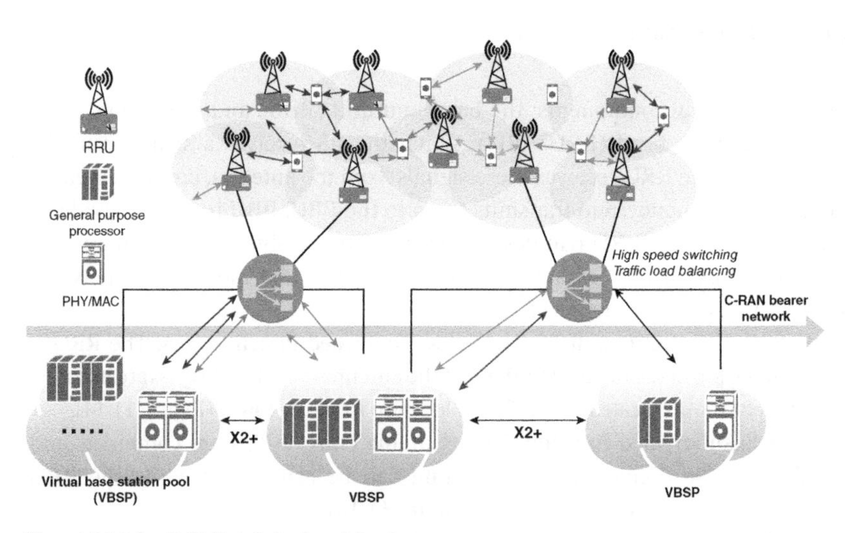

Figure 11.16 C-RAN detailed architecture.

With virtualization technologies, while logical isolation of resources is realized, on the other hand, the sharing of physical resources in a dynamic and scalable manner is facilitated. On C-RAN, these operations are performed in the BBU pool, including network virtualization, networking, computing, and storage.

Each BBU is a virtual node, and communication between these BBUs takes place through virtual connections. In the pool, a machine's shared CPU, memory, and network resources are orchestrated across multiple BBUs. Thanks to this structure, delay times are also reduced.

Centralization of BBUs also provides benefits such as edge point resource virtualization and a more straightforward service setup. On the central side, advanced technologies are used to operate processes that require high processing power. There are also advantages, such as shared processing power and radio sharing with BBU pooling.

In traditional networks, BBUs were installed in remote locations from central offices, on the top of buildings, or in towers. In such an installation, maintenance and installation costs increase. As BBUs get closer to the main office, they can potentially be deployed in more secure data centers. This improves data security while facilitating resource allocation for computationally costly operations. By sharing multiple BBUs, besides lower power consumption, a higher amount of resources can be accessed when needed. By concentrating resources at a central point, communication and user services become available in the edge network, not in the network center. By bringing these services closer to the user, faster services can be established for the users through edge computing. Collecting user traffic on edge is an application that also relieves backbone traffic. This topic will be discussed in detail in the mobile edge computing (MEC) section.

11.7 Holographic MIMO Surfaces

Due to the proliferation of intelligent wireless devices, the widespread use of the Internet of Things, the use of AR/VR applications, etc., in recent years, users' demands for low latency, higher service quality, and low prices have been increasing with increasing momentum as well as high bandwidth. With the increase in the number of users per unit area, the limited spectrum of resources still used has become insufficient to meet these demands [27].

Flexible solutions have been put forward for this bottleneck mentioned by techniques such as mmWave, multiple orthogonal multiplexing methods, and massive MIMO. At this point, solution proposals and standards have already begun to be determined beyond 5G communication systems. These potential solutions focus on nonorthogonal multiple access, OWC, hybrid optical/radio solutions, alternative waveforms, low-cost massive MIMO systems, terahertz communication, and new antenna technologies. Although 6G systems are perceived as an extension of 5G systems, new user requirements, the emergence of entirely new application/usage areas, and new network trends have led to recent paradigm searches for communication infrastructures after 2030, especially in the physical layer [28]. The solution proposals for these searches mainly focus on transceivers (for example, massive MIMO and mmWave), antenna arrays with beamforming features, the use of cognitive spectrum, and adaptive modulation techniques. At this point, the idea that the propagation environment can be transformed into "programmable" environments with the help of new materials has begun to dominate [29]. It is thought that all parts of the device hardware of the future 6G wireless communication systems can be reconfigured with software to adapt to the changes in the wireless environment with intelligent methods [30].

Today's wireless transmission channel models are almost managed by stochastic processes rather than software-controlled deterministic methods. Following recent advances in the fabrication of programmable meta-materials, reconfigurable intelligent surfaces (RISs) offer solutions for 6G networks that have the potential to fulfill this challenging vision [31, 32]. Recently, with the development of RISs and radio systems, it has become possible to control at least some of the wireless communication channels with deterministic methods. Although it is known that reflection and scattering parameters show uncontrollable stochastic behavior by nature, managing or optimizing these parameters (at least some of them) within a system will improve many adverse conditions seen in wireless communication [33].

The reflective surface is a planar aperture synthesized using subwavelength elements (or unit cells). Such a surface can be used to modulate incident waves to the desired wavefront due to reflection. Due to subwavelength unit cell samplings, reflective surfaces can be considered a distinct form of metasurfaces synthesized using a set of metamaterial unit cell elements [34]. Although these structures have been studied in the field of applied electromagnetics, their use in wireless communication networks is still in the interest of researchers. With the recent implementation of 5G technology and upcoming 6G networks, reflective surfaces can significantly reduce the hardware layer's cost and complexity while increasing overall energy efficiency [33].

RISs are considered crucial concepts for intelligent radio systems. These surfaces can be considered thin sheets (surfaces) that operate almost passively and at a low cost. We used the phrase nearly passive because external stimuli are used to manipulate the electromagnetic waves coming into it. The following properties of these surfaces can be listed:

- RIS elements can perform independent reconstructions (for example, phase shifts) on incident electromagnetic waves. They can adapt their responses in real-time.
- RIS surfaces are passive working surfaces produced from low-cost and low-power electronic elements without radio frequency chains.

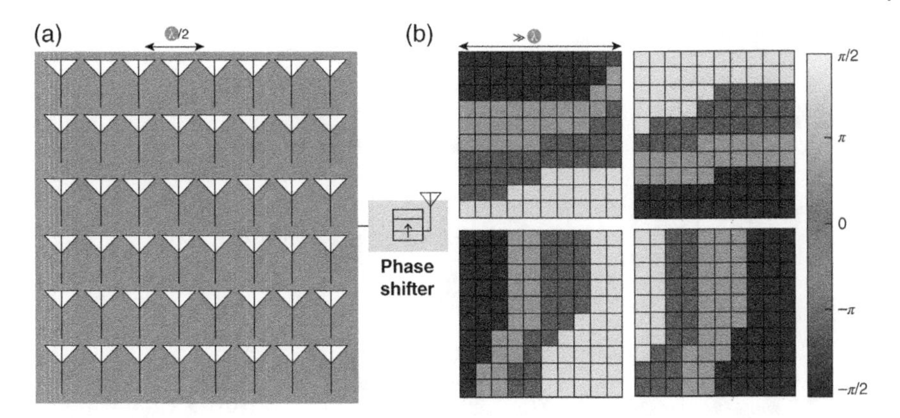

Figure 11.17 RIS: (a) A 48-element reflector array-based RIS. (b) A four-element metasurface-based RIS.

- RIS plates are easily positionable surfaces such as walls, ceilings, and building facades (Figure 11.17).

In the Figure 11.17a a 48-element based reflector array is given. Each element is a conventional antenna connected to a phase shifter. Additionally, in Figure 11.17b a four-element meta-surface-based RIS is given such that each element/slab is a dynamic meta-surface containing many tightly packed metaatoms, and an optional semi-continuous phase gradient can be applied.

In general, the rate of speed that can be achieved in practice in a wireless connection is limited by the modulation order and the number of spatial streams. The current channel realization determines both parameters. The modulation order is adapted according to the perceived power level of the signal at the receiver, which is a result of the channel gain. To keep the error rates low and to avoid retransmission, the user at the cell edge is forced to use low-order modulation, which means that the user connects to the wireless network at a low speed. On the other hand, the number of spatial streams is determined based on the channel's available eigenmode numbers. A direct view link may have high channel gain but will likely operate at low speeds because of the spatially sparse-ranked channel. RISs can be used in these scenarios to change channel realization and significantly improve overall system performance [33].

The visual figure describing outdoor HMIMOS applications is given in Figure 11.18.

The working principle of the indoor HMIMOS application is drawn in Figure 11.19.

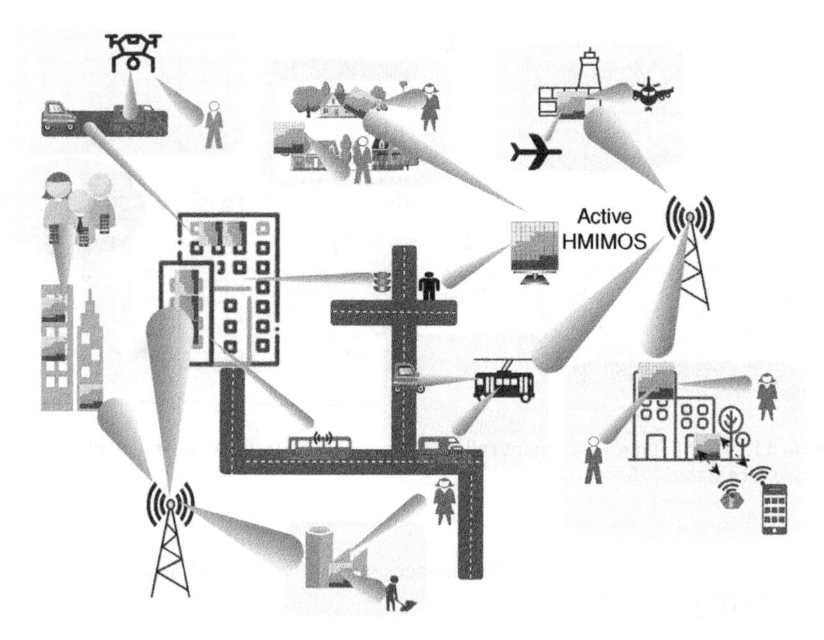

Figure 11.18 Outdoor HMIMOS use cases.

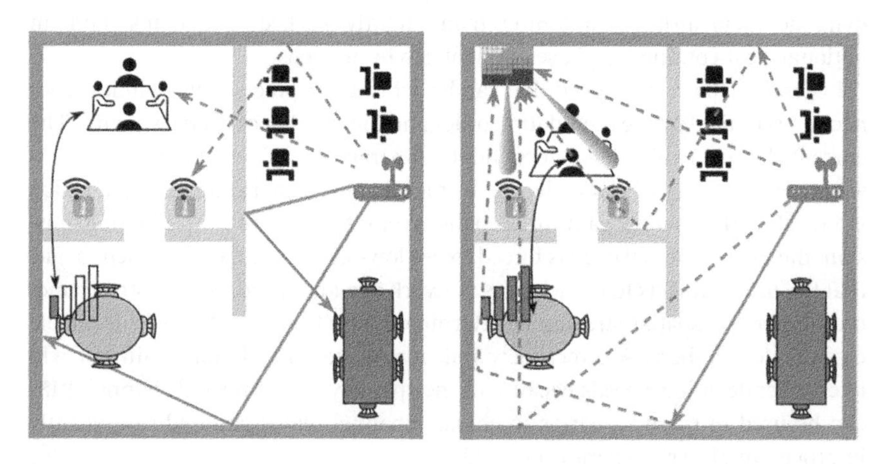

Figure 11.19 Indoor HMIMOS use case.

Holographic surfaces, also called software-defined surfaces, are divided into two categories according to their power consumption and two types according to their hardware structures [35].

- **Classification by power consumption:**
  - **Active HMIMOS:** Surfaces in this class can be used as receivers, transmitters, or reflectors. HMIMOS structures in this class are also called large intelligent surfaces (LIS). A practical application of active HMMO is achieved tightly by integrating an infinite number of small antenna elements. This structure can transmit and receive communication signals across the entire surface, taking advantage of the hologram principle [36, 37]. Another application of active HMIMOS consists of discrete photonic antenna arrays integrating active optical-electric detectors, transducers, and modulators to transmit, receive, and convert optical or RF signals.
  - **Passive HMIMOS:** Passive HMIMOS is also known as RIS. RIS acts as a passive metal mirror or "wave collector" and can be programmed to change an impacting electromagnetic field in a customizable way. Compared to active HMIMOSs, they consist of low-cost passive elements that do not need a power supply. From the point of view of energy consumption, we can say that passive HMIMOS structures shape the waves falling on them and direct these waves without any signal processing or amplification. In addition, these intelligent pads can operate in a full duplex mode without increasing noise levels and causing any obvious self-interference. Another essential advantage of passive MIMOS is that they can be installed on building surfaces, rooms, factory roofs, laptop bags, and even clothes due to their low cost and low energy requirements [31].
- **Classification by hardware structure:**
  - **Adjacent HMIMOS:** In an adjacent HMIMOS, an almost uncountable infinite number of elements are integrated into a finite surface area to form a spatially continuous transceiver aperture. Since continuous aperture takes advantage of the integrated infinite number of antennas with the asymptotic limit of massive MIMO, its potential advantages are to achieve higher spatial resolution and enable the generation and detection of EM waves with arbitrary spatial frequency components without unwanted side lobes [37].
  - **Discrete HMIMOS:** Discrete HMIMOS usually consist of many discrete unit cells made of low-power, software-adjustable metamaterials. It can electronically alter the EM properties of unit cells in more than one way. These range from electronic components to liquid crystals, microelectromechanical systems, electromechanical switches, and other reconfigurable metamaterials. This structure differs significantly from the traditional MIMO antenna array. The arrangement of a discrete surface is based on discrete "meta-atoms" with electronically steerable reflection properties [38].

Figure 11.20 shows the scenario of serving with the help of RIS when the line of sight (LoS) path is blocked by a BS. Figure 11.21 shows the hardware structure of the user device and BS operating in this system [39].

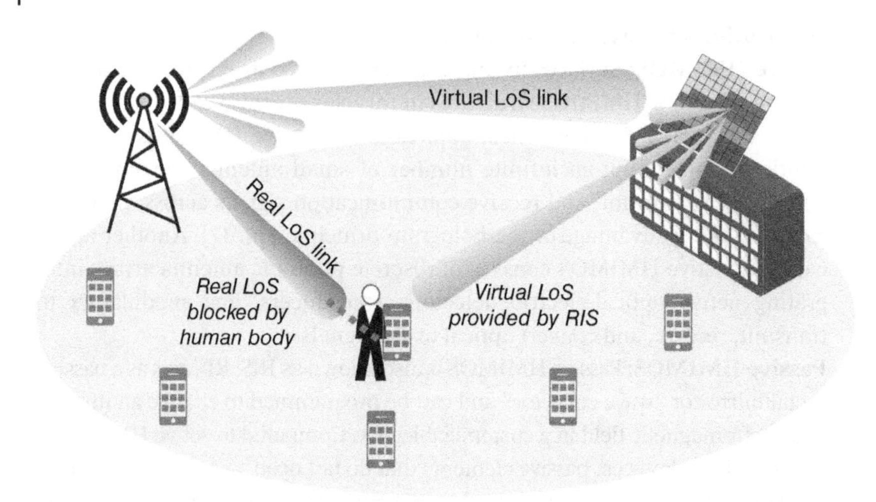

Figure 11.20 LoS RIS operation.

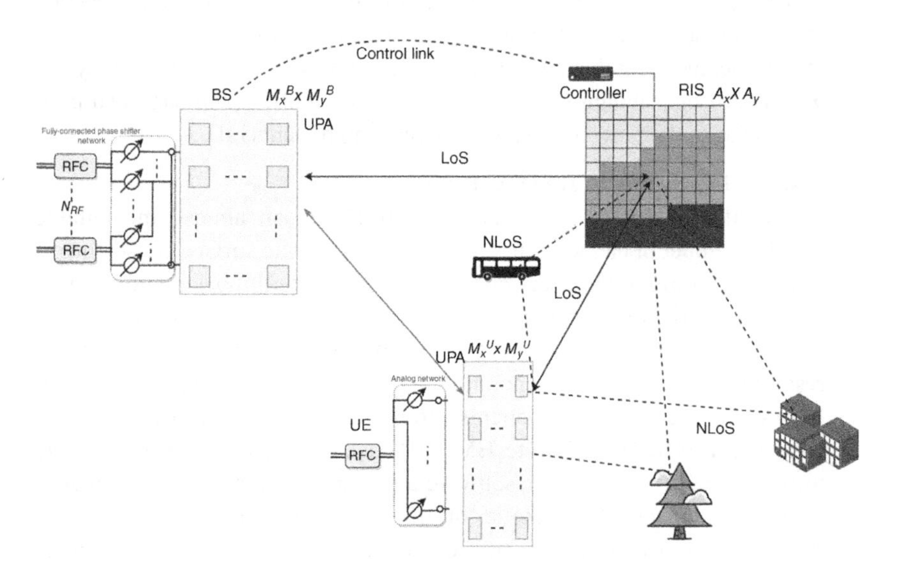

Figure 11.21 BS and user equipment (UE) structure.

11.8 Massive Cell-Free MIMO

Fifth-generation wireless communication systems, which will offer users intense connectivity, ultrareliability, and low latency, have started to be installed in various countries beginning in 2019. This innovative technology can provide high-density beamforming and spatial multiplexing, high spectral efficiency, efficient energy use, and link reliability with multimass MIMO antenna technology centrally positioned on the BSs. Despite all these innovations, the 5G wireless communication infrastructure is lacking in providing better coverage, and a uniform performance graph in this wide coverage area due to the exponentially increasing data need and traffic sizes. The main reason is that mMIMO systems cannot provide an effective solution for the performance drops experienced by users prone to electromagnetic interference between cells and interference occured located at the cell edge ends [40].

Distributed antenna systems (DAS) are recommended as a solution to this low-performance system experienced by users at the edge of the cell. With this solution, the macro-level diversification of MIMO systems to cover dead spots in the cell has been studied [41] (Figure 11.22).

On the other hand, network MIMO and coordinated multipoint (CoMP) are proposed to reduce cell-to-cell interference by adding cooperation between neighboring access points (APs) [42]. This solution divides APs into discrete collaboration clusters to facilitate data sharing (Figure 11.23). However, intercluster signal interference remains a critical issue as it cannot be removed from the cellular structure. Therefore, as long as the cellular structure paradigm remains in practice, there is no method to prevent intercellular interference.

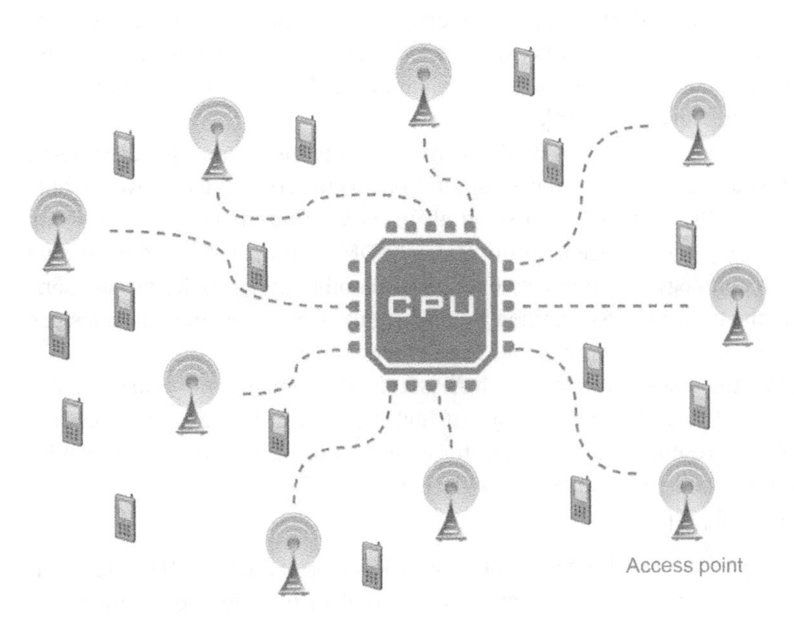

Figure 11.22 Traditional cell-free mMIMO infrastructure.

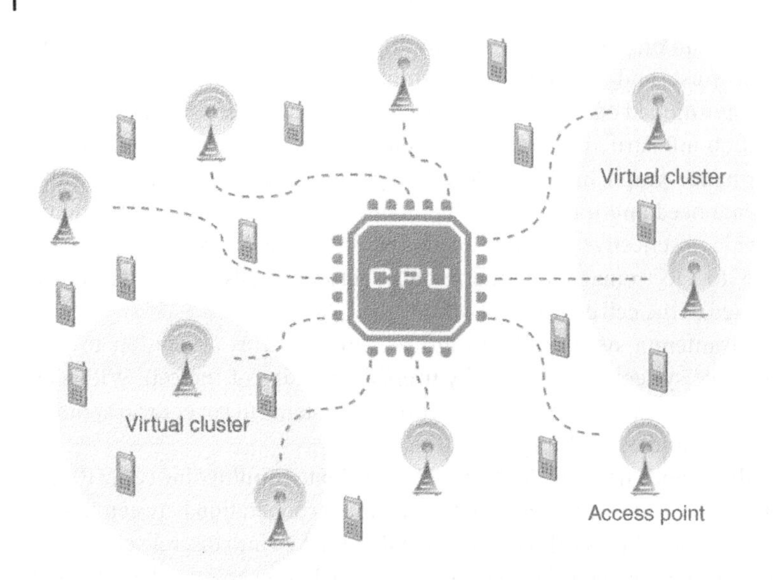

Figure 11.23 Scalable cell-free mMIMO system.

By combining the advantages of mMIMO, DAS, and network MIMO technologies, cell-free mMIMO technology is proposed, with no cell or cell boundaries [43]. Due to its inherent advantages, cell-independent mMIMO is a vital and critical technology for 6G systems. It is expected to bring significant gains such as high throughput, ultralow latency, ultrahigh reliability, high energy efficiency, and uniform coverage everywhere [44].

The basic idea of cell-free mMIMO technology is to install multiple distributed APs connected to a central processing unit (CPU) to serve all users over a wide area. Specifically, each AP is to serve all users via time division multiplexing (TDM) or frequency division multiplexing (FDM). Compared to conventional mMIMO technology, cell-free networks offer smoother connectivity for all users, thanks to the macrodiversity gained from distributed antennas. In addition, since there is no cell concept, there are no in-cell border effects.

However, the assumption that each AP serves all users does not make the system scalable. It has the disadvantage of high power consumption for decoding and increased resource consumption for computation, especially for users with low signal-to-noise-to-noise ratios (SINR) [45].

In a nutshell [44]:

- Cell-independent mMIMO technology was developed based on MIMO technology. With this technology, many APs are used to provide users with many degrees of freedom, high multiplexing, and array gains. These gains can be achieved with simple signal processing techniques due to MIMO technology's favorable propagation and channel-tightening properties. The point to note here is that even if we do have channel tightening in cell-independent MIMO, it is potentially at a lower level compared to side-by-side MIMO.

- In cell-independent MIMO, service antennas (APs) are distributed over the entire network, thereby achieving macrodiversity gains. As a result, a quality network connection with cell-independent MIMO and, accordingly, good services can be provided to all users in the network. Unlike a colocated MIMO structure, where the BS is equipped with giant antennas, cell-independent MIMO is expected to consist of low-cost, low-power components and simple signal processing APs.

11.9 Mobile Cloud Computing (MCC)-Mobile Edge Computing (MEC)

It is planned to integrate mobile cloud computing (MCC) and mobile devices into an integrated structure and develop capacities such as information processing power and storage. This way the user experience is improved by using cloud computing and related services of information processing power and storage-sensitive user applications. The MCC architecture is given in Figure 11.24. As can be seen

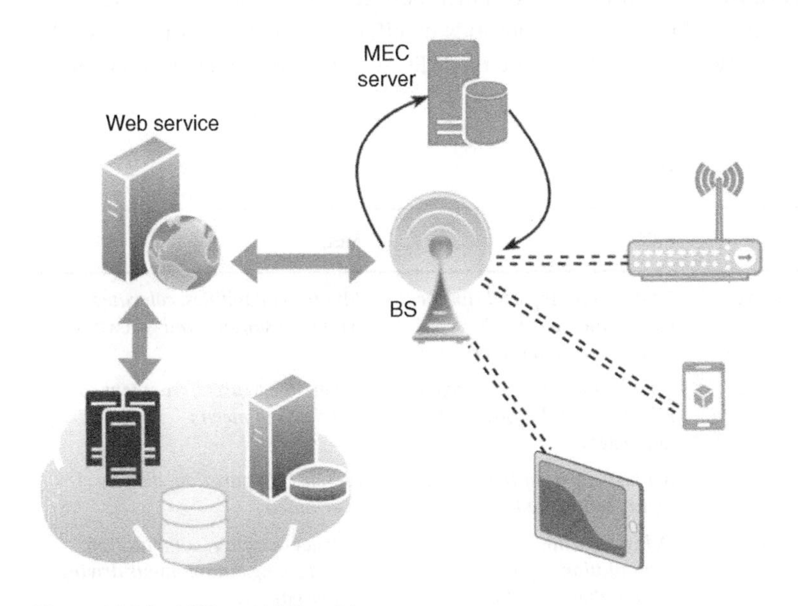

Figure 11.24 MCC and MEC architecture.

in the figure, mobile devices connect to web servers via the nearest BS. Web services work like an application programming interface (API) between mobile devices and the cloud and distribute cloud applications to mobile devices. In the current architecture, mobile devices access cloud services via BSs or WiFi APs in the mobile network. MCC enables resource-limited mobile devices to run latency-insensitive but compute-intensive applications [46].

However, the inherent limitation of MCC is the long distances between mobile devices and the cloud. Due to these distances, long execution delays can occur, and the time constraints of delay-critical applications cannot be satisfied. There are significant differences between MEC and MCC systems regarding computing power and storage. MEC integrates cloud computing into the mobile network to deliver computing power and storage capacity to edge end users. The differences between MEC and MCC are summarized in Table 11.4:

The topics given in the table are explained as follows:

- **Physical server:** Physical servers are deployed in large-scale data centers in the MCC structure. Data center buildings are large and unique buildings. MCC servers have high information processing and storage capacities and are set up as server farms. They have extra protection such as security and redundancy. However, MEC servers are located in smaller buildings where wireless routers, BSs, or gateways are located. MEC servers have limited processing power and storage capacities.
- **Transmission distance:** The distance between users and MCC servers can vary from one kilometer to thousands of kilometers. On the other hand, the distance between the end user and the MEC server varies from tens of meters to hundreds of meters.

Table 11.4 MCC vs. MEC.

	MCC	MEC
Physical server	*High computing and storage capabilities, located in large-scale data centers*	*Limited capabilities, colocated with base stations and gateways*
Transmission distance	*Usually far from users, from kilometers to thousands of kilometers*	*Quiet dose to users, from tens to hundreds of meters*
System architecture	*Sophisticated configuration, highly centralized*	*Simple configuration, densely distributed*
Application characteristics	*Delay-tolerant, computation-intensive, e.g. Facebook, Twitter*	*Latency-sensitive, computation-intensive, e.g. autonomous driving, online gaming*

- **System architecture:** MCC systems are built and operated by colossal information technology companies such as Google and Amazon. The architectures of MCC systems are often very complex and highly centralized. Telecommunication companies, businesses, and communities usually install servers in MEC systems. These servers are widely distributed over the network and have simple configurations. MEC systems are hierarchically controlled, either centrally or distributed.
- **Application characteristics:** Applications in MCC systems generally tolerate a certain degree of latency but require large amounts of computational resources. Therefore, calculation data can be transmitted from end users to MCC servers for computing. Examples of typical MCC applications are online social networks such as Facebook and Twitter. On the other hand, MEC applications are sensitive to latency and calculations in cases such as autonomous driving, image recognition, and online games. The computing processes of MEC applications are run at the network edge to mitigate the long delays between the end user and the cloud (Figure 11.25).

Due to its different distribution architectures, MEC's performance is more critical than MCC in terms of latency, energy consumption, context-aware computing, security, and privacy. Today, the term multiaccess edge computing (MAEC) is also used for MEC. It is an architecture defined by European Telecommunications Standards Institute (ETSI) (Figure 11.26).

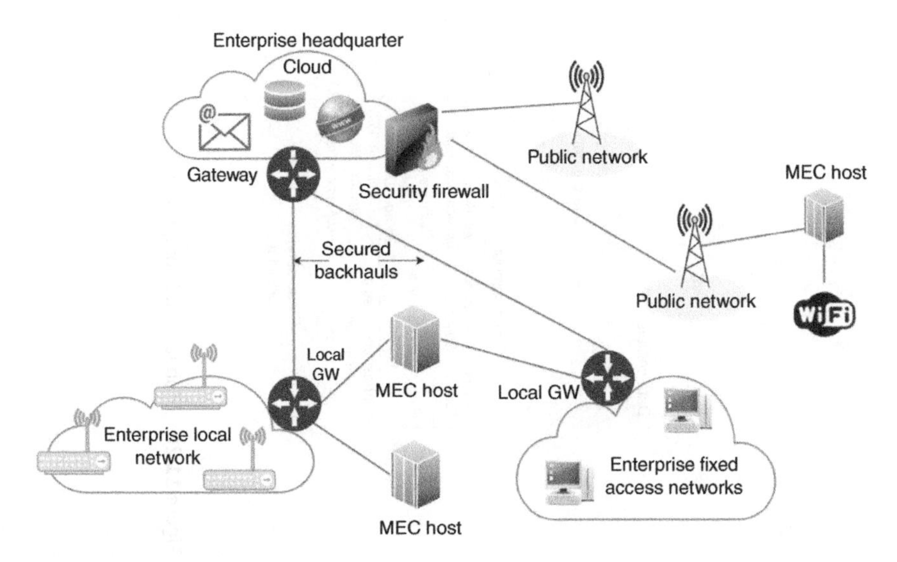

Figure 11.25 MEC application.

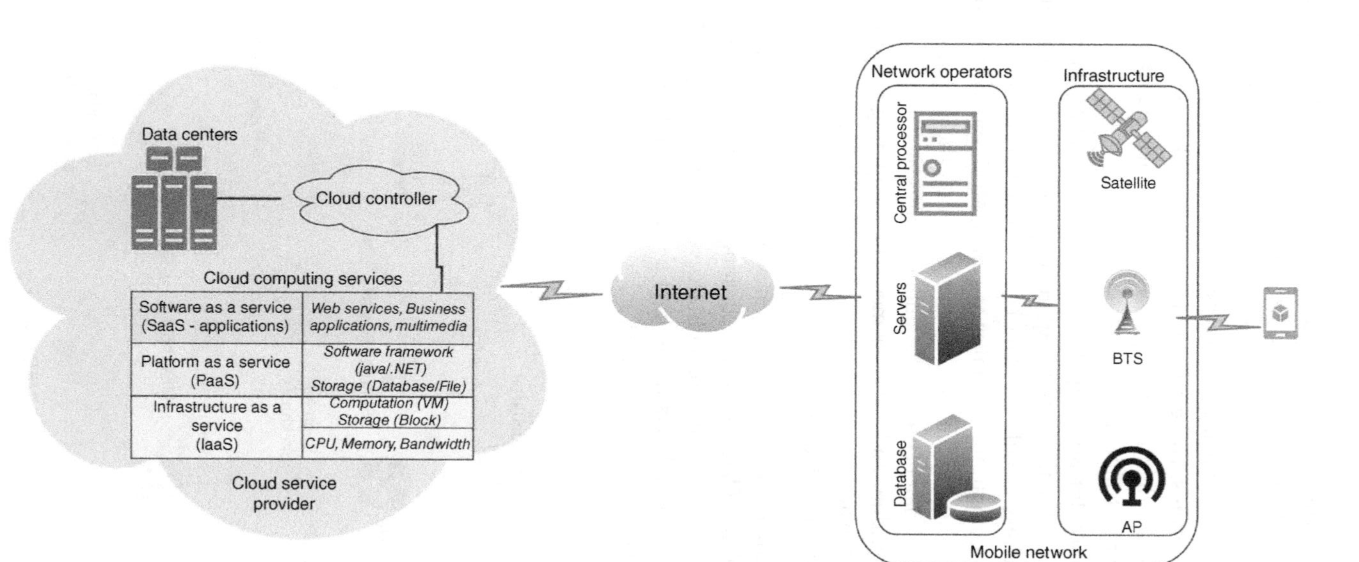

Figure 11.26 Mobile cloud computing.

By shifting the computing power of 5G and beyond systems, especially to the points close to the end user, the improvements in latency will also enable the development of latency-sensitive applications. The implementation of revolutionary applications in the 5G and 6G ecosystems will be realized mainly with MEC systems. MEC has many advantages over traditional cloud computing [47]. To give a real-life example, a study showed that the response time decreased from 900 to 169 ms with the migration of the face recognition application from the cloud to the edge [48]. This improvement was not only limited to response time but up to 30–40% improvements were observed in power consumption (Figure 11.27).

IoT devices are connected to the cloud system via wired/wireless environments. This connection architecture is given in Figure 11.28. The data received through the sensors are delivered to the cloud center via the core Internet to process the information. Due to delays, data transmission means a crash for real-time applications such as health monitoring, autonomous vehicles, production systems/assembly lines, and video surveillance. As we mentioned above, the concept of edge computing has emerged with the idea of extending computing, communication/networking, and control functions to extreme points. After edge computing, Fog, Mist, and Dew concepts have also been put into practice [49]. In terms of latency, it can be said that the cloud has the highest latency and power consumption, while dew computing has the lowest latency and power consumption.

The IoT structure running on the cloud system described in Figure 11.28 is given in Figure 11.29.

Edge runs on the wireless network's micro-/pico-/femtocell structure. This multilayer communication network overlays the layered cell structure using the shared spectrum. In edge applications like the IoT, data traffic flows in predictable patterns [50]. Therefore, dynamic channel allocation can be done by looking at historical data traffic. This way, it is possible to allocate the available channels to the users optimally [51]. It is natural to use cognitive radio technology to increase

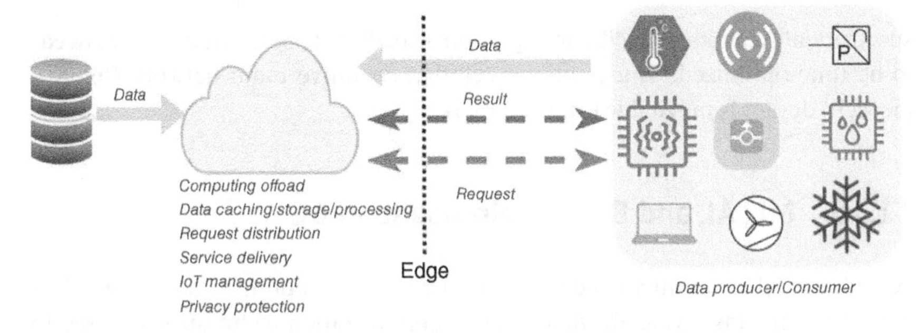

Figure 11.27 Edge computing paradigm.

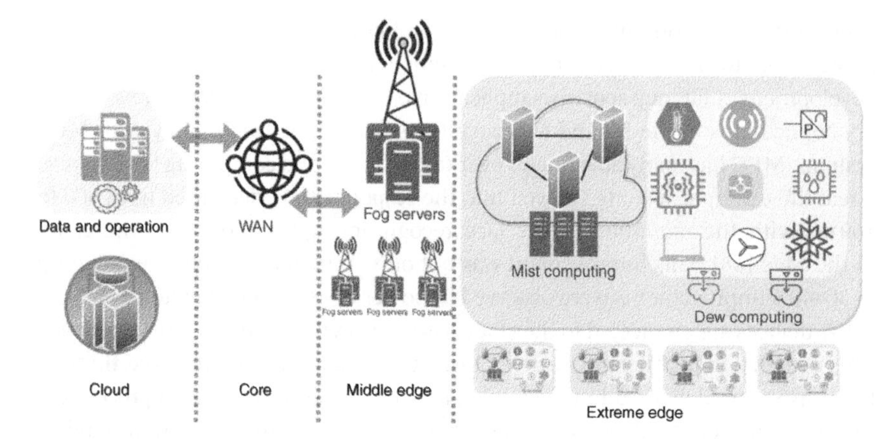

Figure 11.28 Cloud hierarchy.

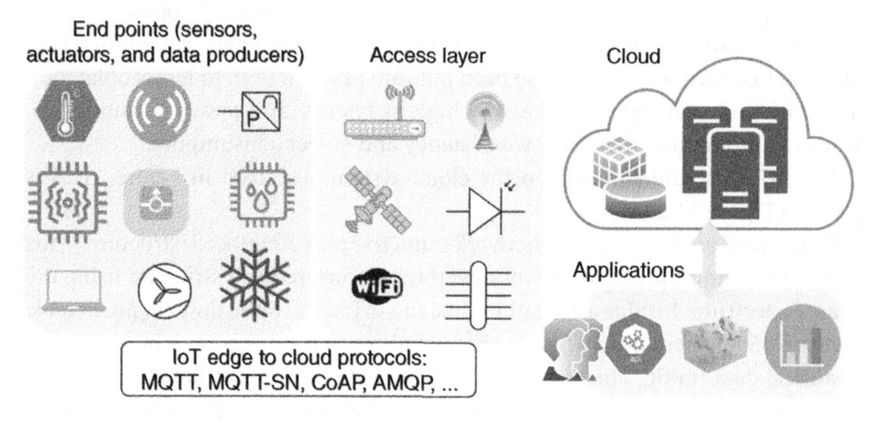

Figure 11.29 IoT cloud structure.

spectrum utilization. To facilitate cognition, broadband spectrum detection needs to be time-optimized. Thus, optimal cellular, cognitive radio network (femto-/picocell) design is inevitable (Figure 11.30).

11.10 ML, AI, and Blockchain Usage in 6G

6G will use AI as an integrated part with the ability to optimize various wireless network problems. Typically, mathematical optimization techniques are used to optimize wireless network problems. We can use convex optimization schemes, matching theory, game theory, heuristics, and brute force algorithms to solve

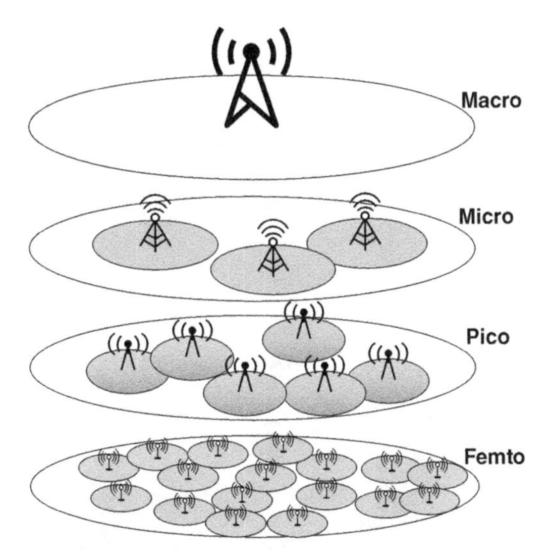

Figure 11.30 Multilayer wireless communication network.

these mathematical optimization problems. However, all these solution approaches may not yield the desired results due to the high complexity that significantly reduces the system capacity. On the other hand, ML can optimize various complex mathematical problems, including problems that cannot be modeled using mathematical equations [9–52].

6G offers a broad-scale, multilayered, high-complexity, dynamic, and heterogeneous network infrastructure. In addition, 6G networks must meet the requirements of a seamless connection environment and quality of service (QoS) of a massive number of connected devices, as well as processing vast volumes of datasets generated by millions of physical devices. At this point, since performance optimization, knowledge discovery, sophisticated learning, structure organization, and complex decision-making will need to be executed intelligently, AI techniques with strong analysis ability, learning ability, optimization ability, and intelligent recognition ability come into play in 6G networks. AI-powered 6G networks can be divided into four main layers: intelligent detection layer, data mining and analytics layer, intelligent control layer, and intelligent application layer [11] (Figure 11.31).

This section will explain the usage areas of ML and AI techniques in 6G systems.

11.10.1 Machine Learning

The applications of today's wireless communication technologies also give us clues for the applications to be developed shortly. Holographic telepresence, eHealth, continuous and widespread connection in intelligent environments,

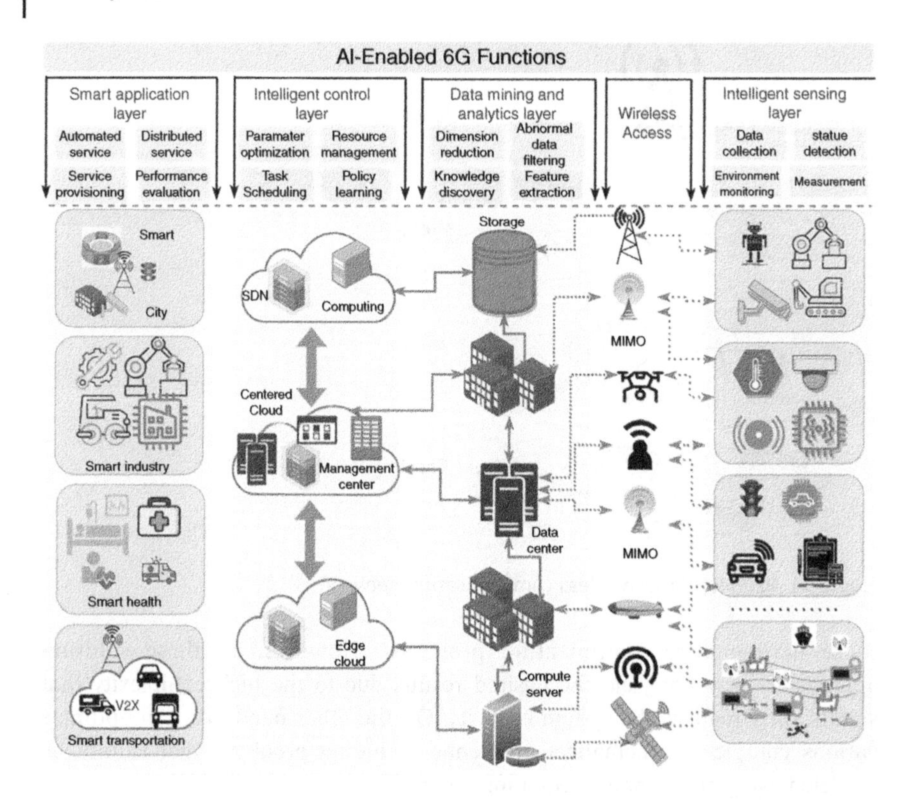

Figure 11.31 AI-enabled 6G functions.

virtual and augmented reality applications, three-dimensional unmanned vehicles, and intense robotic applications are just a few. The development of more intelligent IoT devices increases this need even more. The need for a more efficient, flexible, and robust wireless communication infrastructure is growing daily for applications that continue this development. This is where the 6G vision comes into play. The 6G infrastructure offers solution points for each of these needs.

The size of the data produced by the devices that will be connected at all times and everywhere has started to reach incredible levels. From the beginning, text-based data exchange has now turned into a gigantic echo system that includes image and voice. This change simultaneously forms information stacks that carry knowledge, experience, and past-future knowledge. ML seems to be an essential solution at the point this continuous evolution will reach.

The increasing complexity of both core network devices and terminal devices and the complexity of the modulation and multiplexing techniques used led to a search for new approaches to physical components (channels, antenna patterns, data traffic, mobility, interference, scattering, etc.) and model-based mathematical methods. For example, in systems beyond 5G, network slicing, multiple service classes, and signal interference became a nonstationary and non-Gaussian structure. Similarly, low latency requirements require multilayered, highly structured network architecture. These challenges have led researchers to data-based solutions as well as model-based solutions. This is because data-based approaches are based on optimum learning rather than very complex models, and optimizing wireless communication channels requires computational time and cumbersome mathematical solutions [53].

Of course, questions about ML-based algorithms processing high volumes of data, where ML agents' learning and training processes will be (mobile device, cloud, or edge cloud), and how they will be orchestrated (distributed, centralized, or hybrid) are important challenges. It is important whether the results obtained while performing all processes will be energy efficient.

In wireless communications, the amount of training data (the dataset needed to train the algorithm) is still far from being compared to the vast datasets used by major industry players for core deep learning applications such as computer vision and speech recognition. Due to the multidimensional dataset requirement, deep learning models require massive training datasets to achieve significant performance gains compared to simpler models. Another limiting factor is network diversity due to the coexistence of different mobile network operators. Even if standardized dataset formats ensure interoperability, network application management functions can differ significantly from other operators. Also, due to business-oriented policies, operators may keep the data obtained confidential. These factors can conclude deep learning alone will not be the optimal solution for all data analysis tasks in 6G networks. Instead, various application and platform-dependent models will be required to enable cognitive decision-making even on resource-constrained platforms such as ultralow-power microcontrollers [54].

ML application roles in the multilayered structures are given in Figure 11.32. ML algorithms must be set up and trained at different network layers. The management layer, core, radio BSs, and mobile devices constitute these layers [52].

Since ML algorithms are not in the book's scope, they will not be explained separately, but which algorithms are used will be briefly mentioned.

Artificial neural networks (ANNs) are mainly used to solve the problems seen in wireless networks. If we explain in terms of general usage, multilayer perceptrons (MLPs) are a basic ANN model used in many learning tasks. Convolutional neural networks (CNNs) are another ANN algorithm that reduces data input size and is mainly used in image recognition applications. Recurrent neural networks (RNNs) are the most suitable algorithm for learning topics that require sequential modeling. Auto-coding-based deep learning models used in dimension reduction

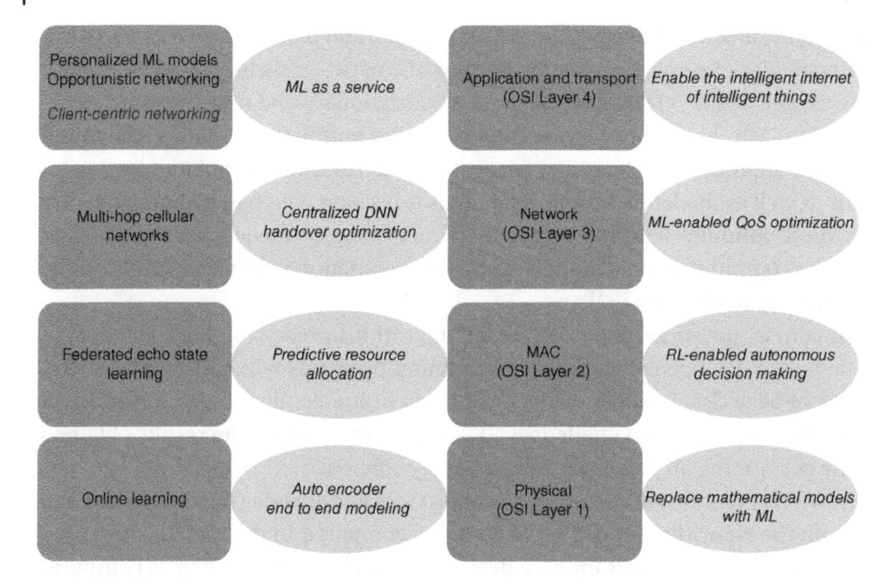

Figure 11.32 Machine learning (ML) in communication.

and generative adversarial networks (GANs) are used to generate samples similar to the dataset.

Some areas where deep learning algorithms can be used for wireless communication networks are given as follows:

- **Physical layer:** Electromagnetic interference detection, uplink/downlink reciprocity on FDD, channel estimation, channel coding, synchronization, location services, signal beaming for smart antenna systems, and physical layer optimization are performed in this layer.
- **MAC (Medium Access Control) layer:** Orientation and mobility prediction in virtual reality (VR) networks, predictive resource allocation in machine-type communication, predictive power management, and asymmetric traffic matching are the responsibility of this layer.
- **Security:** Detection of legal and illegal traffic, classification of network traffic data, spectrum sharing, and secure routing sharing.
- **Application layer:** Network performance management organization, ML-supported UAV control, and appropriate state data transfers for vehicle networks are performed in this layer.

11.10.2 Blockchain

In recent years, blockchain and (as it is generically known) distributed ledger technology has gained momentum and been adopted by the industry and research communities worldwide. Technically, we can think of blockchain as an ever-growing list that keeps growing lists of technology transaction records. If we count the technologies that blockchain technology has added to our lives [55]:

- Evolving the central security structure to a distributed system by eliminating the need for mechanisms that require centrally trusted third parties and intermediaries.
- Transparency through anonymity.
- Unable to deny the changes made to the system with technology or the center that took the action.
- Immutability and tamper resistance of distributed ledger content.
- Preventing all systems due to single-point failure (for example, flexibility and resistance to attacks such as DoS or DDoS).
- Comparatively less transaction delay and transaction fee.

For the reasons listed here, blockchain technology will be an inevitable technology for establishing security in next-generation networks [56].

Usage areas of blockchain technology on 6G (also 5G) are given as follows:

- **Smart resource management:** Network resource management and sharing will be one of the most critical milestones for 6G. Spectrum sharing, orchestration, and decentralized computing should be considered in this area [57].
- **Privacy protection:** Privacy is one of the essential topics in terms of security. Blockchain technologies are recommended for the confidentiality of data being transmitted or stored, especially in substantial data stacks [58].
- **Authentication, authorization, and accounting – AAA:** The authentication process checks whether the user is defined on the system. The authorization control performs the user's authorization level passing through the authentication step (which operations will be allowed). In the first two steps, the transactions conducted by the user logging into the system should be recorded. This job is called accounting.

  Since 6G networks have large-scale connectivity with heterogeneous and fragmented network elements, AAA functions need to be built in a much stronger structure for these decentralized systems to ensure service continuity [58]. For example, (group) key management and access control mechanisms can be installed on blockchain platforms for better scalability (especially for resource-constrained endpoints) and transparency. The network's security, surveillance, and management can be implemented through distributed ledger technology. The distributed ledger is a viable technology in this field as it remains an immutable and transparent log for each event that can be used to audit events.
- **Integrity:** Since integrity is an essential concept in large volumes of data that will emerge with following generation calculations, blockchain technology will also be applied in this area.

- **Availability:** Continuing the availability of the service provided in the case of any attack is an essential topic regarding network metrics. There are proposed blockchain studies against DDoS attacks [59].
- **Accountability:** These records can be stored with blockchain technology to ensure that the list of transactions performed by the user on the system cannot be changed.
- **Scalability:** The scalability limitations of centralized systems will be overcome with blockchain and smart contracts to enable large-scale connectivity in the future.

Applications and services where blockchain technology is used on 6G (also 5G) are given as follows:

- **Industrial applications beyond Industry 4.0:** Holographic communication can be used for industrial use cases such as remote maintenance or dense connectivity of industrial production equipment. Such forms of connectivity require reliable decentralized architectures. Blockchain can provide these capabilities when integrated into these applications or use cases [60].
- **Seamless environmental monitoring and protection:** Blockchain technology can also be used in decentralized and collaborative environmental sensing applications that can be realized globally with 6G. These capabilities, provided by a blockchain, can serve use cases such as smart cities or transportation and environmental protection for the green economy.
- **Smart health apps:** Smart healthcare in 6G will need to go further to address the built-in problems in 5G networks. With blockchain technology's further and omnipresent integration into future networks, existing healthcare systems can be further enhanced, and decentralized infrastructures' security and privacy performance can be improved.
- **Service level agreement – SLA:** 6G wireless networks (similar to 5G networks) are virtualized and sliced but built into a much larger scale network architecture. These networks are set up to serve a broad spectrum of usage scenarios that require different levels of service level agreements. As such, SLA management emerges as an essential system requirement. Blockchain technology provides a decentralized and secure SLA management platform in a situation that requires these complex settings.
- **Spectrum sharing:** Capacity expansion and spectrum variability (change from MHz to THz) for 6G access networks are not manageable by centralized management structures and uncoordinated sharing schemes. Blockchain and smart contracts can be used as a solution for spectrum sharing with collaboration and transparency features [61].
- **Extreme edge:** For 6G networks to provide extremely poor communication and dynamic network usage, it needs to support many primary services from

cloud systems to edge networks. Reliable coordination and transparent resource accounting can be achieved with blockchains in these systems [62]. The usage areas and paradigms of ML in 6G systems were explained in the previous sections. As described, ML can be used in autonomous management and service classification areas and in addressing reconfigurability needs in next-generation systems. These data-driven learning and quantum-assisted computing methods have a significant potential for realizing service-based and fully intelligent 6G wireless communication systems. In this context, significant increases are expected in human-machine connectivity, network nodes, and data traffic [63, 64].

11.11 Quantum Computing in Future Wireless Networks

So, how will quantum calculations take place in these revolutionary changes and developments expected in the near future? In this section, we will try to answer this question. Quantum technology uses features of quantum mechanics, such as the interaction of molecules, atoms, and even photons and electrons, to create devices and systems such as ultraprecise clocks, medical imaging, and quantum computers. However, the full potential of quantum computing (QC) remains to be explored, as research on this topic is still in its infancy. The Quantum Internet is designed as an infrastructure to connect quantum computers, simulators, and sensors through quantum networks and securely distribute information and resources around the world [65].

In seeking to meet the rapidly increasing demands of fast, reliable, secure, intelligent, and green communication, the need for the high computational capability of systems has also grown rapidly. The natural parallelism offered by the fundamental concepts of quantum mechanics and the prospects set by the latest QC technology has a definite potential to outperform conventional computing systems [66]. This immense power of QC derives from fundamental QC concepts such as quantum superpositions, quantum entanglement, or the cloning theorem [67].

In applications that require high computational capacities, such as power field access supported by sequential interference cancellation, optimal routing of data packets in a multipass network, load balancing, channel estimation decoding, and multiuser transmission, QC is seen as a potential solution with its inherent parallel computing capability [65]. This similar computing power comes from the principles of quantum physics, quantum bit (qubit), entanglement, and superposition concepts.

In classical computers, only one of the $2n$ states can be encoded with n "bits" of 1s and 0s at a given time. However, all these $2n$ states, which we talk about with

n "qubits," can be encoded simultaneously in quantum computers. This magic comes from the qubit's superpositions (being 0 and 1 simultaneously!). In classical binary base mathematics, a bit is either 1 or 0 at a given time. In other words, a bit can only be in one of two states at a given time. At any given time, when you look at the state of a bit, it doesn't change a bit if you assume there is nothing to change that bit's state. In the quantum case, however, the notation changes slightly. When we read information from the qubit through a process called "measurement," the qubit always goes into a state. However, when calculating with the qubit before measurement, it is possible to move it to an infinite number of other states and change from one to the other (Figure 11.33).

We can then read this as a bit value of 0 or 1. We represent all states that the qubit can be in as points on the unit sphere |0> at the north pole and |1> at the south pole. Notice that all points on the sphere are equal to quantum states. Mathematically, a qubit always takes a value from one of the "superpositions" whose state is 0 or 1 or in between. In other words, superposition refers to all states except 0 or 1. If a bit is not in one of the poles, we are talking about a non-trivial superposition. Moving to super usually means moving the state to a point above the equator [68].

This enormous speed and parallelism are why quantum principles want to be used in communication. It is inevitable to use quantum physics principles for complex algorithms that require high processing power and speed and for the vast datasets that these algorithms process. These quantum mechanics concepts can generate intuitive statistical data patterns that classical computers cannot produce effectively [69].

Classical ML methods also produce data with the same characters to determine statistical data characteristics on a given dataset. Classical ML algorithms deal

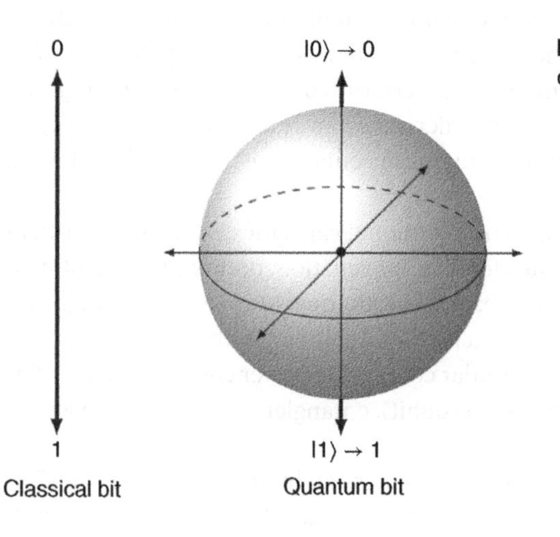

Figure 11.33 Classical bit and a quantum bit (qubit).

with the manipulation and classification of large datasets in large vector forms. Here the polynomial time required for computation is proportional to the data size. On the other hand, QC has excellent potential to properly manipulate such large-size data vectors in large tensor product spaces. In addition, with the combination of QC and classical ML features, the production and determination of statistical data patterns that cannot be performed effectively with classical computer and classical ML methods can be performed effectively. With this combination, the concept of quantum machine learning (QML) has been introduced into our lives.

Quantum principles can be used in many areas, from terrestrial communications to satellite and maritime communications. The most debated issue in this regard is whether quantum communication can be used in optical fiber communication based on classical electromagnetic fields and affected by undesired waves. Also, the noise ratio of quantum mechanical origin can be limited by the performances of photodetectors. To address all these issues, optical communication can be designed under a quantum mechanical framework [70].

One of the most promising issues with quantum communication is security. Infinitely secure communication environments can be realized with the quantum key distribution (QKD) protocol, in which any third parties cannot obtain security keys due to quantum mechanics principles being used. The Quantum Internet, which is about the communication of qubits from one computer to another, is considered a potential infrastructure of the near future [71]. Another important application of quantum communication is transferring a particular quantum state to another place with quantum devices using classical bits instead of quantum bits. The quantum entanglement principle is used for this application [72].

The opportunities that QC will bring to 6G are summarized as follows [73]:

- **Quantum-Assisted Radio Access Networks (qRAN):** Both QC and quantum communication can be leveraged to improve RAN efficiency and security. For example, radio resource allocation and cell planning can be realized using quantum searching algorithms, providing higher energy and spectrum efficiency [74]. In addition, in open RAN (O-RAN) applications, high-security environments can be realized thanks to quantum communication.
- **Quantum Space Information Networks (qSIN):** Thousands of altitude satellites (LEO, MEO, and GEO) are installed in Earth orbit to realize the 3D wireless networks we mentioned about satellite integration with 6G. Space information networks (SINs) in this mesh can often be interconnected by free-space optical links and a powerful satellite hosting a quantum computer. Thus, quantum communication can be established between two satellite nodes, and satellite-ground communication can be realized using free-space optical channels as quantum channels. In addition, powerful satellite nodes with quantum

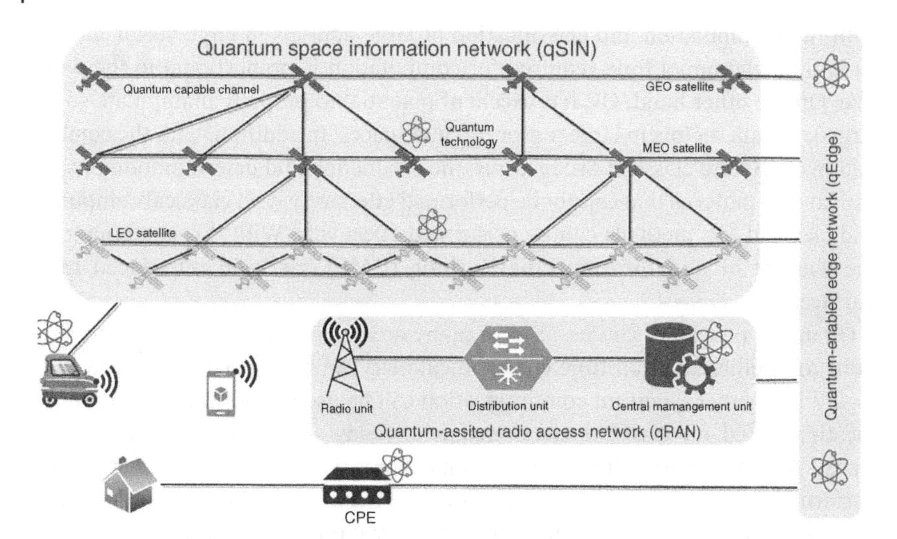

Figure 11.34 Quantum capable access network.

computers can provide QC services to other satellites and ground stations. On the other hand, a satellite node can be used as a trusted node or quantum repeater to assist and enhance quantum communications, such as satellite-based QKD [75] (Figure 11.34).

- **Quantum-Assisted Edge Networks (qEdge):** With 6G, an increasing number of edge nodes are expected to be installed to provide general edge computing services. With such a distributed structure, security, task offloading, and edge resource allocation also arise. Using QC to establish secure quantum communication between nodes and to find the optimum solution for edge resource allocation and task offloading are recommended solutions for these problems [73].
- **Quantum-Assisted Data Center (qDC):** In modern data centers, optical fiber and free-space optic connections are established between racks to increase data rates and avoid interference. These optical links can also be used as quantum channels. As a result, quantum communication and quantum encryption can be used to improve inter-rack security [76]. QC can also be used to solve the optimal data flow and energy consumption management problems for data centers.
- **Quantum-Assisted Blockchain (qChain):** Blockchain, or distributed ledger technology, elegantly combines several mechanisms (for example, distributed consensus protocols, distributed database, cryptography, and hashing) to realize a decentralized system with multiple advantages such as transparency and immutability. Blockchain technology can be used in many applications, such as decentralized authentication and distributed wireless resource sharing between parties that do not trust each other. However, blockchain technology also

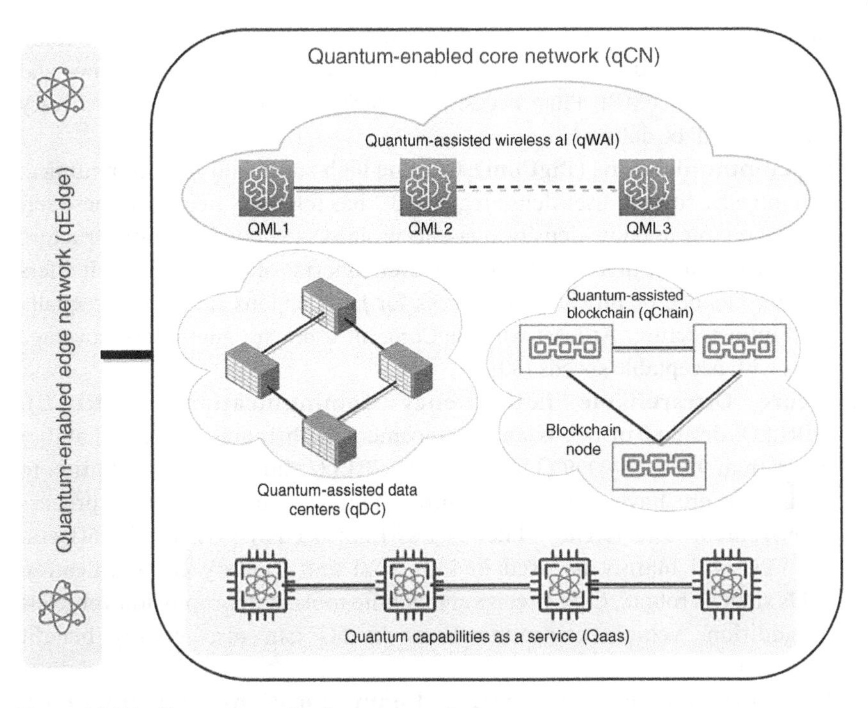

Figure 11.35 Quantum-enabled core network.

inherits potential issues such as security attacks from malicious nodes, slow transaction speed due to consensus protocols, and privacy breaches due to transparent data contained in blocks. These problems can be mitigated or resolved using quantum information technologies (QIT) called qChain [77].

- **Quantum-Assisted Wireless Artificial Intelligence (qWAI):** The 6G system will be more intelligent and autonomous thanks to the availability of big data from ubiquitous devices and network nodes and the application of AI algorithms such as deep learning, deep reinforcement learning, unified learning, and transfer learning [78]. Here, QIT will be helpful in the features of QC, such as security and optimization (Figure 11.35).

11.12 5G Concepts in 6G (eMBB, uRLLC, and mMTC)

We mentioned that 5G communication systems have three focal points: eMBB, uRLLC, and mMTC. 6G systems will come with developing these three concepts and additional components. These topics are reviewed as follows [79]:

- **Enhanced Mobile Broadband Plus (eMBB-Plus):** It is an enhanced version of eMBB in 5G. It offers much higher requirements and standards than eMBB. eMBB-Plus promises a higher capacity for extensive data transmission and processing and mobile network optimization in terms of interference and handover. With eMBB-Plus, necessary security, confidentiality, and privacy standards will be defined.

- **Big Communications (BigCom):** Offering high-speed and good communication infrastructure in user-dense regions, 5G has relatively neglected these services in remote and low-density areas. 6G promises to provide an infrastructure that will eliminate this distinction. However, in terms of being feasible, it offers a better balance of providing resources for both regions rather than equally good infrastructure. At least with BigCom, all users are guaranteed coverage areas with acceptable speeds [80].
- **Secure Ultrareliable Low-Latency Communications (SURLLC):** SURLLC, defined in 6G, is an enhancement with much higher reliability (more than 99.999999999%) based on 5G URLLC and mMTC. In addition to security, there have also been improvements in quality control, process improvement, and latency (latency less than 0.1 ms) [81]. In the 6G era, SURLLC will mainly be used in industrial and military communication fields such as robots, high-precision machine tools, and propulsion vehicles. In addition, vehicle communications in 6G can also greatly benefit from SURLLC.
- **Three-Dimensional Integrated Communications (3D-InteCom):** 3D-InteCom in 6G emphasizes that network analysis, planning, and optimization should be increased from two dimensions to three dimensions, thus taking into account the number of communication nodes. UAVs and underwater communications are examples of this three-dimensional scenario. In both structures, three-dimensional analysis, planning, and optimization can be carried out. Accordingly, the analytical framework created for two-dimensional wireless communication arising from stochastic geometry and graph theory needs to be updated in the 6G era [82]. Considering the number of nodes also enables the implementation of height beamforming with full-size MIMO architectures, providing another aspect for network optimization [83].
- **Unconventional Data Communications (UCDC):** UCDC is probably the most open-ended application scenario in 6G communication. Currently, the definition and regulation of UCDC still await further research, but it is emphasized that it should at least cover holographic, tactile, and human-connected communications.

In light of the explanations given here, the comparison between 5G and 6G is shown in Figure 11.36.

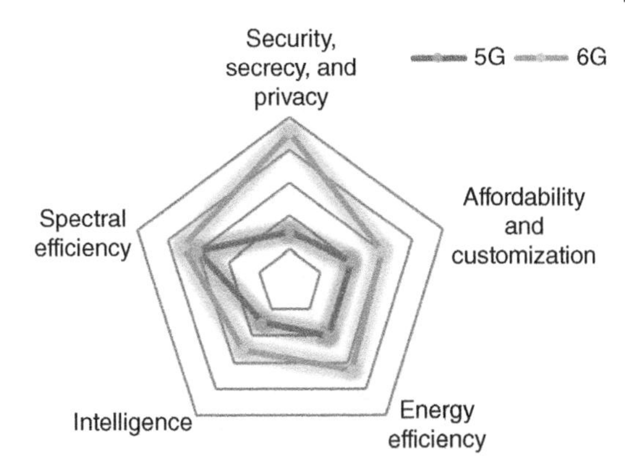

Figure 11.36 Qualitative comparison of 5G–6G systems.

11.13 6G Use Cases

Up to this section, the evolution from 5G to 6G and the 6G concepts that will emerge with this evolution have been explained in detail. This section will give the services offered to the users with the 6G infrastructure. These service and usage scenarios are summarized in Figure 11.37 as application areas [9].

Before explaining each title, it is helpful to summarize the content and application developments throughout communication generations. This journey, starting from 1G communication and extending to 6G, which applications have entered and will enter our lives, is summarized in Figure 11.38. In light of this summary, 6G application scenarios will be explained in detail in the subtitles.

11.13.1 Virtual, Augmented, and Mixed Reality

Virtual reality (VR) services are simulated experiences that will allow the user to experience a virtual and immersive environment from a first-person perspective. VR technology can potentially enable geographically separated people to communicate effectively in groups. They can make eye contact and manipulate common virtual objects. It will require the real-time movement of extremely high-resolution electromagnetic signals to geographically distant locations to convey various thoughts and emotions. Extended reality (XR) with high-definition imaging and 4K/8K high-resolution, entertainment services (including video games and 3D cameras), education and training, meetings with physical and social experience, workplace communication, etc., will be used appropriately in innovative applications [84]. These new applications will likely fill the existing 5G spectrum requiring data rates over 1 Tbps. Additionally, real-time user interaction in an immersive environment will require minimal latency and ultrahigh reliability (Figure 11.39).

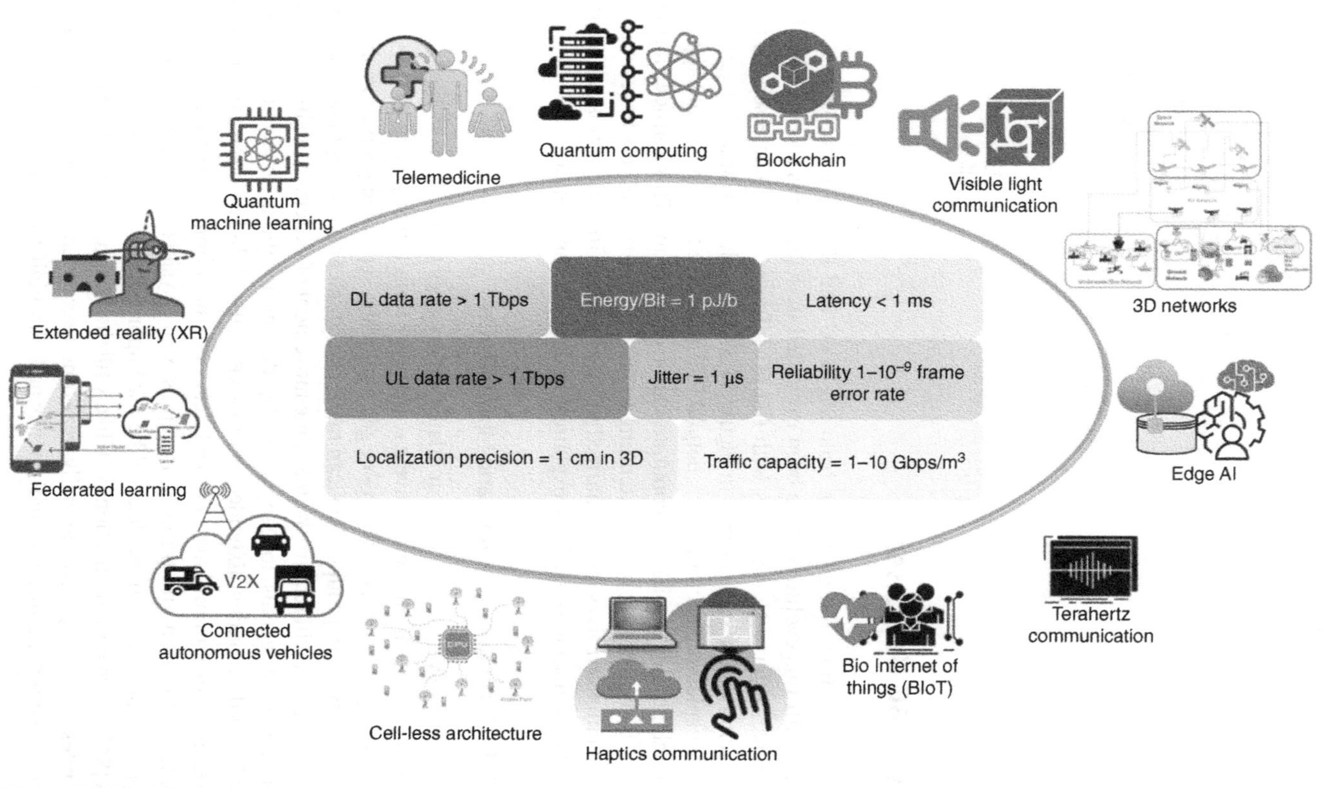

Figure 11.37 6G use cases.

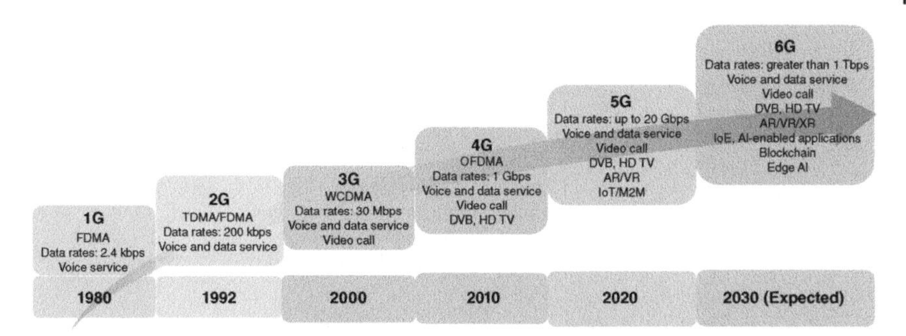

Figure 11.38 From 1G to 6G.

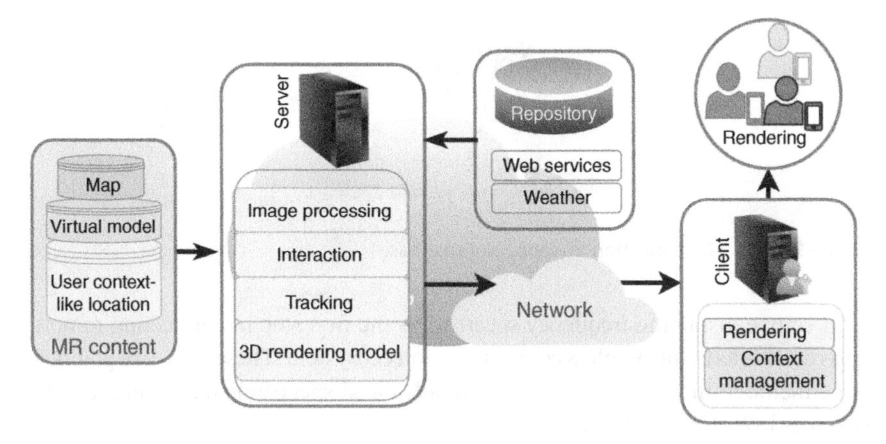

Figure 11.39 General architecture of mixed reality system.

11.13.2 Rural Areas/Depopulated Areas

Approximately, 50% of the global population lives in rural areas and villages. 67% of the world's population has a mobile subscription, but 3.7 million do not have an Internet connection, and most live in remote and rural areas [85]. As such, people in these regions cannot benefit from many advantages (education, health, online shopping, etc.) brought by high technology. The importance of these advantages has become more evident in the worldwide COVID-19 epidemic in 2020. Wireless connections in rural areas are expected to have significant economic implications.

At this point, connection scenarios with 6G have started to be produced for regions where a relatively radical solution was not proposed in previous generations [86]. From a commercial point of view, an important marketing segment will also join the digital economy.

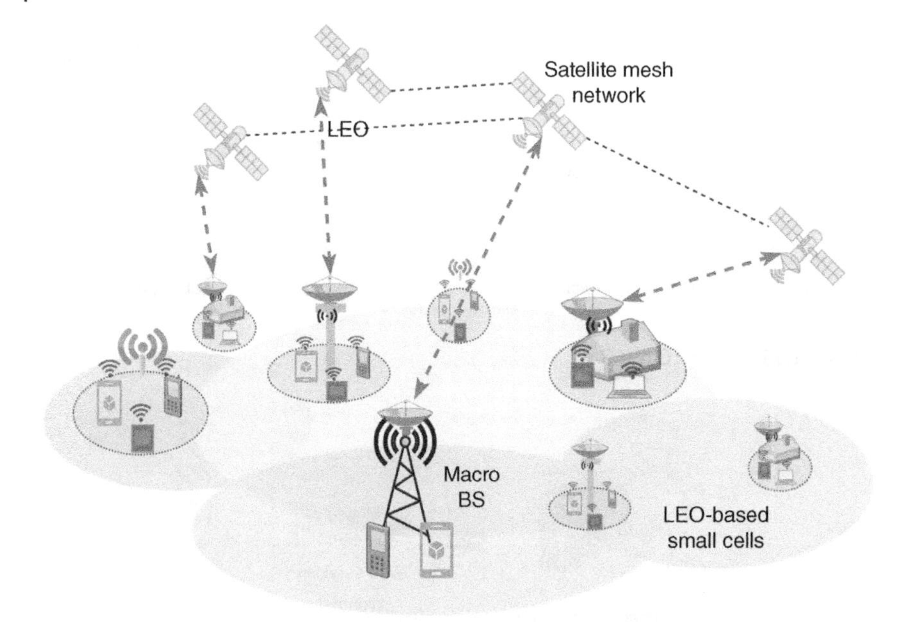

Figure 11.40 6G connection in depopulated areas.

Of course, using the frequency spectrum is the first step in connecting remote and rural areas to the wireless communication ecosystem. These areas are planned to be included in the coverage area with advanced spectrum sharing and co-use schemes [87] (Figure 11.40).

11.13.3 Nonterrestrial Communication

Disconnection during natural disasters causes significant damage to human life, property, and business. Nonterrestrial communication (NTC) will also be explored to support ubiquitous coverage and large-capacity global connectivity with 6G technology. To overcome the coverage limitations of 5G, 6G technology efforts focus on exploring nonterrestrial networks (NTNs) to support global, ubiquitous, and continuous connectivity (Figure 11.41).

NTN can assist in dynamically offloading traffic from terrestrial components and reaching unserved areas. Therefore, nonterrestrial stations such as UAVs, high-altitude platform stations (HAPSs), drones, and satellites are likely to complement terrestrial networks. This will provide many benefits, such as cost-effective coverage in congested areas, support for high-speed mobility, and high-throughput services. It can be considered that NTNs support applications including meteorology, surveillance, broadcast information, remote sensing, and

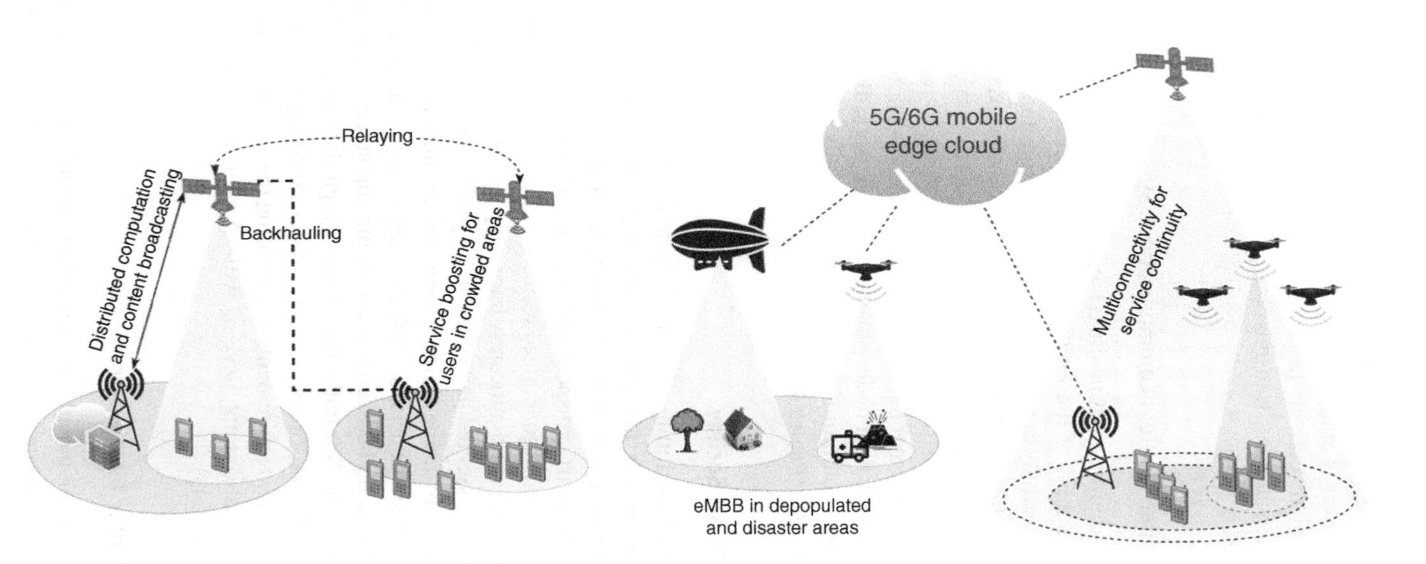

Figure 11.41 Integration of terrestrial and nonterrestrial networks.

navigation. It can supplement terrestrial towers, serving as an alternative route in case the terrestrial tower is out of service. Similarly, NTC can assist in delivering alert and entertainment content to large static and mobile audiences. Long-distance and intersatellite transmission can be achieved with laser communication [88].

11.13.4 Underwater Wireless Communications Systems

The underwater communications medium is crucial for providing worldwide connectivity as it covers most of the Earth's surface. Underwater networks will be used to provide connectivity as well as to observe and monitor various ocean and deep sea activities. Unlike land, water exhibits different propagation properties. Thus, bidirectional underwater communication requires more underwater hubs and also needs to use acoustic and laser communications to realize high-speed data transmission. Underwater networks establish the connection between underwater BSs and communication nodes of submarines, sensors, divers, etc. In addition, this underwater communication network can also be coordinated with terrestrial networks [89] (Figure 11.42).

11.13.5 Super Smart Society

The 6G technology of the future will contribute to creating a smart city environment. This environment is expected to have tight connectivity requirements connecting millions of applications, including utilities (electricity, water, and waste management), smart transportation, smart grid, residential environment, telemedicine, shopping with guaranteed security, etc. Smart devices' seamless and ubiquitous connectivity will significantly improve people's quality of life.

Smart homes are an essential part of the intelligent lifestyle. In the beginning, the concept of the smart home mainly progressed through the development of electricity meters and smart devices. Still, this trend is being renewed because it facilitates close integration between IoT and home appliances, enabling connectivity anytime, anywhere. Naturally, this requires high data rates and extremely high security of users' data. The 6G system envisions meeting this stringent data rate, latency, and security requirements by providing the necessary infrastructure at home and fully integrating devices with AI for autonomous decision-making [90] (Figure 11.43).

11.13.6 Holographic Telepresence

With the ongoing developments in technology, people have become highly dependent on the innovations supported by technology. One such innovation is

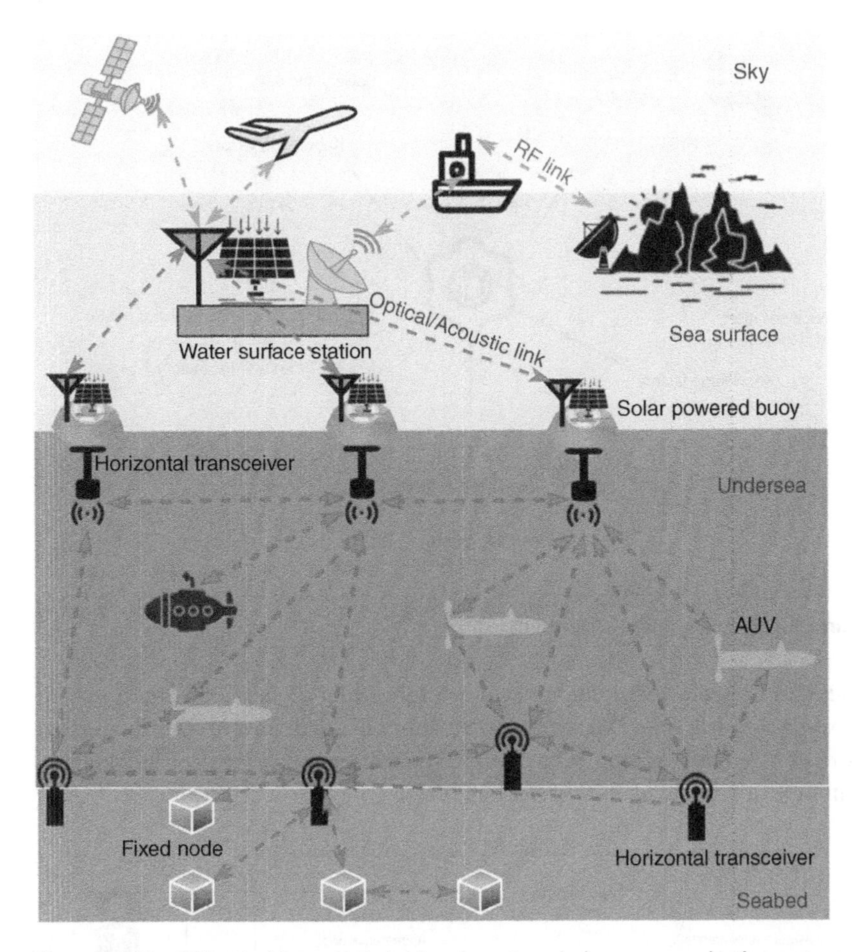

Figure 11.42 RF/optical/acoustic hybrid underwater wireless communication systems.

holographic telepresence, which allows people or objects in different positions to appear directly in front of another person. It can potentially bring about a tremendous change in how we communicate and is slowly becoming a part of mainstream communication systems. Some interesting applications of telepresence include enhancing the experience of watching movies and television, playing games, controlling robots, remote surgery, etc. People's tendency to connect remotely may experience a shift from a traditional video conference to a virtual face-to-face meeting, thus reducing the need for travel for work. A three-dimensional image combined with stereo sound is required to capture the physical entity. The three-dimensional holographic display would require a massive data rate of about 4.32 Tbps over the ultrareliable communication network.

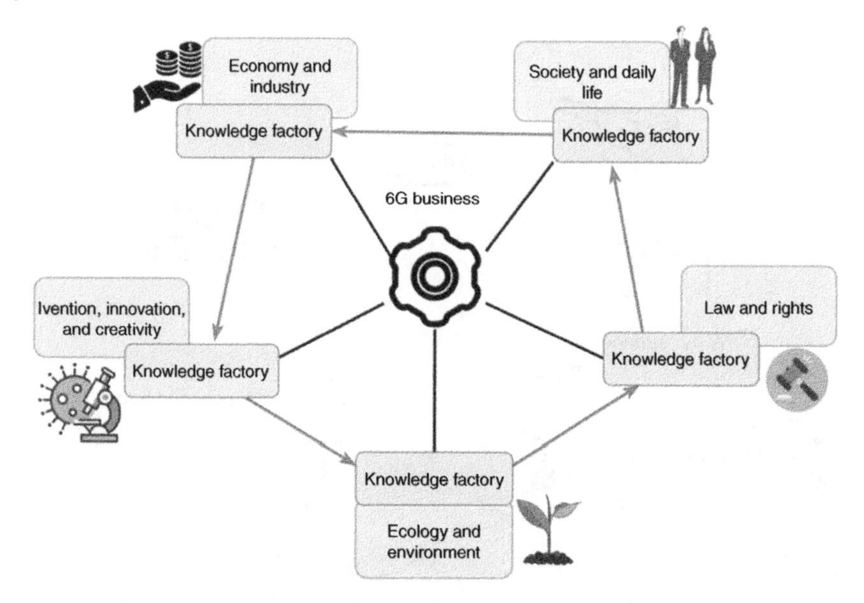

Figure 11.43 6G business ecosystem.

Similarly, latency requirements will be in the order of submilliseconds, which helps synchronize many viewing angles. This will create severe communication restrictions on the existing 5G network. The conceptual scheme of holographic communication is given in Figure 11.44 [91].

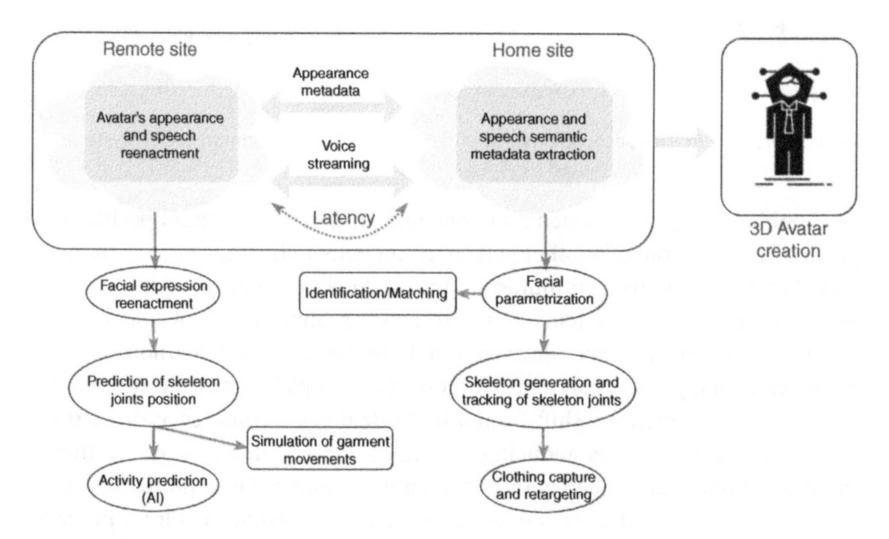

Figure 11.44 Holographic communication system architecture.

11.14 Comparison of 5G and 6G Network Architectures

5G and 6G systems have been explained in detail up to this section. To facilitate understanding, the sources of both generations are given in Figure 11.45. As can be seen from the figure, 5G layers have evolved into intelligent structures by integrating technologies such as AI, ML, blockchain, and QC [92].

With this intelligent structure, the changes seen in the network will be managed flexibly by predicting techniques and transmission infrastructures that will provide maximum benefit to users with intelligent traffic management. The backbone traffic has been alleviated with the dew and fog computing infrastructures installed at the edge points. With 6G, there will be no area that is not covered worldwide. Again, with 6G, wireless communication systems will become a kind of cyber organism. This massive change in the wireless network structure seems to change business and social life.

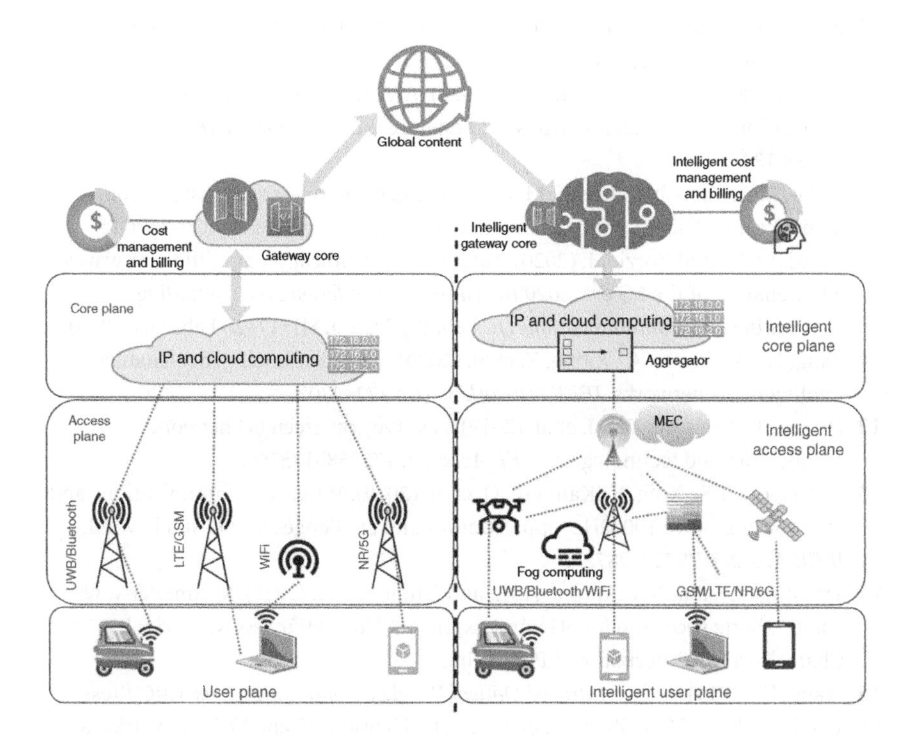

Figure 11.45 Comparison of 5G and 6G architectures.

References

1 Maier, M. (2021). Toward 6G: a new era of convergence. *2021 Optical Fiber Communication Conference (OFC)*, San Francisco, CA (6–10 June 2021).

2 Dang, S., Amin, O., Shihada, B., and Alouini, M.-S. (2020). What should 6G be? *Nature Electronics* 3 (1): 20–29.

3 Prasad, R. (2022). *6G: The Road to the Future Wireless Technologies 2030*. CRC Press.

4 Wu, Y., Singh, S., Taleb, T. et al. (2021). *6G Mobile Wireless Networks*. Springer Nature.

5 Khan, F., Pi, Z., and Rajagopal, S. (2012). Millimeter-wave mobile broadband with large scale spatial processing for 5G mobile communication. *2012 50th Annual Allerton Conference on Communication, Control, and Computing (Allerton)*, Monticello, IL (1–5 October 2012), pp. 1517–1523. IEEE.

6 Chowdhury, M.Z., Shahjalal, M., Ahmed, S., and Jang, Y.M. (2020). 6G wireless communication systems: applications, requirements, technologies, challenges, and research directions. *IEEE Open Journal of the Communications Society* 1: 957–975.

7 Ziegler, V., Viswanathan, H., Flinck, H. et al. (2020). 6G architecture to connect the worlds. *IEEE Access* 8: 173508–173520.

8 Saad, W., Bennis, M., and Chen, M. (2020). A vision of 6G wireless systems: applications, trends, technologies, and open research problems. *IEEE Network* 34 (3): 134–142.

9 Khan, L.U., Yaqoob, I., Imran, M. et al. (2020). 6G wireless systems: a vision, architectural elements, and future directions. *IEEE Access* 8: 147029–147044.

10 Giordani, M. and Zorzi, M. (2020). Satellite communication at millimeter waves: a key enabler of the 6G era. *2020 International Conference on Computing, Networking and Communications (ICNC)*, Big Island, HI (17–20 February 2020).

11 Yang, H., Alphones, A., Xiong, Z. et al. (2020). Artificial-intelligence-enabled intelligent 6G networks. *IEEE Network* 34 (6): 272–280.

12 Huang, T., Yang, W., Wu, J. et al. (2019). A survey on green 6G network: architecture and technologies. *IEEE Access* 7: 175758–175756.

13 Rappaport, T.S., Xing, Y., Kanhere, O. et al. (2019). Wireless communications and applications above 100 GHz: opportunities and challenges for 6G and beyond. *IEEE Access* 7: 78729–78757.

14 Tripathi, S., Sabu, N.V., Gupta, A.K., and Dhillon, H.S. (2021). Millimeter-wave and terahertz spectrum for 6G wireless. In: *6G Mobile Wireless Networks*, 83–121. Cham: Springer International Publishing.

15 Wang, K. (2019). *Indoor Infrared Optical Wireless Communications*. CRC Press.

16 Wu, X., Soltani, M.D., Zhou, L. et al. (2021). Hybrid LiFi and WiFi networks: a survey. *IEEE Communications Surveys and Tutorials* 23 (2): 1398–1420.

17 Plank, T., Leitgeb, E., Pezzei, P., and Ghassemlooy, Z. (2012). Wavelength-selection for high data rate Free Space Optics (FSO) in next generation wireless communications. *2012 17th European Conference on Networks and Optical Communications,* Vilanova i la Geltru, Spain (20–22 June 2012).

18 Khalighi, M.-A., Xu, F., Jaafar, Y., and Bourennane, S. (2011). Double-laser differential signaling for reducing the effect of background radiation in free-space optical systems. *Journal of Optical Communications and Networking* 3 (2): 145.

19 Koepf, G.A., Marshalek, R.G., and Begley, D.L. (2002). Space laser communications: a review of major programs in the United States. *AEU – International Journal of Electronics and Communications* 56 (4): 232–234.

20 Djordjevic, I.B. (2022). *Advanced Optical and Wireless Communications Systems.* Springer Nature.

21 Fang, X., Feng, W., Wei, T. et al. (2021). 5G embraces satellites for 6G ubiquitous IoT: basic models for integrated satellite terrestrial networks. *IEEE Internet of Things Journal* 8 (18): 14399–14417.

22 Sharma, S.K., Chatzinotas, S., and Arapoglou, P.-D. (ed.) (2018). *Satellite Communications in the 5G Era*. Institution of Engineering & Technology.

23 Evans, B., Werner, M., Lutz, E. et al. (2005). Integration of satellite and terrestrial systems in future multimedia communications. *IEEE Wireless Communications* 12 (5): 72–80.

24 Xu, C., Ishikawa, N., Rajashekar, R. et al. (2019). Sixty years of coherent versus non-coherent tradeoffs and the road from 5G to wireless futures. *IEEE Access* 7: 178246–178299.

25 Chen, Y., Liu, W., Niu, Z. et al. (2020). Pervasive intelligent endogenous 6G wireless systems: prospects, theories and key technologies. *Digital Communications and Networks* 6 (3): 312–320.

26 Prasad, R. and Henrique, P.S.R. (2022). *6G The Road to the Future Wireless Technologies 2030*. River Publishers.

27 Akyildiz, I.F., Nie, S., Lin, S.-C., and Chandrasekaran, M. (2016). 5G roadmap: 10 key enabling technologies. *Computer Networks* 106: 17–18.

28 Basar, E. (2019). Transmission through large intelligent surfaces: a new frontier in wireless communications. *2019 European Conference on Networks and Communications (EuCNC),* Valencia, Spain (18–21 June 2019).

29 Liaskos, C., Tsioliaridou, A., Nie, S. et al. (2019). An interpretable neural network for configuring programmable wireless environments. *2019 IEEE 20th International Workshop on Signal Processing Advances in Wireless Communications (SPAWC),* Cannes, France (2–5 July 2019).

30 Renzo, M.D., Debbah, M., Phan-Huy, D.-T. et al. (2019). Smart radio environments empowered by reconfigurable AI meta-surfaces: an idea whose time has come. *EURASIP Journal on Wireless Communications and Networking* 2019 (1): 129.

31 Huang, C., Zappone, A., Alexandropoulos, G.C. et al. (2019). Reconfigurable intelligent surfaces for energy efficiency in wireless communication. *IEEE Transactions on Wireless Communications* 18 (8): 4157–4170.

32 Wu, Q. and Zhang, R. (2019). Intelligent reflecting surface enhanced wireless network via joint active and passive beamforming. *IEEE Transactions on Wireless Communications* 18 (11): 5394–5409.

33 ElMossallamy, M.A., Zhang, H., Song, L. et al. (2020). Reconfigurable intelligent surfaces for wireless communications: principles, challenges, and opportunities. *IEEE Transactions on Cognitive Communications and Networking* 6 (3): 990–1002.

34 Chen, H.-T., Taylor, A.J., and Yu, N. (2016). A review of metasurfaces: physics and applications. *Reports on Progress in Physics* 79 (7): 076401.

35 Huang, C., Hu, S., Alexandropoulos, G.C. et al. (2020). Holographic MIMO surfaces for 6G wireless networks: opportunities, challenges, and trends. *IEEE Wireless Communications* 27 (5): 118–125.

36 Yurduseven, O., Marks, D.L., Fromenteze, T., and Smith, D.R. (2018). Dynamically reconfigurable holographic metasurface aperture for a Mills-Cross monochromatic microwave camera. *Optics Express* 26 (5): 5281.

37 Pizzo, A., Sanguinetti, L., and Marzetta, T.L. (2022). Spatial characterization of electromagnetic random channels. *IEEE Open Journal of the Communications Society* 3: 847–866.

38 Liaskos, C., Nie, S., Tsioliaridou, A. et al. (2018). A new wireless communication paradigm through software-controlled metasurfaces. *IEEE Communications Magazine* 56 (9): 162–169.

39 Wan, Z., Gao, Z., and Alouini, M.-S. (2020). Broadband channel estimation for intelligent reflecting surface aided mmWave massive MIMO systems. *ICC 2020 – 2020 IEEE International Conference on Communications (ICC)*, Dublin, Ireland (7–11 June 2020).

40 He, H., Yu, X., Zhang, J. et al. (2021). Cell-free massive MIMO for 6G wireless communication networks. *Journal of Communications and Information Networks* 6 (4): 321–335.

41 Choi, W. and Andrews, J. (2007). Downlink performance and capacity of distributed antenna systems in a multicell environment. *IEEE Transactions on Wireless Communications* 6 (1): 69–63.

42 Zhang, J., Chen, R., Andrews, J. et al. (2009). Networked MIMO with clustered linear precoding. *IEEE Transactions on Wireless Communications* 8 (4): 1910–1921.

43 Ngo, H.Q., Ashikhmin, A., Yang, H. et al. (2017). Cell-free massive MIMO versus small cells. *IEEE Transactions on Wireless Communications* 16 (3): 1834–1850.

44 Interdonato, G., Björnson, E., Quoc Ngo, H. et al. (2019). Ubiquitous cell-free massive MIMO communications. *EURASIP Journal on Wireless Communications and Networking* 2019 (1): 197.

45 Bjornson, E. and Sanguinetti, L. (2020). Scalable cell-free massive MIMO systems. *IEEE Transactions on Communications* 68 (7): 4247–4261.

46 Abbas, N., Zhang, Y., Taherkordi, A., and Skeie, T. (2018). Mobile edge computing: a survey. *IEEE Internet of Things Journal* 5 (1): 450–465.

47 Jabbarpour, M.R., Marefat, A., Jalooli, A., and Zarrabi, H. (2017). Could-based vehicular networks: a taxonomy, survey, and conceptual hybrid architecture. *Wireless Networks* 25 (1): 335–354.

48 Caffe (2022). Deep learning framework. *Berkeley Vision*. https://caffe.berkeleyvision.org/ (accessed 16 February 2023).

49 Cao, J., Zhang, Q., and Shi, W. (2018). *Edge Computing: A Primer*. Springer.

50 Al-Turjman, F. (2018). *Edge Computing*. Springer.

51 Grover, J. and Garimella, R.M. (2018). Optimization in edge computing and small-cell networks. In: *Edge Computing* (ed. F. Al-Turjman), 17–31. Springer Nature.

52 Jagannath, J., Polosky, N., Jagannath, A. et al. (2019). Machine learning for wireless communications in the Internet of Things: a comprehensive survey. *Ad Hoc Networks* 93: 101913.

53 Gunduz, D., De Kerret, P., Sidiropoulos, N.D. et al. (2019). Machine learning in the air. *IEEE Journal on Selected Areas in Communications* 37 (10): 2184–2199.

54 Zappone, A., Di Renzo, M., and Debbah, M. (2019). Wireless networks design in the era of deep learning: model-based, AI-based, or both? *IEEE Transactions on Communications* 67 (10): 7331–7376.

55 Hewa, T., Gur, G., Kalla, A. et al. (2020). The role of blockchain in 6G: challenges, opportunities and research directions. *2020 2nd 6G Wireless Summit (6G SUMMIT)*, Levi, Finland (17–20 March 2020).

56 Nguyen, T., Tran, N., Loven, L. et al. (2020). Privacy-aware blockchain innovation for 6G: challenges and opportunities. *2020 2nd 6G Wireless Summit (6G SUMMIT)*, Levi, Finland (17–20 March 2020).

57 Maksymyuk, T., Gazda, J., Han, L., and Jo, M. (2019). Blockchain-based intelligent network management for 5G and beyond. *2019 3rd International Conference on Advanced Information and Communications Technologies (AICT)*, Lviv, Ukraine (2–6 July 2019).

58 Yang, H., Zheng, H., Zhang, J. et al. (2017). Blockchain-based trusted authentication in cloud radio over fiber network for 5G. *2017 16th International Conference on Optical Communications and Networks (ICOCN)*, Wuzhen, China (7–10 August 2017).

59 Sharma, P.K., Singh, S., Jeong, Y.-S., and Park, J.H. (2017). DistBlockNet: a distributed blockchains-based secure SDN architecture for IoT networks. *IEEE Communications Magazine* 55 (9): 78–85.

60 Mahmood, N.H., Alves, H., Lopez, O.A. et al. (2020). Six key features of machine type communication in 6G. *2020 2nd 6G Wireless Summit (6G SUMMIT)*, Levi, Finland (17–20 March 2020).

61 Nguyen, D.C., Pathirana, P.N., Ding, M., and Seneviratne, A. (2020). Blockchain for 5G and beyond networks: a state of the art survey. *Journal of Network and Computer Applications* 166: 102693.

62 Alhosani, H., ur Rehman, M.H., Salah, K. et al. (2020). Blockchain-based solution for multiple operator spectrum sharing (MOSS) in 5G networks. *2020 IEEE Globecom Workshops (GC Wkshps)*, Taipei, Taiwan (7–11 December 2020).

63 Bockelmann, C., Pratas, N.K., Wunder, G. et al. (2018). Towards massive connectivity support for scalable mMTC communications in 5G networks. *IEEE Access* 6: 28969–28992.

64 Sharma, S.K. and Wang, X. (2020). Toward massive machine type communications in ultra-dense cellular IoT networks: current issues and machine learning-assisted solutions. *IEEE Communications Surveys & Tutorials* 22 (1): 426–471.

65 Nawaz, S.J., Sharma, S.K., Wyne, S. et al. (2019). Quantum machine learning for 6G communication networks: state-of-the-art and vision for the future. *IEEE Access* 7: 46317–46350.

66 Day, C. (2007). Quantum computing is exciting and important--really. *Computing in Science and Engineering* 9 (2): 104.

67 Gyongyosi, L. and Imre, S. (2019). A survey on quantum computing technology. *Computer Science Review* 31: 51.

68 Kietzmann, J., Demetis, D.S., Eriksson, T., and Dabirian, A. (2021). Hello quantum! How quantum computing will change the world. *IT Professional* 23 (4): 106–111.

69 Biamonte, J., Wittek, P., Pancotti, N. et al. (2017). Quantum machine learning. *Nature* 549 (7671): 195–202.

70 Shapiro, J. (2009). The quantum theory of optical communications. *IEEE Journal of Selected Topics in Quantum Electronics* 15 (6): 1547–1569.

71 Pinto, A.N., Silva, N.A., Muga, N.J. et al. (2017). Quantum communications: an engineering approach. *2017 19th International Conference on Transparent Optical Networks (ICTON)*, Girona, Spain (2–6 July 2017 2017).

72 Bennett, C.H., Brassard, G., Crépeau, C. et al. (1993). Teleporting an unknown quantum state via dual classical and Einstein-Podolsky-Rosen channels. *Physical Review Letters* 70 (13): 1895–1899.

73 Wang, C. and Rahman, A. (2021). Quantum-enabled 6G wireless networks: opportunities and challenges. *IEEE Wireless Communications* 29 (1): 58–69.

74 Botsinis, P., Alanis, D., Babar, Z. et al. (2019). Quantum search algorithms for wireless communications. *IEEE Communications Surveys and Tutorials* 21 (2): 1209–1242.

75 Khan, I., Heim, B., Neuzner, A., and Marquardt, C. (2018). Satellite-based QKD. *Optics and Photonics News* 29 (2): 26.

76 Hamedazimi, N., Qazi, Z., Gupta, H. et al. (2014). FireFly: a reconfigurable wireless data center fabric using free-space optics. *Proceedings of the 2014 ACM Conference on SIGCOMM*, Chicago, IL, 319–330 (August 2014).

77 Edwards, M., Mashatan, A., and Ghose, S. (2020). A review of quantum and hybrid quantum/classical blockchain protocols. *Quantum Information Processing* 19 (6): 184.

78 Letaief, K.B., Chen, W., Shi, Y. et al. (2019). The roadmap to 6G: AI empowered wireless networks. *IEEE Communications Magazine* 57 (8): 84–90.

79 Dang, S., Amin, O., Shihada, B. et al. (2020). What should 6G be? *Nature Electronics* 3: 20–29. https://doi.org/10.1038/s41928-019-0355-6.

80 Shi, H., Prasad, R.V., Onur, E., and Niemegeers, I.G.M.M. (2014). Fairness in wireless networks: issues, measures and challenges. *IEEE Communications Surveys and Tutorials* 16 (1): 5–24.

81 Hong, T. and Qi, F. (2021). Applications and implementations of 6G Internet of Things. In: *6G Wireless Communications and Mobile Networking* (ed. X. Xie, B. Rong, and M. Kadoch), 208–223. Bentham Science Publishers.

82 Haenggi, M., Andrews, J., Baccelli, F. et al. (2009). Stochastic geometry and random graphs for the analysis and design of wireless networks. *IEEE Journal on Selected Areas in Communications* 27 (7): 1029–1046.

83 Nadeem, Q.-U.-A., Kammoun, A., and Alouini, M.-S. (2019). Elevation beamforming with full dimension MIMO architectures in 5G systems: a tutorial. *IEEE Communications Surveys & Tutorials* 21 (4): 3238–3273.

84 Rokhsaritalemi, S., Sadeghi-Niaraki, A., and Choi, S.-M. (2020). A review on mixed reality: current trends, challenges and prospects. *Applied Sciences* 10 (2): 636.

85 Pirinen, P., Saarnisaari, H., Van de Beek, J. et al. (2019). Wireless connectivity for remote and arctic areas – food for thought. *2019 16th International Symposium on Wireless Communication Systems (ISWCS)*, Oulu, Finland (27–30 August 2019).

86 Yaacoub, E. and Alouini, M.-S. (2020). A key 6G challenge and opportunity—connecting the base of the pyramid: a survey on rural connectivity. *Proceedings of the IEEE* 108 (4): 533–582.

87 Chaoub, A., Giordani, M., Lall, B. et al. (2022). 6G for bridging the digital divide: wireless connectivity to remote areas. *IEEE Wireless Communications* 29 (1): 160–168.

88 Giordani, M. and Zorzi, M. (2021). Non-terrestrial networks in the 6G era: challenges and opportunities. *IEEE Network* 35 (2): 244–251.

89 Ali, M.F., Jayakody, D.N.K., Chursin, Y.A. et al. (2019). Recent advances and future directions on underwater wireless communications. *Archives of Computational Methods in Engineering* 27 (5): 1379–1412.

90 Imoize, A.L., Adedeji, O., Tandiya, N., and Shetty, S. (2021). 6G enabled smart infrastructure for sustainable society: opportunities, challenges, and research roadmap. *Sensors* 21 (5): 1709.

91 Manolova, A., Tonchev, K., Poulkov, V. et al. (2021). Context-aware holographic communication based on semantic knowledge extraction. *Wireless Personal Communications* 120 (3): 2307–2319.

92 Akhtar, M.W., Hassan, S.A., Ghaffar, R. et al. (2020). The shift to 6G communications: vision and requirements. *Human-Centric Computing and Information Sciences* 10 (1): 53.

12

Internet of Things (IoT)

12.1 Introduction

The Internet of Things (IoT) or the Internet of Everything (IoE) can be defined as connecting all possible objects to the network to collect and share data. The vision of this ecosystem, which we call IoT or IoE, is to analyze the vast data generated from these terminals by connecting ubiquitous electronic devices to the network/Internet, thereby developing intelligent applications for the advancement of society.

If we look at its historical development, Kevin Ashton used the phrase Internet of Things in 1999, although it took many years for the technology to catch up with the vision. In 1999, Auto-ID labs and MIT started developing Electronic Product Code (EPC) and using Radio Frequency Identification (RFID) to identify objects on the network. From 2003 to 2004, with the development of projects that serve the idea of the IoT, such as Cooltown, Internet, and Disappearing Computer Initiative, the concept of IoT began to appear in books for the first time. RFID was established and widely deployed by the US Department of Defense. When a report was first published by the International Telecommunication Union (ITU) in 2005, IoT entered a new level. In 2008, a well-known group of companies such as Cisco, Intel, SAP, and more than 50 company members came together to form the IPSO Alliance, promoting the use of Internet Protocol (IP) in something called "smart object" communication and enabling the concept of IoT. In 2008–2009, the IoT was invented, or in other words, we can say it was "BORN" by the hand of the Cisco Internet Business Solutions Group (IBSG). Figure 12.1 shows the number of connected devices that will reach gigantic dimensions [1].

Imagine a world where almost anything you can think of is online and communicating with other things and people to enable new services that improve our lives. From self-driving drones that deliver your grocery orders to sensors that

Evolution of Wireless Communication Ecosystems, First Edition. Suat Seçgin.

Published 2023 by John Wiley & Sons, Inc.

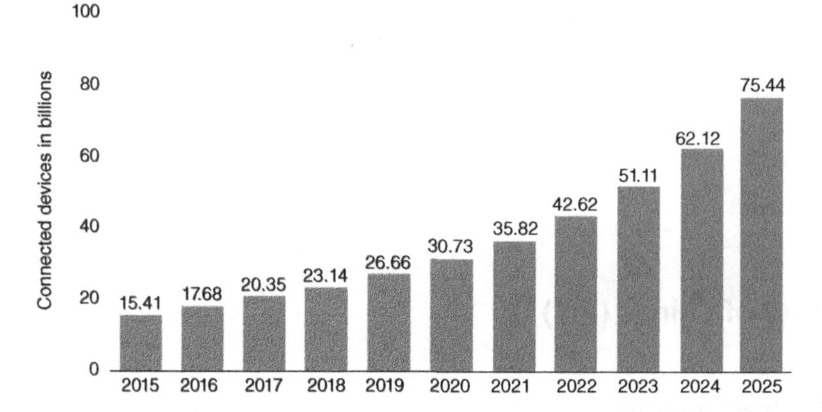

Figure 12.1 Number of devices connected to the IoT system by year.

monitor your health in your clothes, the world as you know it is poised to undergo a significant technological shift.

12.2 IoT Vision

This system, where devices are connected to the network by going one step ahead of M2M communication, is known as the Internet of Things (IoT). Tighter integration between the physical world and computers is achieved when objects and machines can be remotely sensed and controlled over a network. This allows for efficiency, accuracy, and automation improvements in addition to enabling advanced applications [2]. This intelligent connection system will have many effects on business and social life. At the last point, we can say that the world will be "digitized" with the interconnection of people, processes, data, and objects over IoT (Figure 12.2).

IoE is expected to meet three basic expectations to realize this imagined digital world. Three expectations of IoE [3] (Figure 12.3):

- **Scalability:** Scalability means building a scalable network for IoE to cover anywhere and anything flexibly. In this sense, IoE can meet various communication requirements for different geographic scenarios, including urban, rural, underwater, terrestrial, air, and space. To achieve this goal, scalable IoE networking requires broad coverage, massive reach, and ubiquitous connectivity. In this connection, mobile cellular network (MCN), wireless local area network, wireless sensor network (WSN), satellite network, and mobile ad hoc network (MAHN) infrastructures are utilized.

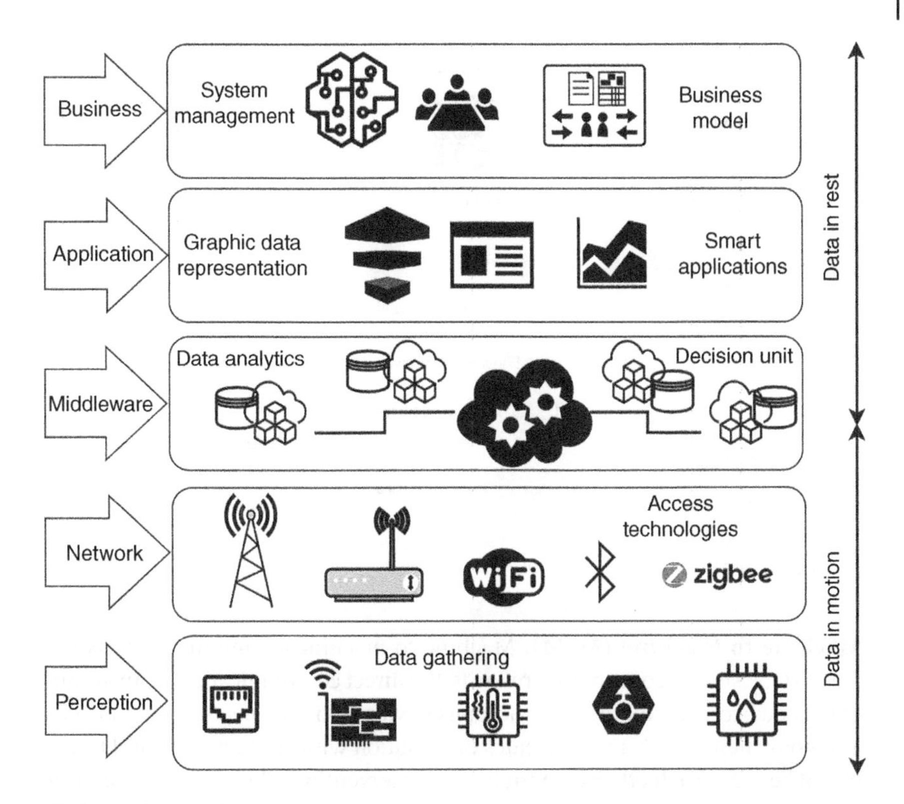

Figure 12.2 Layered IoT architecture.

- **Intelligence:** This step aims to obtain the meaningful information using the techniques such as artificial intelligence data mining, etc. from the vast data produced by IoT devices. Intelligent decision systems are created using this information. From this point of view, it can be thought that there will be a global computing ecosystem with distributed databases and storage areas and big data algorithms running at the top. Descriptive, diagnostic, predictive, and prescriptive analyzes can be made through big data processing algorithms. From the point of view of the IoT cloud structure, intelligence can be divided into three parts: local, edge, and cloud intelligence.
- **Diversity:** It means supporting different applications. The diversity here can be categorized as geographic, stereoscopic, business, and technology diversity. Sometimes IoT and M2M communication are concepts that are confused with each other. Although they are similar technologies, there are essential differences between them.

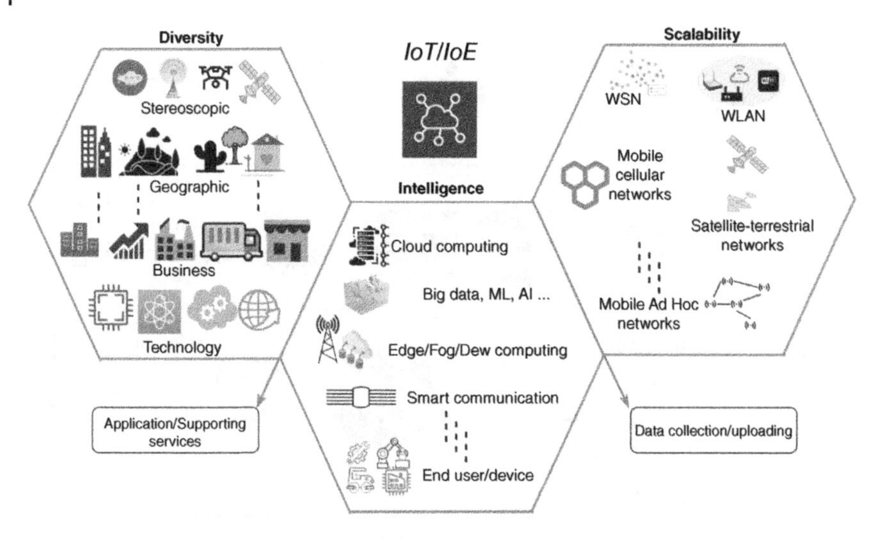

Figure 12.3 Three expectations of IoE/IoT.

- **Machine to Machine (M2M):** Machine-to-machine communication, as the name suggests, is a concept that refers to the direct communication of an autonomous device with another autonomous device. What is meant by the idea of autonomy here is that a node creates information with another node and transmits it between each other without human intervention. The form of communication is left open to the application. Indeed, it has been beneficial in that M2M devices do not use inherent services or topologies for communication, leaving these applications to typical Internet devices that are regularly used for cloud services and storage. In addition, the M2M system can work on non-IP-based channels such as serial ports or custom protocol communication.
- **IoT:** IoT systems may be incorporated with some M2M nodes, such as Bluetooth mesh non-IP-based communication. However, data is aggregated in an edge router or gateway in IoT. An edge device such as a router or gateway serves as an entry point to the Internet. Alternatively, some sensors with high computational capacity can be connected directly to the cloud network and the Internet. In other words, the ability to connect to the Internet fabric defines the IoT. It allows the data produced by sensors to provide an Internet connection for edge processors and intelligent devices to be processed on cloud-based systems. Without this capacity and opportunity, we would still stay in the M2M World [4] (Figure 12.4).
- **Internet of Everything (IoE):** The IoE brings people, processes, data, and things together to make networked connections more relevant and valuable than ever. It transforms knowledge into actions that create new capabilities,

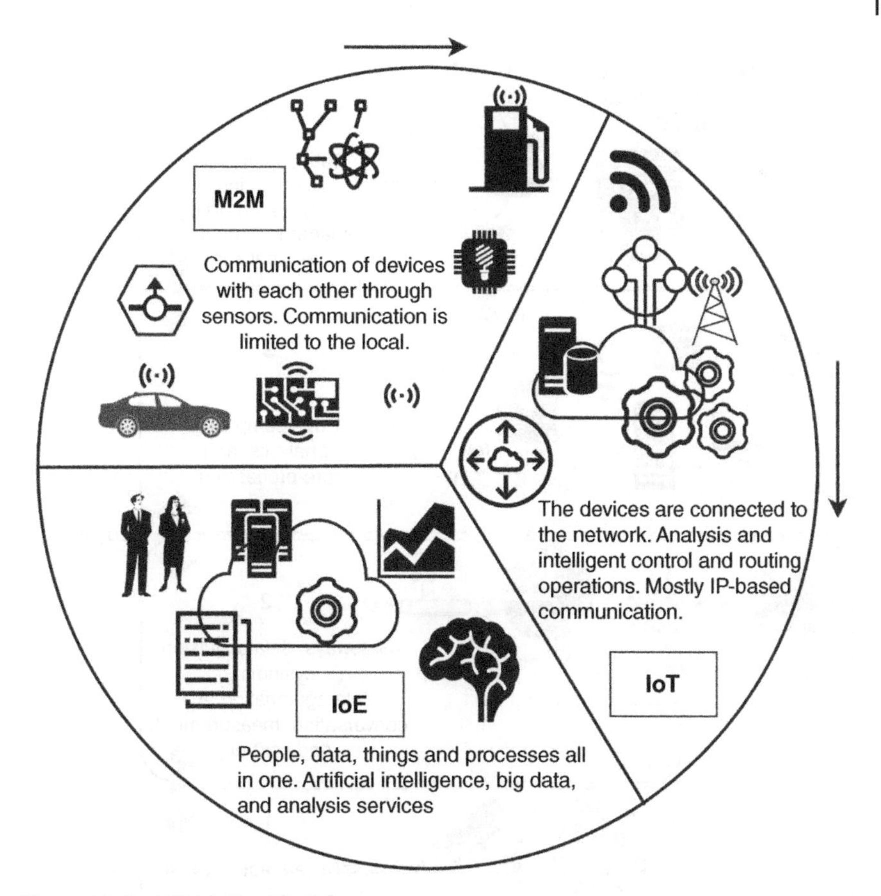

Figure 12.4 M2M, IoT, and IoE.

richer experiences, and unprecedented economic opportunities for businesses, individuals, and countries. In other words, IoE can be defined as IoT's extended version or new phase. Every component in the IoE loop is interrelated, creating a closed loop that starts with people and ends with people.

12.3 Architecture and Communication Model

The architecture that will provide communication between IoT and machines (machine-to-machine communication [M2M]) can be examined at four levels: The "sensing" layer, where the connections of sensors and end devices are provided (data is collected at this layer), the network layer where data transmission is

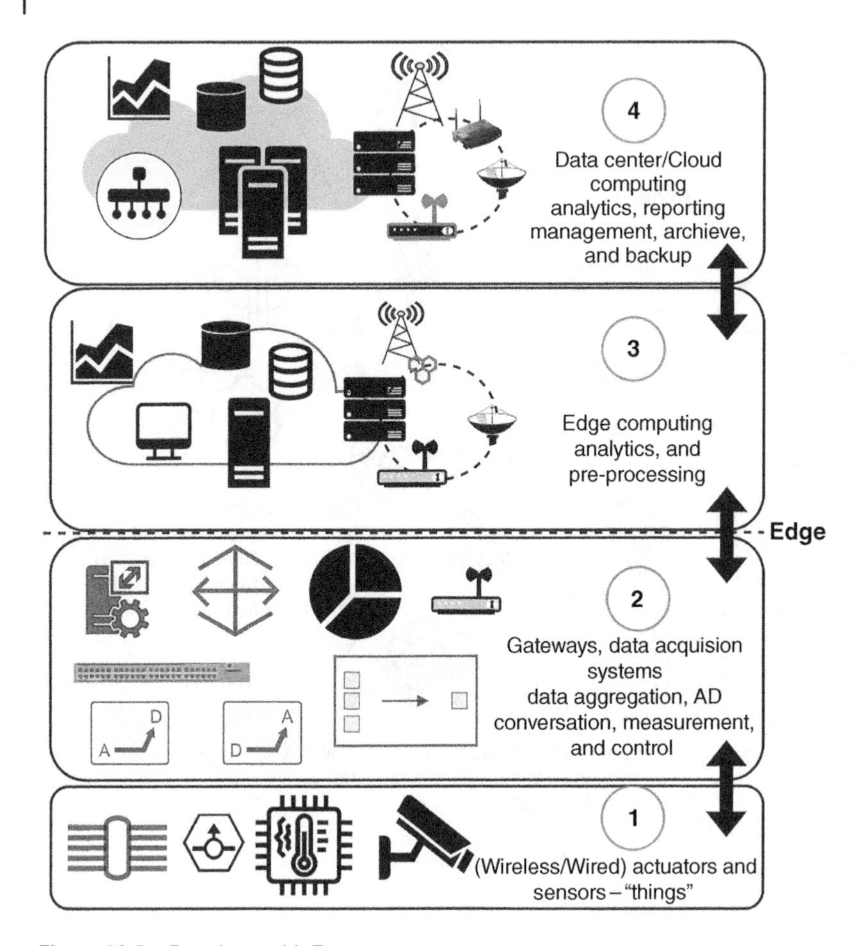

Figure 12.5 Four-layered IoT structure.

made, gateways and various network technologies are used, the data processing layer (Edge IT processing) where data is processed, process information is compiled, and the application layer (data center and cloud applications) with intelligent applications and management consoles at the top (Figure 12.5).

In Figure 12.6, the interaction block diagram of the actors in the IoT infrastructure is given. As you can see, nodes at the "edge" point, which we can describe as data collectors, send the data they collect to the cloud over gateways. Higher-level structures and applications are also connected to the cloud, performing data processing and monitoring functions [5].

Gateways located at the critical connection point in this structure are software or physical device. Gateways are a translator for the cloud and controllers,

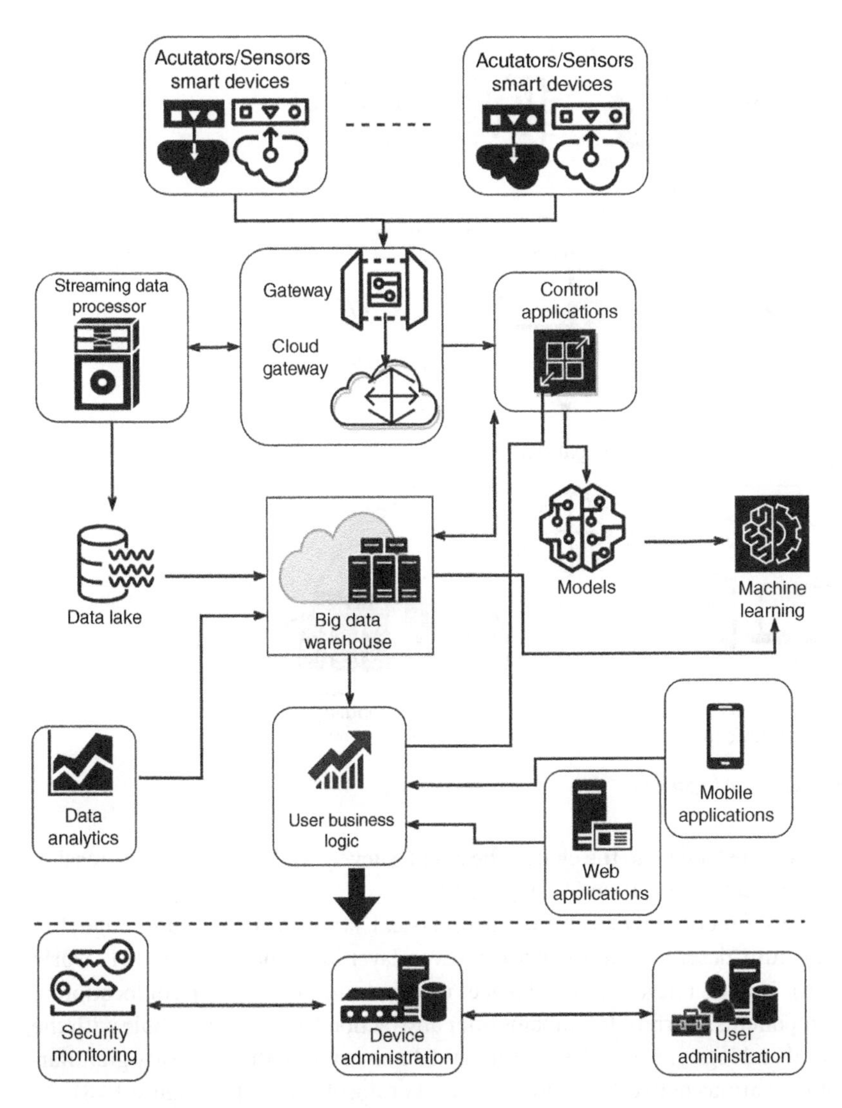

Figure 12.6 IoT conceptual architecture.

sensors, and intelligent devices. Its most important function (which we will explain in the following sections) is to translate different protocols into each other. An IoT gateway software or device is called an intelligent gateway or handle tier. However, quality and reliability become the system's most critical components when the network has hundreds or thousands. The protocols translated here are low-profile protocols with limited bandwidth, battery power, speed, and the

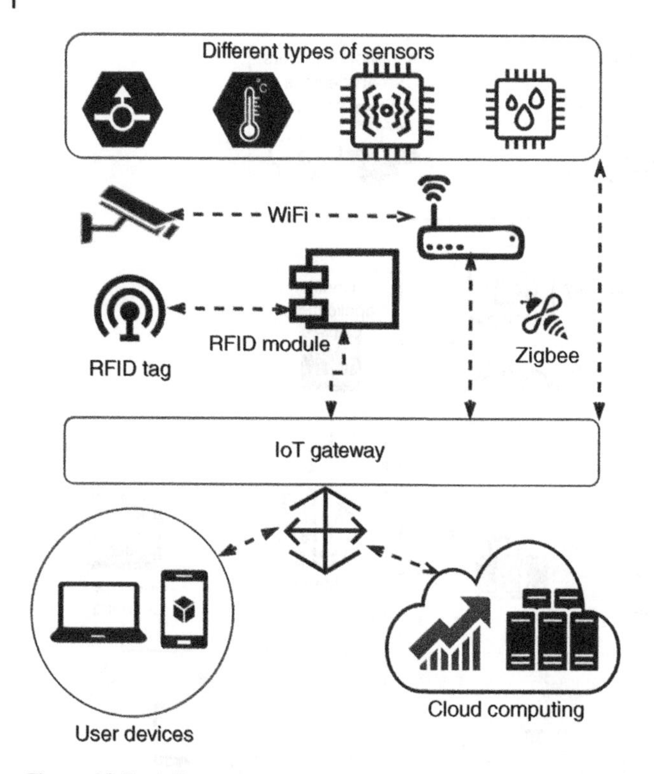

Figure 12.7 IoT gateway.

ability to transfer data to the cloud. The cloud gateway facilitates data compression and secure communication [6] (Figure 12.7).

We can functionally examine the physical structures and four-stage IoT architecture described above with a seven-layer reference model (IoT World Forum). The IoT reference model places no constraints on the scope or location of its components. Current IoT models offer abstraction levels incompatible with the physical and logical network structure and do not capture the necessary granularity of the various network architectures and protocols in use [7] (Figure 12.8).

The layers are described below [8];

- **Physical devices and controllers:** The IoT reference model starts at the bottom layer with physical devices and controls that control different devices. These nodes are objects within the IoT fabric and refer to endpoint devices that exchange data. These devices have a wide variety and do not require regulation on their location, size, form factors, and beginnings. These devices can carry out the analog-to-digital conversion, data generation, interrogation, and control over the Internet when necessary.

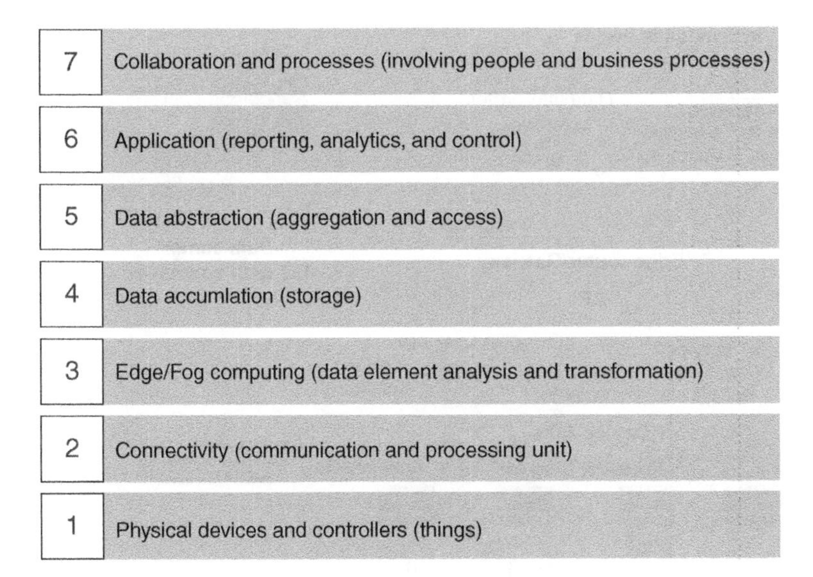

Figure 12.8 IoT world forum reference model.

- **Connectivity:** Connectivity is the most basic function for the next stage and provides a reliable and timely transmission environment. The IoT reference model provides the necessary definitions for the communication and processing performed by existing networks. The IoT reference model does not need to create another network based on existing networks. As first-stage devices increase, their interaction with second-stage connectivity equipment may change. Without attention to detail, the first-stage devices interact with the second-stage connection equipment and communicate over the IoT system. This level covers transmission between devices and the network. In summary, this layer performs functions such as communication with and between level 1 devices, reliable delivery across the networks, implementation of various protocols, switching/routing, translation between protocols, security at the network level, and intelligent networking analysis (Figure 12.9).
- **Edge computing:** The miniature data processing function on the sensor nodes and gateway is performed by edge computing. The functions of the third stage result from the need to transform the network data streams into information suitable for storage and higher stage processing in the fourth stage [9]. The third phase activities focus on high-volume data analysis and transformation. Since data is usually sent to second-stage network equipment by devices in miniature units, third-stage processing is performed on a packet basis.

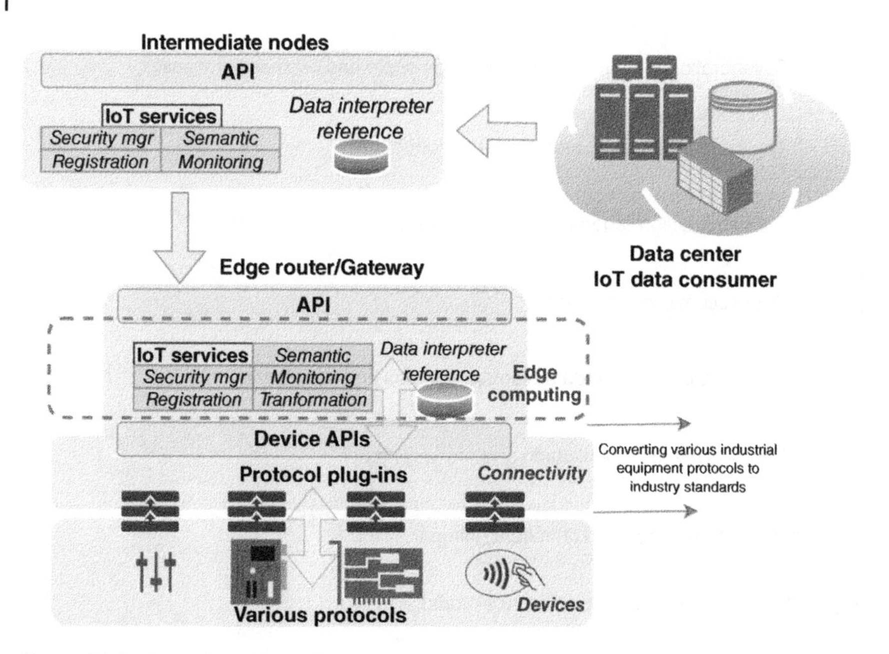

Figure 12.9 Level 2 and level 3 interaction.

Level 3 processes work on a package-by-package basis. Processing is limited here because it works on data units, not sessions or transactions. Evaluation of data units, formatting (reformatting of data for a higher level of consistency processing), expanding/decoding (processing encrypted data with additional context, distillation/reduction (data effect, network effect, etc.) functions such as reducing and summarizing data to minimize traffic and higher-level processing systems, and assessment (determining whether the data represents a threshold or alarm) are carried out in this stage. In this process, operations such as redirecting the data to additional destinations are also performed. Figure 12.9 shows a connectivity and data element analysis.

- **Data accumulation:** The data sent by the sensor nodes over the Internet via gateways are kept in a database on the cloud. Transmission media in the system are responsible for transporting data securely [10]. However, several computational activities may take place in the second phase, such as protocol translation or application of network security policy. In the third stage, computational tasks such as packet inspection can be performed. Computing methods that are as close as possible to the edge of the IoT system can be given as an example of fog computing. In the fourth stage, event-based data is converted into query-based processes. This phase bridges the gaps between the nonreal-time

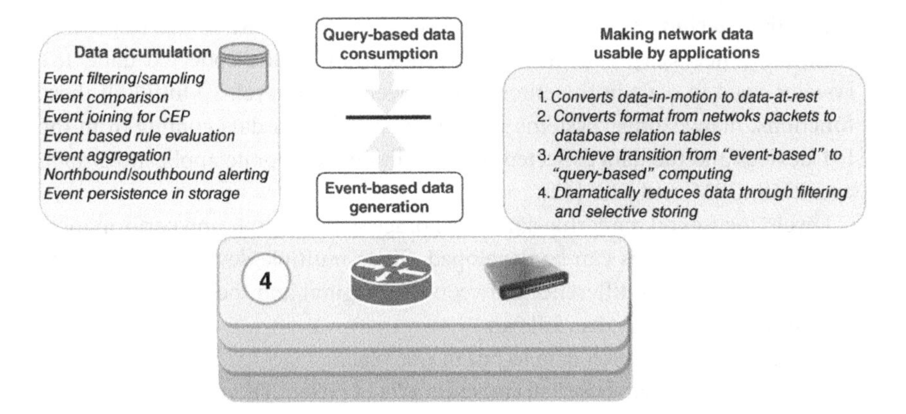

Figure 12.10 IoT reference model level 4.

application world and the real-time networking world. Before level 4, data flows over the network, and data organization is determined by the device that produces the data. The data in motion up to the fourth level goes into "at-rest" state on the data memory or a disk with the fourth level (Figure 12.10).

At this stage, it is checked whether the data is of interest to higher levels. When necessary, data is served in line with the needs of higher levels. If data is persistent, it is given whether it should be kept on a nonvolatile disk or stored in memory for short-term use. The type of storage needed depends on the persistence requirement for a file system, big data system, or relational database. It is checked whether the data is organized according to the required storage type. In the final stage, data from some non-IoT sources is combined and recomputed. In stage 4, data is captured and put "at rest" state. Thus, the data is now available for applications operating on a nonreal-time basis. In other words, event-based data processing is converted to query-based data processing.

- **Collaboration and processes:** The seventh and top layer of the IoT reference model. Human interaction and participation are the most neglected parts of IoT application scenarios [11]. People use apps and app-related data for their specific needs. Frequently, multiple people use the same app for different purposes. In this case, the objective is people's ability to do their job better than the application itself. Applications at level 6 must give business people the correct data at the right time and take the right action. But frequently, axion requires more than one people. People use the traditional Internet to communicate and gather. In this sense, communication and collaboration often require multiple steps. And this process is often used in multiple applications. This is why level 7 represents a higher level than a single application.

- **Data abstraction:** The primary purpose of the data abstraction stage is to extract the necessary and meaningful ones from all the collected data. IoT systems need to scale at the enterprise or even global level. To fulfill all these functions, many storage systems should be available for data coming from the IoT devices and the data collected from traditional corporate applications such as HRMS, CRM, and ERP [12].

  Level 5 focuses on rendering data and storage so that simple and performance-enhanced applications can be developed. Since multiple devices generate the data, there may be a difference between the original and the stored data. The reasons for this are listed as follows:
  - There may be too much data to put in one place.
  - Moving data to a database can consume a lot of power. Therefore, the retrieval process should be separated from the data generation process. This is done by online transaction processing (OLTP) of databases and data warehouses.
  - Devices are geographically separated, and processing is locally optimized.
  - Levels 3 and 4 might distinguish raw data from data representing events of a continuous stream. The data storage for streaming data can be significant in a data management system like Hadoop. The event data storage can be a relational database with a faster query time (RDMS).

  For the reasons described here, the data abstraction level must handle many different things. Reconcile multiple data in various formats from other sources, ensure consistency of data semantics across sources, confirm that data is compatible for higher-level applications, consolidate data into a single domain (with ETL, ELT, or replication), or access multiple data through data virtualization. It is also necessary to index, normalize, or denormalize the data to ensure data protection, protect it with appropriate authentication and authorization, and finally provide fast application access.
- **Application:** Responding to data by controlling actuators in analytical and sensor nodes is one of the primary applications of IoT architecture. The sixth layer is the application layer, where information interpretation takes place. The software in this layer interacts with layer 5 and other data. Therefore, it does not need to operate at network speed. The IoT reference model we have provided does not strictly define applications. Applications may vary according to vertical markets, nature of device data, and business requirements. Some apps monitor device data while others collect device and nondevice data. Some apps focus on device controls. Monitor and control applications describe various application models, servers, hypervisors, multithreading, multitenancy, programming patterns, and software stacks.

  Implementation complexity can vary. Applications can be mission-critical applications such as specialized industry solutions or generalized ERP, mobile applications that handle simple interactions, business intelligence applications, analytical applications for decision support systems, or system management/control center applications.

References

1 Alam, T. (2018). A reliable communication framework and its use in internet of things (IoT). *CSEIT1835111*|Received, 2018 [10], pp. 450–456.
2 Hanes, D., Salgueiro, G., Grossetete, P. et al. (2017). *IoT Fundamentals: Networking Technologies, Protocols, and Use Cases for the Internet of Things*. Cisco Press.
3 Liu, Y., Dai, H.-N., Wang, Q. et al. (2020). Unmanned aerial vehicle for internet of everything: opportunities and challenges. *Computer Communications* 155: 66–63.
4 Cloud Credential Council (2022). Knowledge Byte: The Differences Between IoT, M2M and IoE. *Cloud Credential Council*. https://www.cloudcredential.org/blog/knowledge-byte-the-differences-between-iot-m2m-and-ioe/ (accessed 16 February 2023).
5 Wu, M., Lu, T.-J., Ling, F.-Y. et al. (2010). Research on the architecture of Internet of Things. *2010 3rd International Conference on Advanced Computer Theory and Engineering (ICACTE)*, Chengdu (20–22 August 2010).
6 Datta, S.K., Bonnet, C, and Nikaein, N. (2014). An IoT gateway centric architecture to provide novel M2M services. 2014 IEEE World Forum on Internet of Things (WF-IoT), Seoul, South Korea (6–8 March 2014).
7 Gordillo, R.X., Romero, C.G., Abasolo, S.E., and Carrera, M.A. (2014). Testbed for evaluating reference models of Internet of Things (IoT). *2014 IEEE Colombian Conference on Communications and Computing (COLCOM)*, Bogota, Colombia (4–6 June 2014).
8 Perwej, Y., Ahmed, M., Kerim, B., and Ali, H. (2019). An Extended Review on Internet of Things (IoT) and its Promising Applications. *Communications on Applied Electronics*. 7 (26): 8–22.
9 Da Cruz, M.A.A., Rodrigues, J.J.P.C., Al-Muhtadi, J. et al. (2018). A reference model for Internet of Things middleware. *IEEE Internet of Things Journal* 5 (2): 871–883.
10 Yi, S., Qin, Z., and Li, Q. (2015). Security and privacy issues of fog computing: a survey. In: *Wireless Algorithms, Systems, and Applications* (ed. K. Xu and H. Zhu), 685–689. Cham: Springer.
11 Bin, S., Yuan, L., and Xiaoyi, W. (2010). Research on data mining models for the internet of things. *2010 International Conference on Image Analysis and Signal Processing*, Zhejiang (9–11 April 2010), pp. 127–132.
12 Boyi, X., Li Da, X., Cai, H. et al. (2014). Ubiquitous data accessing method in IoT-based information system for emergency medical services. *IEEE Transactions on Industrial Informatics* 10 (2): 1578–1586.

13

Non-IP-Based WPAN Technologies

13.1 Introduction

Wireless personal area networks (WPANs) are short-range networks formed by interconnected devices with humans at the center. The methods to exchange information with end-node devices such as sensors and actuators create WPAN communication. WPAN connectivity is becoming increasingly common in commercial, industrial, or consumer Internet of Things (IoT) connections. Connections between endpoints and Internet or enterprise networks are divided into IP-based and non-IP-based. Connection types are divided into four topics: device-to-device, device-to-cloud, device-to-gateway, and back-end data-sharing models [1].

- **Device-to-device communication model:** In M2M communication, the device-to-device communication model symbolizes paired devices without any proxy. In this connection type, devices connect and communicate directly with each other. To communicate directly, devices use Bluetooth, ZigBee, etc., protocols. This communication model follows a one-to-one communication protocol to exchange information to achieve its respective functions. Home-based IoT devices such as light bulb switches, door locks, washing machines, microwave ovens, refrigerators, and air conditioners often use device-to-device communication via tiny data packets and low data rates. One main disadvantage of device-to-device communication is that they must deploy/select similar category devices because device-to-device statement protocols are well-suited (Figure 13.1).
- **Device-to-cloud communication model:** In the device-to-cloud communication model, IoT devices are directly connected to the cloud service provider, which monitors data interchange and imposes restrictions on message traffic. This model uses existing wired or wireless connections to create a connection between devices with the Internet network protocol connected to the cloud

Evolution of Wireless Communication Ecosystems, First Edition. Suat Seçgin.

Published 2023 by John Wiley & Sons, Inc.

Figure 13.1 Device-to-device communication model.

Figure 13.2 Device-to-cloud communication model.

service at the endpoint. Here, the device and cloud service must originate from the same manufacturer (Figure 13.2).

- **Device-to-gateway communication model:** In this model, an IoT/M2M device is connected to the cloud via an application layer gateway (ALG). The application software runs on a local gateway and acts as an intermediary between the device and the cloud. This application software also provides security-related functions while transmitting data. For example, smartphones have become gateway devices that run applications to communicate with another device and distribute data to the cloud service. Home-based automation applications using hub devices are device-to-gateway type models. Here, hubs act as a gateway between independent IoT devices alongside the cloud service (Figure 13.3).
- **Back-end data sharing model:** This communication model facilitates the importing, exporting, and evaluating smart/small device records from a cloud service. These models allow the merging and later examination of records or data generated by the individual IoT device. This type of communication model allows data portability as needed [2] (Figure 13.4).

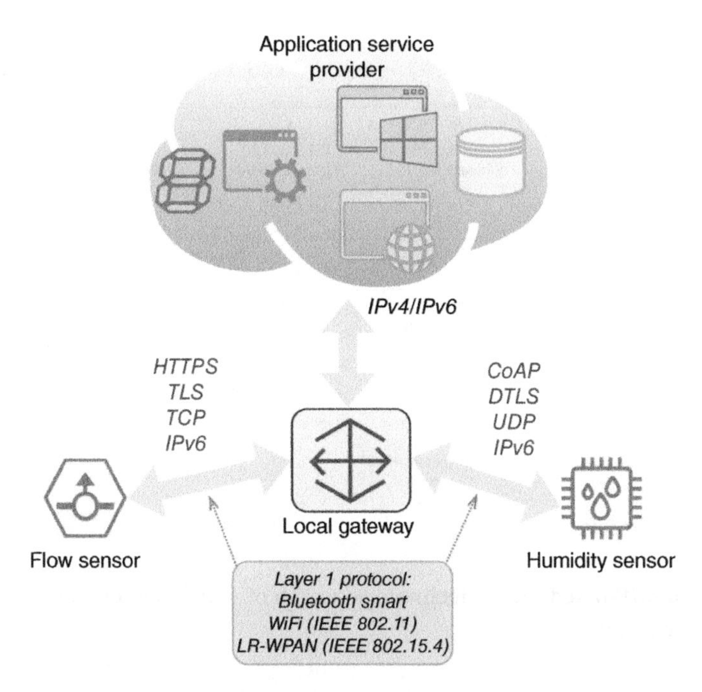

Figure 13.3 Device-to-gateway communication model.

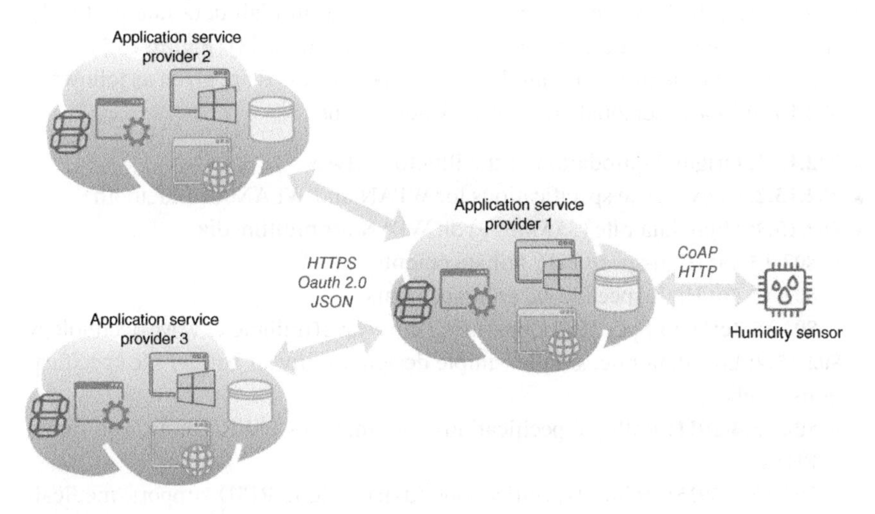

Figure 13.4 Back-end data sharing model.

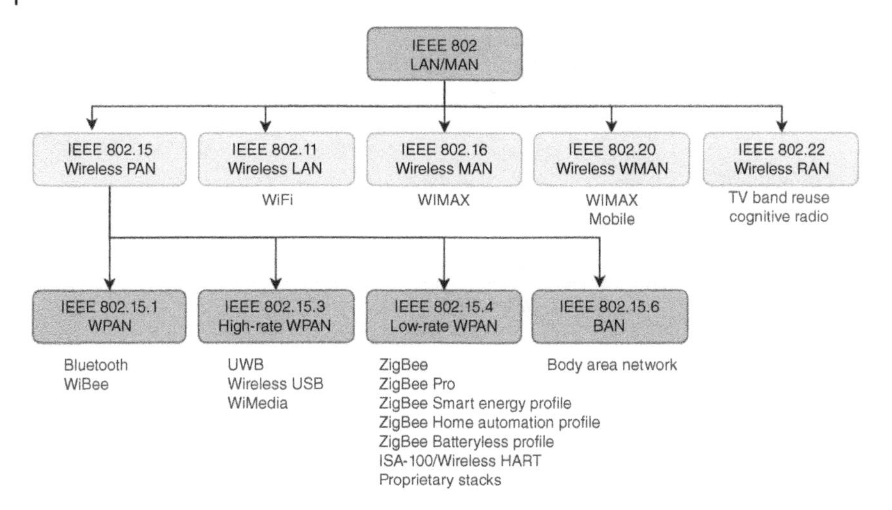

Figure 13.5 802.15 protocols.

In this section, non-IP-based WPAN technologies, each of which has a specific usage area, will be explained.

13.2 802.15 Standards

Many of the technologies described in this section are based on the IEEE 802.15 standard specifications. The 802.15 group was initially established to set wearable device standards. However, their work has focused on high data rate protocols, meter-to-kilometer ranges, and specialty communications [3] (Figure 13.5).

The details of the protocol, standards, and specifications are given as follows:

802.15: Wireless personal area network definitions

- **802.15.1:** Original foundation of the Bluetooth PAN
- **802.15.2:** Coexistence specifications for WPAN and WLAN for Bluetooth
- **802.15.3:** High data rate (55 Mbps+) on WPAN for multimedia
  - **802.15.3a:** High-speed PHY enhancements
  - **802.15.3b:** High-speed MAC enhancements
  - **802.15.3c:** High-Speed (>1 Gbps) using mm-wave (millimeter wave) technology
- **802.15.4:** Low data rate, simple, simple design, multiyear battery life specifications (Zigbee)
  - **802.15.4-2011:** Rollup (specifications a–c) includes UWB, China, and Japan PHYs
  - **802.15.4-2015:** Rollup (specifications (d–p) include RFID support, medical-band PHY, low energy (LE), TV white spaces, and rail communications

  - **802.15.4r (on hold):** Ranging protocol
  - **802.15.4s:** Spectrum Resource Utilization (SRU)
  - **802.15.t:** High rate PHY of 2 Mbps
- **802.15.5:** Mesh networking
- **802.15.6:** Body area networking for medical and entertainment
- **802.15.7:** Visible light communications using structured lighting
  - **802.15.7a:** Extends range to UV and near-IR, changed name to optical wireless
- **802.15.8:** Peer Aware Communications (PAC) infrastructure-less peer-to-peer (P2P) at 10 kbps to 55 Mbps
- **802.15.9:** Key Management Protocol (KMP), management standard for critical security
- **802.15.10:** Layer 2 mesh routing, recommend mesh routing for 802.15.4, multi-PAN
- **802.15.12:** Specifically, a comparison of the most well-known non-IP-base WPAN technologies is given in Table 13.1.

13.3 Radio Frequency Identification

Radio Frequency Identification (RFID) is one of the latest technologies developed for IoT and M2M. In this technology, small reading devices read the message, a radio device, and frequency transponders known as RF tags. This tag is vital because it contains the programmed information that enables the RFID to read the signals. There are two different tag systems in RFID, one known as the active reader tag and the other passive reader tag. An active tag's most important feature is that it operates at a higher frequency than a passive tag. An RFID system in IoT is used primarily in healthcare applications, agriculture, and national security systems [4].

A typical RFID system consists of three main components shown in Figure 13.6: RFID tags, a reader, and an application system [5]. RFID uses electromagnetic fields to track and identify tags attached to objects. RFID tags are known as transponders (transmitter/responder) attached to objects for counting and identification purposes. Tags can be active or passive. Active tags are partially or fully battery-powered tags that can communicate with other tags. Active tags can initiate their dialogs with tag readers. Passive tags can harvest energy from a nearby RFID reader that emits interrogating radio waves. Active tags have a local power source (such as a battery) and can operate hundreds of meters from the RFID reader. Tags mainly consist of a helical antenna and a microchip, which is the primary purpose of data storage. The reader is called a transceiver (transmitter/receiver), which consists of a radio frequency interface (RFI) module and a control unit. Its main functions are to activate tags, configure the communication order with the tag, and transfer data between the application software and the

Table 13.1 Comparison of IoT connection technologies [10].

Technology	RFID	NFC	BLE	ZigBee	6LoWPAN	LoRa	SigFox	NB-IoT	MIOTY
Range	*1–5 m*	*1–10 cm*	*1–10 m*	*75–100 m*	*100 m*	*2–15 km*	*3–50 km*	*10–15 km*	*15–20 km*
Bandwidth	*2–26 MHz*	*14 kHz*	*1–2 MHz*	*2 MHz*	*3 MHz*	*<500 kHz*	*100 Hz*	*180 kHz*	*200 kHz*
Band	*125–134.2 kHz* *13.56–433 MHz* *860–960 MHz*	*13.56 MHz*	*2.4 GHz*	*868 MHz* *915 MHz* *2.4 GHz*	*2.4 GHz* *Sub 1 GHz*	*868 MHz* *915 MHz* *Sub 1 GHz*	*915–928 MHz* *Sub 1 GHz*	*700 MHz* *800 MHz* *900 MHz*	*133–966 MHz*
Data rate	*4–640 kbps*	*100–424 kbps*	*1 Mbps*	*250 kbps*	*250 kbps*	*50 kbps (FSK)*	*>100 bps*	*200 kbps*	*Sub 1 kbps*
Latency	*1–10 ms*	*100 ms*	*6 ms*	*~15 ms*	*2–6 ms*	*1–10 ms*	*10–30 ms*	*40 ms to 10 s*	*10 ms to 10 s*
Modulation	*OOK, FSK, PSK*	*ASK*	*DQPSK/DPSK*	*O-QPSK*	*QPSK/BPSK*	*FSK*	*GFSK/DBPSK*	*BPSK, QPSK*	*BPSK, MSK*
Battery lifetime	*Battery free*	*Battery free*	*1–5 days*	*1–5 years*	*1–2 years*	*<10 years*	*<10 years*	*>10 years*	*Up to 20 years*
Application	*Materials management attendee tracking*	*Payment, ticketing*	*IoT device authentication localization*	*Smart home healthcare*	*Smart city infrastructure*	*Air pollution monitoring fire detection*	*Smart meter, pet tracking*	*Street lightning, agriculture*	*Dense IoT network scenarios*
Advantages	*High speed and convenience very low cost*	*Convenient to use*	*Support by most smartphones*	*Highly reliable, scalable*	*Massive connection without complex routing*	*High immune to interference, adaptive data rate*	*High reliability low device complexity*	*Better coverage range and coverage*	*Low packet error rate, high energy efficiency*
Limitation	*Bring security and privacy concerns*	*Limited to data rates*	*Limited range and battery life*	*Short range, communication security issues*	*Limited range and data rates*	*Longer latency, not acknowledges all packets*	*Multiple transmissions suffer from interference*	*No hand-off support, low interference immunity lacks in ACK*	*Low data rates*
Standard body	*ISO/IEC*	*ISO/IEC*	*Bluetooth SIG*	*ZigBee Alliance*	*IETF*	*LoRa Alliance*	*Sigfox*	*3GPP*	*MIOTY Alliance*

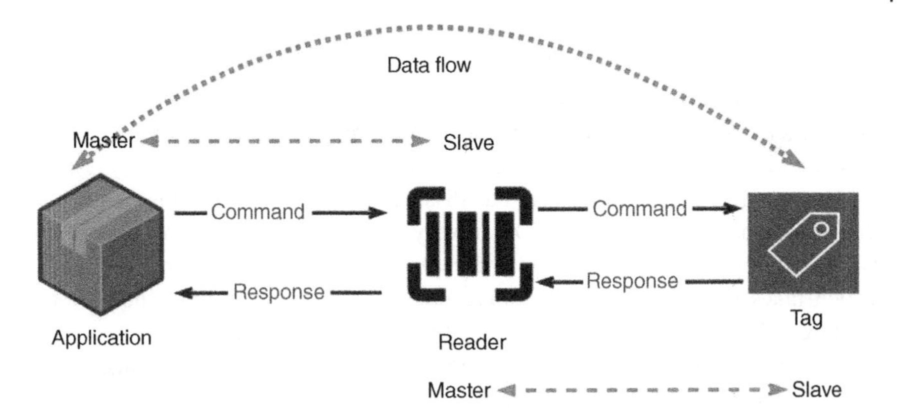

Figure 13.6 RFID system architecture.

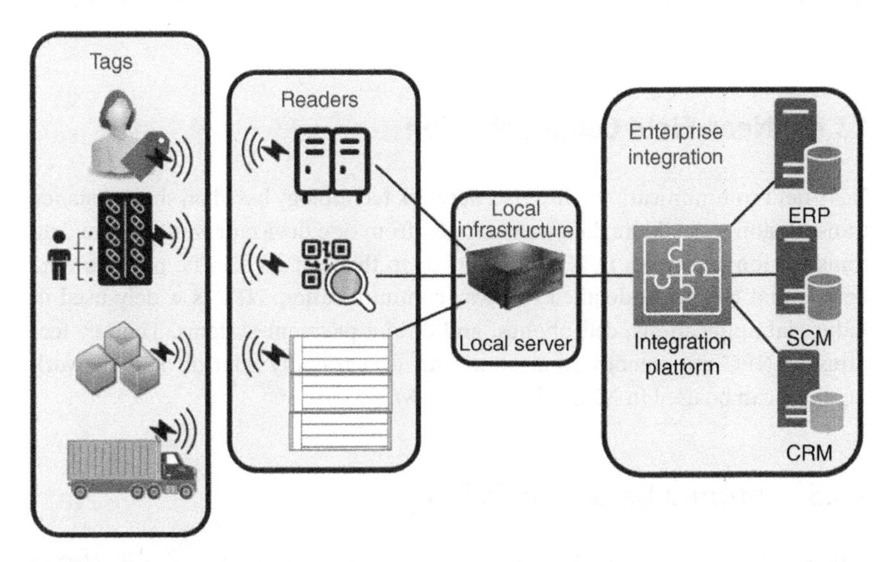

Figure 13.7 RFID structure.

tags. An application system is a data processing system that can be an application or a database, depending on the application. The application software initiates all reader and tag activities. RFID provides a fast, flexible, and reliable way to detect, track, and control various items electronically (Figure 13.7).

RFID systems use radio transmissions to energize an RFID tag, while the tag sends a unique identification code to a data acquisition reader connected to an information management system. The data collected from the tag can then be sent directly to a host or stored in a portable reader and then uploaded to the host. RFID technology has many advantages. The RFID tag and reader do not need a line of sight for the system to work, and an RFID reader can scan multiple tags simultaneously. Unlike barcodes, RFID tags can store more information. It can follow the commands/directions of the reader and transmit its location along with its ID to the reader. RFID technology is inherently versatile. Therefore, smaller and larger RFID devices are available depending on the application. Unlike barcodes, tags can be read only as well as read/write.

However, RFID technology also has disadvantages. Active RFID is expensive due to the use of batteries. RFID's privacy and security issues are associated because they can be easily read and intercepted. RFID devices need programming, which takes time. External electromagnetic interference may limit the remote reading of RFID tags. Passive tags have a limited range of around 3 m.

RFID finds its market in many different areas such as inventory management, personnel and asset tracking, controlling access to restricted areas, supply chain management, and fraud prevention. The adoption of RFID promotes the innovation and development of IoT, widespread in households, offices, warehouses, parks, and many other places.

13.4 Near-Field Communication

Near-field communication (NFC) is a network technology based on short-distance transmission to facilitate data transmission from one device to another. The data transmission principles in NFC are similar to those of RFID. The point to note here is that NFC uses detailed two-way communication. NFC is widely used in industrial applications, cell phones, and online payment systems. The key features of NFC are smooth connection and user-friendly control. P2P network topology can be used in NFC [6] (Figure 13.8).

13.5 Infrared Data Association

Infrared data is a standard defined by the Infrared Data Association (IrDA) Consortium. It specifies the wireless data stream via infrared radiation. IrDA's initial specifications supported data rates of up to 115.2 kbps. The specification describes the IrDA protocol stack. Stack has layers such as serial infrared (SIR), infrared link access protocol (IrLAP), and infrared link management protocol (IrLMP) places device. IrDA range is around 2 m. A point-to-point line-of-sight connection is established between two IrDA devices. This technology uses light wavelengths from 850 to 900 nm for transmission. IrDA devices communicate using infrared light-emitting diodes (LEDs).

IrDA does not provide link-level security. Since there is no authorization and authentication, the transmitted information is sent without encryption. However,

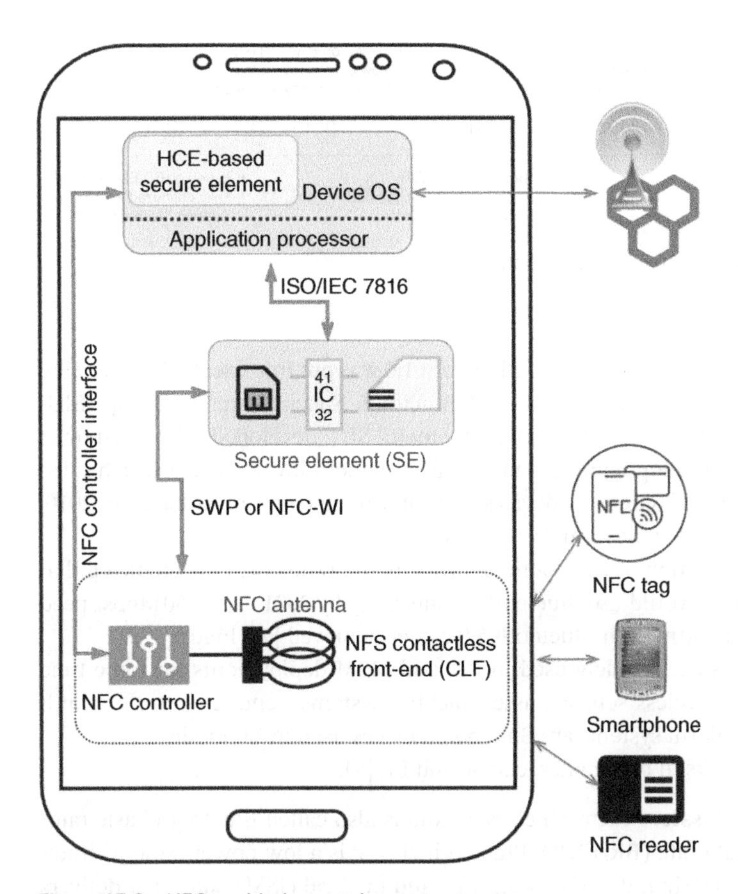

Figure 13.8 NFC-enabled smartphone structure.

although security is not defined, it is considered a relatively secure technology. IrDA PDAs have many uses, such as mobile phones, printers, cameras, and laptop computers. Point-to-point data transmission has advantages such as direct sight communication, security, low power consumption, and low cost, as well as disadvantages such as the necessity of direct sight and the capacity to connect to a single device simultaneously.

13.6 Bluetooth

Ericsson originally designed Bluetooth technology in 1994 to replace cables and cords of computer peripherals with RF connections. Subsequently, Intel and Nokia also joined the initiative for the wireless connection of cell phones to

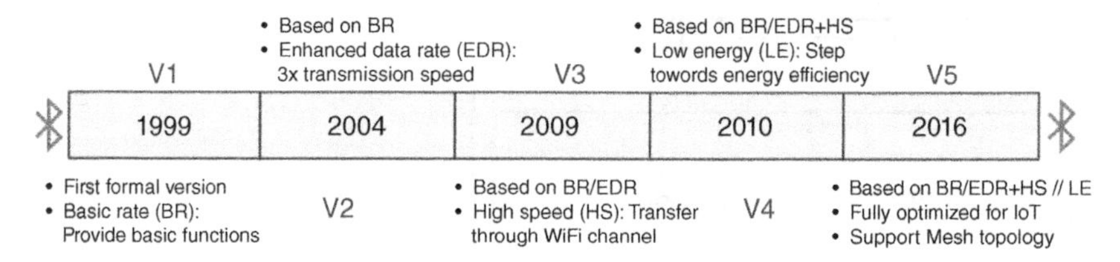

Figure 13.9 The evolution of Bluetooth.

computers. In 1998, the version of Bluetooth 1.0 was put into use. In 2005, version 2.0 was approved, and the number of Bluetooth Special Interest Group (SIG) members exceeded 4000. In 2007, Bluetooth SIG developed ultralow-power Bluetooth. The development, also known as Bluetooth low energy (BLE), has led to the production of Bluetooth devices that can run on coin cell batteries. In 2010, the Bluetooth 4.0 specification was approved.

With the ratification of Bluetooth 5.0 in 2016, the Bluetooth SIG predicted that 13 billion devices would use Bluetooth connectivity by 2021. Up to 50 Mbps speed at a distance of 240 m with Bluetooth 5.0 can be achieved [7] (Figure 13.9).

Bluetooth has been widely used in IoT and M2M deployments for some time now. Beacons, wireless sensors, asset tracking systems, remote controls, health monitors, and alarm systems are Bluetooth devices used in LE mode.

Bluetooth works in two modes, classic and LE [8]:

- **Bluetooth classic:** Bluetooth classic radio is also called Bluetooth basic rate/enhanced data rate (BR/EDR). Bluetooth classic is a low-power radio technology in the 2.4 GHz industrial, scientific, and medical (ISM) band that delivers data streams in over 79 channels. Bluetooth Classic is mainly used for wireless audio streaming to support point-to-point device communication. It has become the standard radio protocol behind wireless speakers, headphones, and in-car entertainment systems. The Bluetooth classic radio also enables data transfer applications, including mobile printing.
- **Bluetooth low energy (BLE):** The BLE radio is designed for low power usage. By transmitting data over 40 channels in the 2.4 GHz license-free ISM frequency band, the Bluetooth LE radio gives developers tremendous flexibility to create products that meet the unique connectivity requirements of their markets. Bluetooth LE supports multiple communication topologies, expanding to point-to-point and finally mesh, enabling Bluetooth technology to support the creation of reliable, large-scale device networks. Although initially known for its device communication capabilities, Bluetooth LE is now widely used as a device positioning technology to meet the growing demand for high-precision indoor location services. Initially supporting simple presence and proximity features,

Bluetooth LE now supports Bluetooth® orientation and soon highly accurate distance measurement.

Whether Bluetooth basic rate (BR) or Bluetooth LE, nodes can work as advertisers or scanners according to this definition. Advertiser devices transmit advertised packets. Devices running in scanner mode are devices that receive advertiser packets with no intention of connecting. Finally, initiator devices are devices that try to establish a connection. Several Bluetooth events are happening in a Bluetooth WPAN.

- **Advertising:** Advertising is initiated by a device to match a message in the advertising packet or broadcast to scanning devices to alert them of the presence of a device that wants to forward.
- **Connecting:** This event involves pairing a device with a host.
- **Periodic advertising (PA):** PA (for Bluetooth 5) allows an advertiser to advertise periodically over 37 nonprimary channels with channel hopping between 7.5 ms and 81.91875 s.
- **Extended advertising (EA):** EA (for Bluetooth 5) allows extended protocol data units (PDUs) to support advertise chaining and large PDU payloads as well as new use cases involving audio or other multimedia.

A device must be compatible with the subset of Bluetooth profiles (often called services or functions) required to use the desired services. A Bluetooth profile is a feature related to one aspect of Bluetooth-based wireless communication between devices. It sits above the Bluetooth core specification and (optionally) additional protocols. While a profile may use certain features of the core specification, specific versions of profiles rarely link to particular versions of the kernel specification, making them independent. For example, hands-free profile (HFP) 1.5 applications use Bluetooth 2.0 and 1.2 core features. These profiles are listed as follows [9]:

- Advanced audio distribution profile (A2DP)
- Attribute profile (ATT)
- Audio/Video remote control profile (AVRCP)
- Basic imaging profile (BIP)
- Basic printing profile (BPP)
- Common ISDN access profile (CIP)
- Cordless telephony profile (CTP)
- Device ID profile (DIP)
- Dial-up networking profile (DUN)
- Fax profile (FAX)
- File transfer profile (FTP)
- Generic audio/video distribution profile (GAVDP)
- Generic access profile (GAP)

- Generic attribute profile (GATT)
- Generic object exchange profile (GOEP)
- Hard copy cable replacement profile (HCRP)
- Health device profile (HDP)
- Hands-free profile (HFP)
- Human interface device (HID) profile
- Headset profile (HSP)
- iPod accessory protocol (iAP)
- Intercom profile (ICP)
- LAN access profile (LAP)
- Mesh profile (MESH)
- Message access profile (MAP)
- OBject EXchange (OBEX)
- Object push profile (OPP)
- Personal area networking profile (PAN)
- Phone book access profile (PBAP, PBA)
- Proximity profile (PXP)
- Serial port profile (SPP)
- Service discovery application profile (SDAP)
- SIM access profile (SAP, SIM, rSAP)
- Discontinued systems
- Synchronization profile (SYNCH)
- Synchronization mark-up language profile (SyncML)
- Video distribution profile (VDP)
- Wireless application protocol bearer (WAPB)

13.7 Zigbee

ZigBee is a low-cost, low-data-rate, and short-range wireless ad hoc networking standard that includes a set of communication protocols. ZigBee is developed by the ZigBee Alliance based on the IEEE 802.15.4 reference stack model and operates in the 2.4 GHz- and 868/915 MHz-frequency bands [10] (Figure 13.10).

ZigBee has many advantages in terms of low data rate, low power consumption, low complexity, high security, and support for different network topologies. Since it generally works in the 2.4 GHz band, its operating distance is between 750 and 100 m due to the difficulty of passing through obstacles such as walls. Although operating at 250 kbps, ZigBee provides small area monitors, security, discovery, profiling, and similar services for industrial control, home automatic control, and where sensor network-based applications are deployed [11]. For example, ZigBee can be used in building energy consumption monitoring systems due to its low

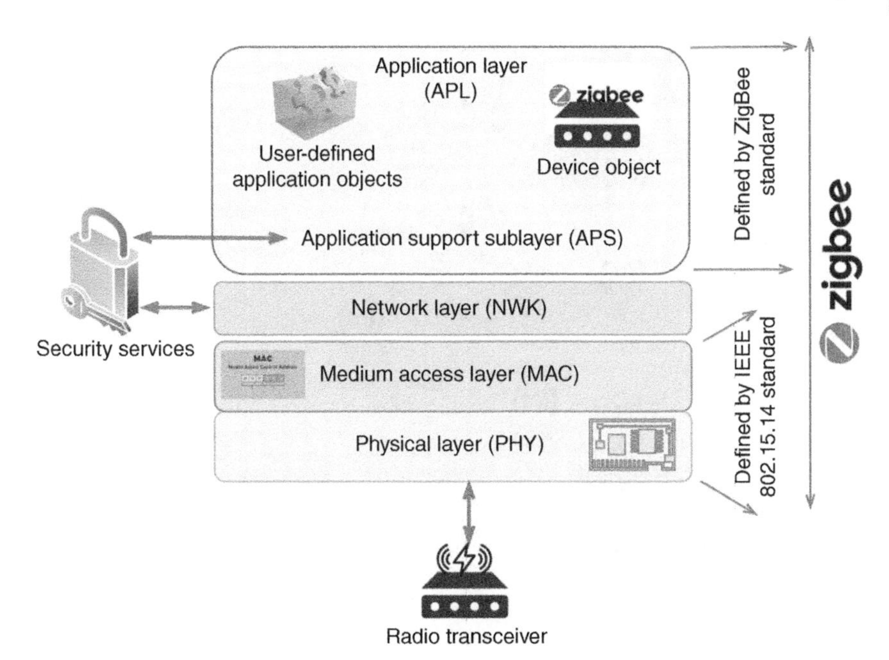

Figure 13.10 ZigBee system architecture.

cost, device sparsity, low energy (LE) consumption, and self-organization features. End devices with different sensors are used to monitor temperature, humidity, voltage, etc. End devices can also collect data from water, gas, and electricity meters. These data collected from various end devices will be sent to the upper computer. The particular system will establish the policy to achieve energy consumption monitoring, temperature control, and energy-saving operation management [10] (Figure 13.11).

ZigBee system structure consists of three main components: ZigBee coordinator, router, and end device. Each ZigBee network has a coordinator working between the network and ZigBee. This coordinator is a hub that receives and stores important information in transmitting data operations. On the other hand, we can say that the coordinator is a particular router. In addition to the router's capabilities, a coordinator is responsible for creating the network. It must select the appropriate channel, PAN ID, and extended network address. It is also responsible for choosing the security mode of the network.

On the other hand, the ZigBee router (ZR) acts as an intermediate between this information hub and end devices. End devices, on the other hand, have limited communication access to save useful power, energy, or the battery itself. In this sense, we can say that the most crucial advantage of ZigBee is power saving. Power

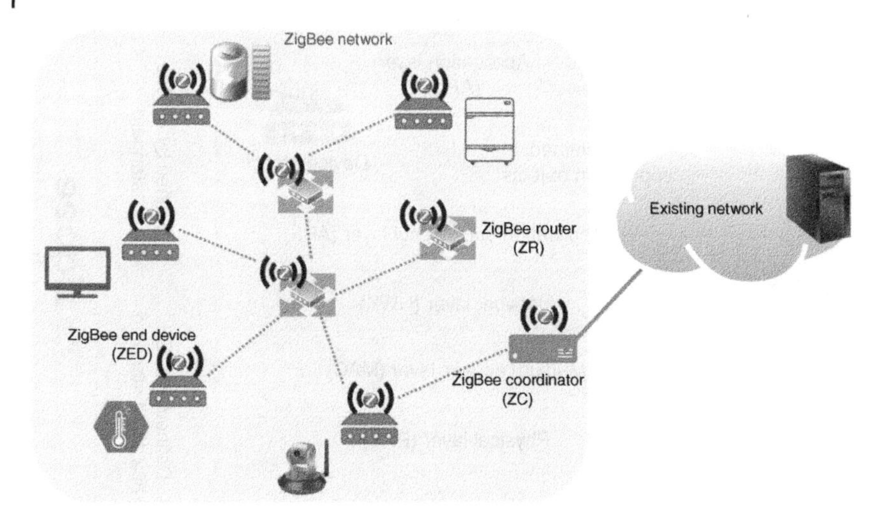

Figure 13.11 ZigBee network.

savings are at high levels due to low data rates and small coverage areas. In addition, using a lightweight protocol stack requires much lower software support than others. It also has a self-healing feature due to its mesh topologies. Since ZigBee devices go into sleep mode at leisure, they have a battery life that lasts for months. These three components are connected to the star, tree, and mesh network topology infrastructure.

To better understand the working principle, it is necessary to know the ZigBee 5 layer.

- **Physical layer (1):** This layer performs the modulation task and demodulates various transmitted and received signals. Depending on locality, different frequencies, data rates, and channels can be used in this layer.
- **MAC layer (2):** The function of this layer is to provide reliable data transfer communication by accessing different networks with carrier sense multiple access (CSMA) and collision avoidance (CA).
- **Network layer (3):** This layer accompanies all network-related operations such as connection, network disconnection, routing, and various device configurations between the router and different end devices.
- **Application support sublayer (4):** This layer is responsible for interfering with ZigBee devices with different object application devices to communicate through the network layer. It is also responsible for matching peripherals to their services, applications, and needs.

- **Application framework (5):** Nonspecific message is a construct characterized by the developer, critical value matching is used to get properties inside application objects. ZDO provides an interface between application elements and the APS layer in ZigBee gadgets. It is responsible for distinguishing, initializing, and restricting different gadgets to the system.

In ZigBee technology, data transmission is carried out in two different modes: Beacon mode and non-beacon mode. In beacon mode, coordinators and routers consume a lot of power as they constantly monitor changes in flowing data. In this mode, routers and coordinators do not go into sleep mode because they can exchange signals to communicate or respond to the node anytime. However, overall power usage is low as most devices are hidden in the system for long periods, even though they require more power supplies.

On the contrary, in beacon mode, the router and coordinators go into sleep mode without data transmission. A cyclical process uses counters that periodically turn routers on and off to transmit data to multiple nodes within a network. These networks work for appropriate time slots, i.e. when correspondence requires results at lower duty cycles and more extended battery usage [12].

The functions of the three critical components of a ZigBee network are summarized as follows [3]:

- **ZigBee controller (ZC):** It is a high-capacity device that initiates and creates network functions. Each ZigBee network has a single ZC that acts as the 802.15.4 2003 PAN coordinator. After the network is formed, the ZC can act as a ZR. It can provide a logical network address and allow nodes to join or leave the network.
- **ZigBee router (ZR):** This component is optional but does handle some mesh network hoping and routing coordination. Simultaneously, a PAN can fulfill the coordinator role and associate with the ZC. A ZR takes its messages through multihop routing and allows logical address assignment and nodes to join or leave the network.
- **ZigBee and device (ZED):** This is usually a simple end device such as a light switch or thermostat. Coordinators have functions that are capable of communicating. It does not have routing logic. Therefore, any message reaching the ZED and not targeting that device is relayed. Theoretically, a ZigBee network could contain 65 536 ZEDs (Figure 13.12).

Lastly, two unique addresses are used per ZigBee node. One of them is the 64-bit long address assigned to the device by the manufacturer and is immutable (the first 24-bit organizational unique identifier and the remaining 40-bit areas are the OEM-managed address). The other is a 16-bit short address that is the same as the 802.15.4 PAN ID.

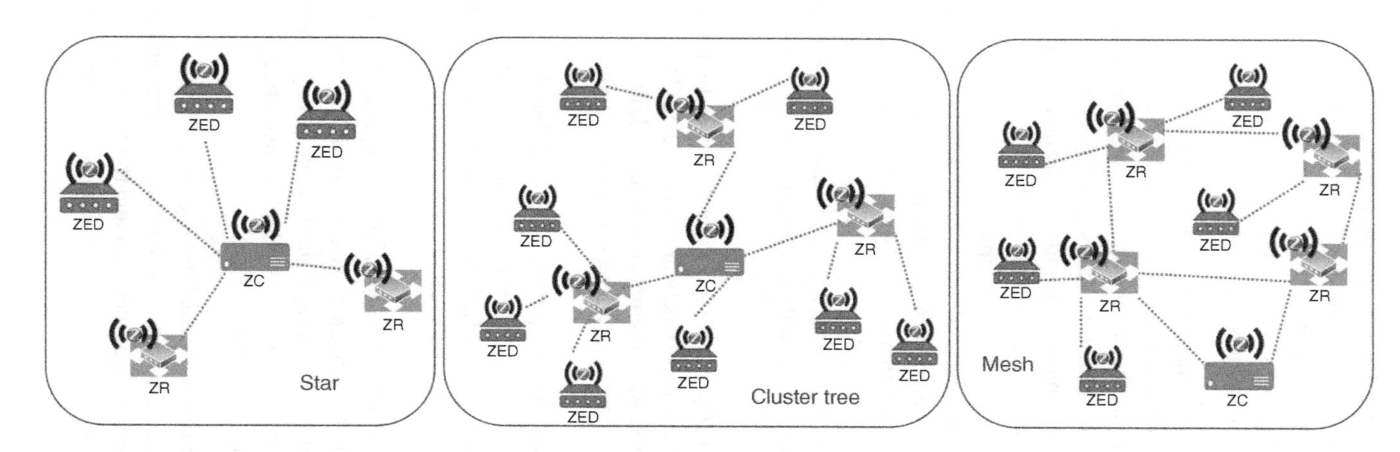

Figure 13.12 ZigBee network topologies.

13.8 Z-Wave

Z-Wave is another mesh PAN technology introduced by a Danish company in 2001 to be used in health, ventilation, and air conditioning (HVAC) and light control, operating at 900 MHz (may vary depending on the country). The main work area is home automation. Companies such as SmartThings, Honeywell, Belkin, Bosch, Carrier, ADT, and LG are members of the Z-Wave Alliance.

It operates at speeds of 100, 40, and 9.6 kbps, depending on the Z-Wave propagation band. It has a high wavelength because it operates at a low frequency. In this way, it can penetrate through obstacles such as walls. Thus, it provides a more reliable and fast communication topology between X-Wave devices (Figure 13.13).

Z-Wave provides a source-routed mesh network topology consisting of sensors (end devices) and home automation with a primary node called a hub or controls [13]. All Z-Wave networks are identified by the primary controller's network ID. Z-Wave devices, called secondary nodes, are identified with their node IDs. Primary controllers assign Network IDs to network end devices. Network IDs of all devices in a particular network are the same. As such, devices on different networks (with different network IDs) cannot communicate with each other.

- **Controller device:** This top-level device provides a routing table for the mesh network and has responsibilities such as including and removing end devices from the network. There are two types of controllers: primary controller and secondary controller.
- **Slave device/node:** They are end devices that respond to the commands they receive. These devices cannot communicate with neighboring end devices

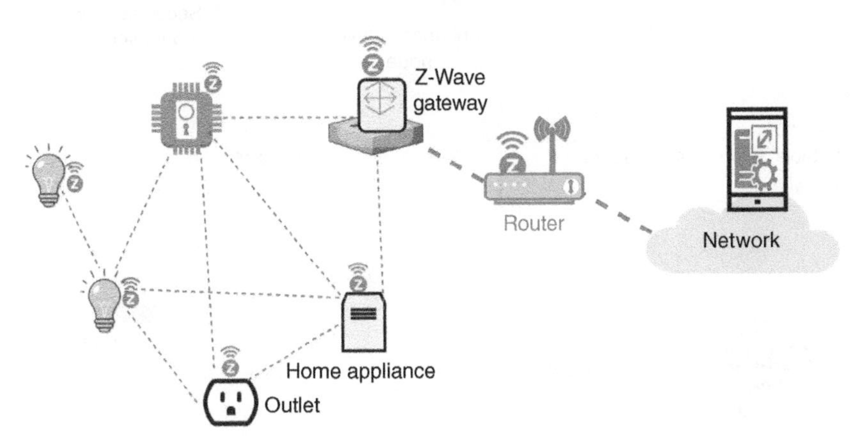

Figure 13.13 Z-Wave architecture.

unless specified with a command. They can store routing information but cannot perform computation or update the routing table. Typically, they act as repeaters in the mesh network.

Due to the mesh network topology, the primary controller can send a message to the target Z-Wave device in remote locations by routing the signal through intermediate devices by finding the shortest possible path. This signal transfer process is often referred to as signal hopping. If devices within the path are busy, they will route signals through other available paths (Figure 13.14).

Z-Wave features are summarized as follows:

- Compared to Wi-Fi, Z-Wave uses much lower power and offers three to five years of battery life.
- Low interference and high-penetration capacity as it operates at 900 MHz.
- Interoperability with backward compatibility.
- High security in 100 kbps with AES128.
- Access to a wide range of household products such as lighting, thermostats, security sensors, and locks.
- A range of up to 100 m.
- Since it works at different frequencies, it does not interfere with the signals emitted by WiFi or any household device.
- Up to 232 devices can be defined in a Network ID.

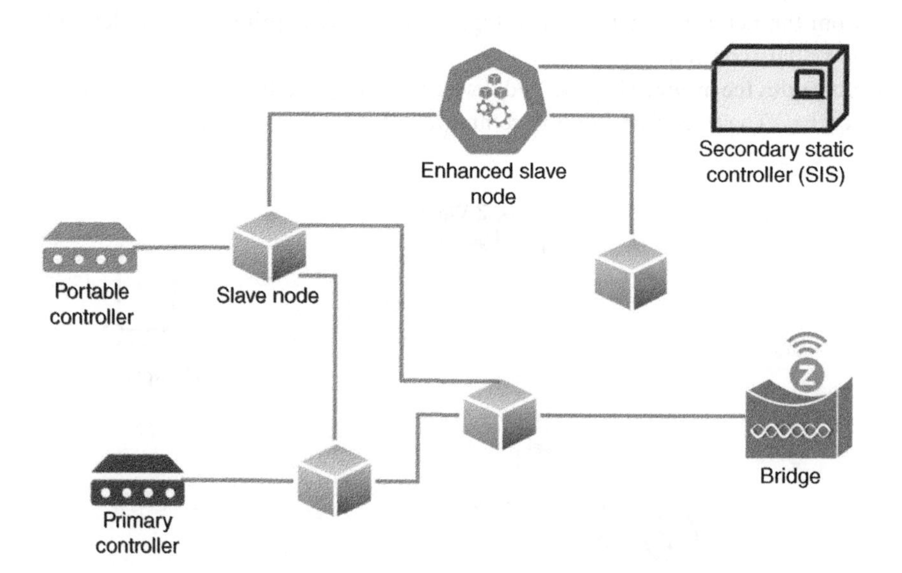

Figure 13.14 Z-Wave topology.

13.9 Power Line Communication

Power line communication (PLC) involves using existing electrical wiring to carry data and is a long-established method. Power grids have been using this technology for many years to send or receive a limited amount of data to the existing power grid. The PLC uses already existing infrastructure in every home and industrial facility, eliminating the unnecessary expense and the hassle of installing new cables to achieve high signal penetration. In this way, it allows the use of plug-and-play-type PLC-enabled systems.

Home and industry automation is an essential part of modern life, helping control and monitor a wide variety of home and industry devices such as air conditioning, refrigeration, or lighting systems. In addition to the functionality and comfort these systems provide, they also help increase the energy efficiency of appliances and other electrical appliances. The rapid development of communication technologies has accelerated the integration of automation systems into cyber-physical networks and enabled these systems to be managed remotely. On the other hand, more measures have emerged to ensure security and privacy. PLC provides a wide range of communication frequencies and systems. These systems are divided into two: Narrowband PLC is mainly used in automation. On the other hand, broadband PLC allows for home network applications. Many PLC-based home and industrial automation solutions are now available [14] (Figure 13.15).

The figure shows that the power line connects appliances, sensors, and controllers. Light, heat, and smoke sensors are installed from all house parts and transmit the signals to the control unit through communication interfaces such as Ethernet and RS232. The task of the control unit, as in other home automation systems, is to record the data, determine the command required for automation, and transmit

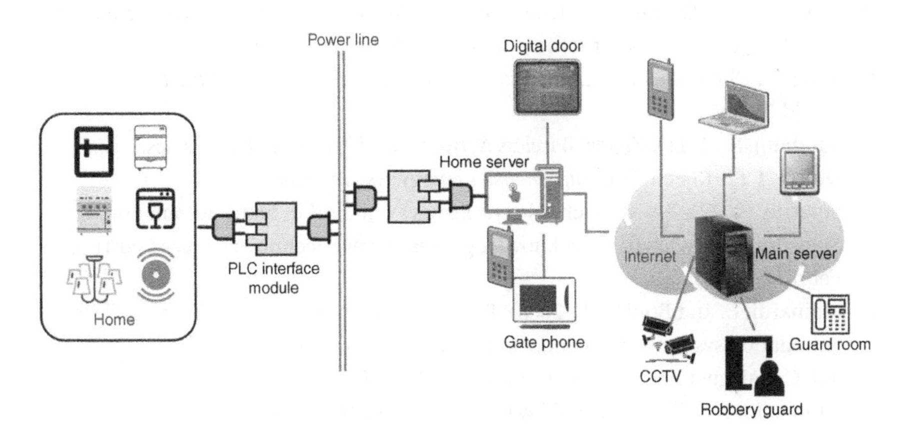

Figure 13.15 PLC application.

these commands to the actuator and regulators. Sensors, control units, devices, and actuators are interconnected via a communication channel with power lines available in the case of PLC [15].

References

1 Peng, S.L., Pal, S., and Huang, L. (ed.) (2020). *Principles of Internet of Things (IoT) Ecosystem: Insight Paradigm*. Springer International Publishing.

2 Bouazza, H., Zohra, L.F., and Said, B. (2019). Integration of Internet of Things and social network. In: *Advances in Data Science, Cyber Security and IT Applications*, Communications in Computer and Information Science (ed. A. Alfaries, H. Mengash, A. Yasar, and E. Shakshuki), 312–324. Cham: Springer.

3 Lea, P. (2018). *Internet of Things for Architects: Architecting IoT Solutions By Implementing Sensors, Communication Infrastructure, Edge Computing, Analytics, and Security*. Packt Publishing Ltd.

4 King, S. and Nadal, S. (2012). Ppcoin: Peer-to-peer crypto-currency with proof-of-stake. Self-published paper, 19 August 2012.

5 Finkenzeller, K. (2010). *RFID Handbook: Fundamentals and Applications in Contactless Smart Cards, Radio Frequency Identification and Near-Field Communication*. Wiley.

6 Coskun, V., Ozdenizci, B., and Ok, K. (2012). A survey on near field communication (NFC) technology. *Wireless Personal Communications* 71 (3): 2259–2294.

7 Cirani, S., Ferrari, G., Picone, M., and Veltri, L. (2018). *Internet of Things*. Wiley.

8 Bluetooth.org (2022). Bluetooth SIG, Inc. https://www.bluetooth.org (accessed 16 February 2023).

9 Wikipedia (2022). List of Bluetooth profiles. *Wikipedia*. https://en.wikipedia.org/wiki/List_of_Bluetooth_profile (accessed 16 February 2023).

10 Yang, Y., Chen, X., Tan, R., and Xiao, Y. (2021). *Intelligent IoT for the Digital World*. Wiley.

11 Farahani, S. (2011). *ZigBee Wireless Networks and Transceivers*. Newnes.

12 Dincer, I. (2018 Feb 7). *Comprehensive energy systems*. Elsevier.

13 Smartify (2022). Z-Wave technology: the new standard in home automation. *Smartify*. https://smartify.in/knowledgebase/z-wave-technology (accessed 16 February 2023).

14 Mainardi, E. and Bonfè, M. (2008). Powerline communication in home-building automation systems. In: *Robotics and Automation in Construction* (ed. C. Balaguer and M. Abderrahim), 53–70. InTech.

15 Lampe, L., Tonello, A.M., and Swart, T.G. (ed.) (2016). *Power Line Communications: Principles, Standards, and Applications from Multimedia to Smart Grid*. Wiley.

14

IP-Based WPAN and WLAN

14.1 Introduction

Flexible and scalable wireless local networks can be quickly established using IP-based wireless connectivity in the local area. This section explains IP-based wireless personal area networks and wireless local area network technologies used in machine-to-machine communication and IoT.

14.2 HaLow WiFi (Low-Power WiFi)

Low-power Wi-Fi is a standard that corresponds to the IEEE 802.11ah standard, which emerged with the amendment of the IEEE 802.11-2007 wireless network standard. It is also called Wi-Fi HaLow by the Wi-Fi Alliance. Wi-Fi HaLow is expected to find wide use in industrial, retail, agricultural, and smart city environments and power-efficient applications such as smart homes, connected cars, and digital healthcare [1].

WiFi HaLow benefits and features are given as follows:

Features:

- Sub-1 GHz spectrum operation
- Narrowband OFDM channels
- Several device power-saving modes
- Native IP support
- Latest Wi-Fi security (IPv6 and WPA)

Benefits:

- Long range: approximately 1 km
- Penetration through walls and other obstacles

Evolution of Wireless Communication Ecosystems, First Edition. Suat Seçgin.

Published 2023 by John Wiley & Sons, Inc.

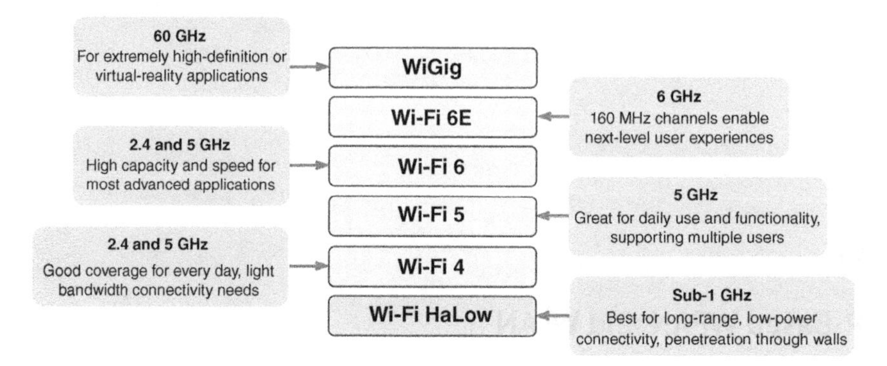

Figure 14.1 WiFi HaLow.

- Supports coin cell battery devices for months or years
- No need for proprietary hubs or gateways.

In Figure 14.1, the place of WiFi Halow among the Wi-Fi generations is given.

WiFi HaLow's ability to penetrate walls and other barriers is a crucial advantage. 802.11ah can solve many problems associated with deploying large-footprint WiFi networks by allowing many devices, providing power-saving services, and allowing long distances to the AP. A typical 802.11ah AP can be related to 8000 devices within a 1 km. This makes WiFi HaLow an ideal solution for installing high-concentration sensors and other devices such as street light controllers and smart parking meters. The 802.11ah standard also includes new PHY and MAC layers that group devices into traffic indicator maps to host small units (such as sensors) and M2M communications [2] (Figure 14.2).

14.3 ISA 100.11a Wireless

ISA 100.11a is a wireless networking protocol developed by the International Society of Automation (ISA). Its official full definition is "Wireless Systems for Industrial Automation: Process Control and Related Applications." ISA 100 is an industrial wireless network standard that addresses all the features a factory may need, such as process control, personnel and asset tracking, and identification convergence of networks [3].

ISA 100 is designed to address a key header: security, reliable wireless communication, good power management, multifunctional, multispeed monitoring, open, scalable, global usability, control ready, multiprotocol, and quality of service. In addition, it includes three different service classes: safety (emergency action), control (closed loop, supervisory control, and open-loop control), and monitoring (alerting, logging, and downloading/uploading).

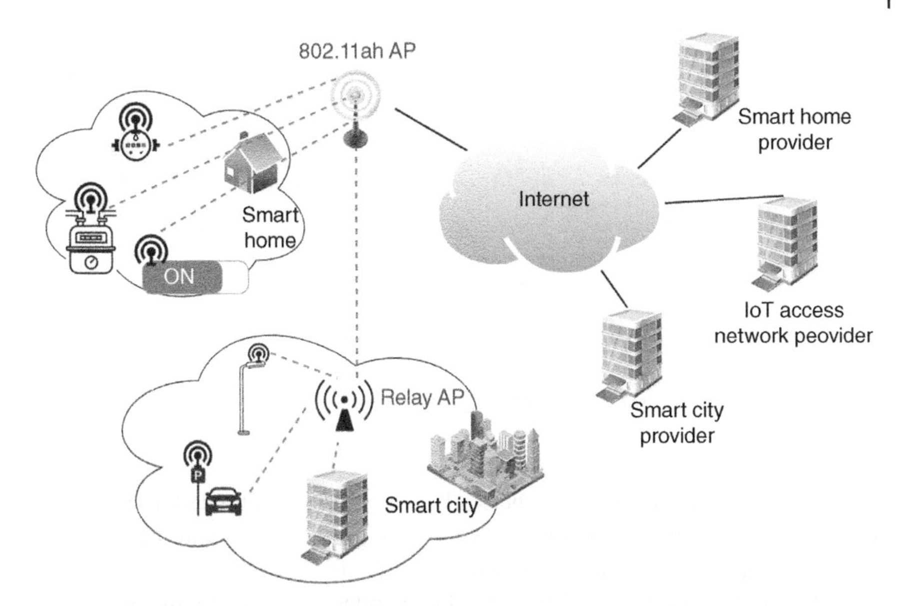

Figure 14.2 IEEE 802.11ah network model.

Table 14.1 ISA-100 on ISO/OSI.

ISO/OSI layer	ISA 100 wireless functionality
Application	*Object-oriented, pub/sub, client/server, native/non-native, tunneling*
Presentation	
Session	
Transport	*Connectionless, optional security (symmetric/asymmetric)*
Network	*6LoWPAN, graph/source/superframe routing, fragmentation*
Data link	*CSMA/CA, TSMP, channel hoping (slotted, frequency), black listing*
Physical	*IEEE 802.15.4 (2.4 GHz)*

ISA 100 runs on the other five tiers of the ISO/OSI model, excluding the presentation and session tiers [4]. The ISA-100 wireless stack for these layers is given in Table 14.1.

As can be seen from Table 14.1, ISA 100 uses the IEEE 802.15.4 standard at the physical layer. There are 27 channels in total, numbered from 0 to 26 in this physical layer. The ISA 100 uses the 2.4 Ghz frequency band (almost unlicensed

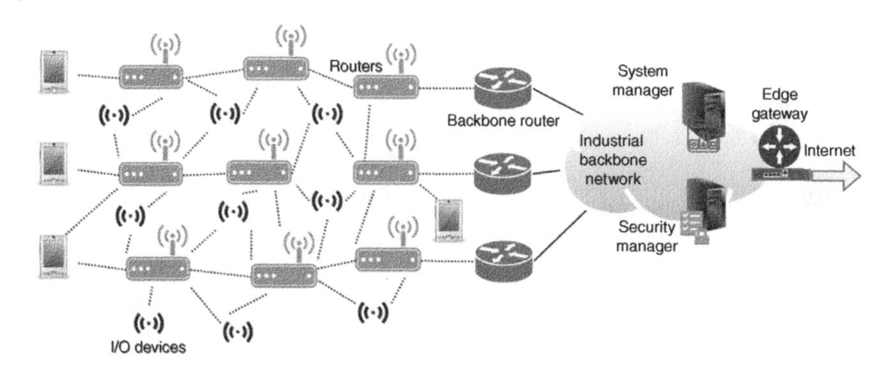

Figure 14.3 ISA 100 wireless network architecture.

worldwide). When we look at the network layer, we see that the ISA 100 wireless communication standard also supports the IPv6 protocol. This way, the Internet traffic is supported and compatible with the industrial Internet and provides an infrastructure solution for industry 4.0. ISA wireless networking architecture is given in Figure 14.3 [5].

ISA 100 defines wireless devices in two main classes: field and backbone devices and system and security management devices. The system administrator is responsible for network resource management and communication, while the security administrator implements the security design based on the adopted security policy. The field devices class can support wireless devices with or without routing capability. For example, a mobile, handheld device such as a smartphone may be classified as a field device without routing capability. The decisive purpose of such a device in the network is to associate with a routing device and transmit essential data or monitor and analyze network traffic. The roaming of mobile devices is outside the scope of ISA100 Wireless. Backbone devices in the network are supplied with high energy and constantly powered, while field devices typically have limited battery capacity. The system administrator acts as the global network clock and manages time and time synchronization information. It is assumed that ISA100 Wireless implements a time-division multiple access (TDMA) network topology targeting relevant Industry 4.0 applications, resulting in a strongly centralized and relatively stable deployment. Although the standard can support the star topology, the network topology is preferred as it offers excellent reliability and a greater capacity for external interference management. ISA100 Wireless centralized data management (via system manager) performs dynamic topological changes in response to faults based on selected routing mechanisms and predetermined time-synchronized message scheduling [3].

14.4 Wireless Highway Addressable Remote Transducer Protocol (HART)

It is an industrial wireless sensor structure used in the industrial IoT segment. The classification of industrial wireless sensor networks (IWSNs) is given hereunder [6] (Figure 14.4).

Based on the Highway Addressable Remote Transducer (HART) Protocol, WirelessHART has a wide range of applications as it is a suitable solution for industrial requirements. With its latest version in 2016 by the International Electrotechnical Commission, HART has the following updated features:

- Wireless mesh networking
- Time synchronization and stamping
- Security encryption and decryption
- Enhanced burst mode messaging
- Pipes for high-speed file transfer

The WirelessHART communication protocol uses five of the seven layers of the OSI model, excluding presentation and session. These five layers are physical, data-link, network, transport, and application layers. The central network manager handles routing, communication scheduling, and corresponding signal generation.

WLAN, Bluetooth, Zigbee, and IPv6 have not been widely accepted for use in the industry due to the limits in their controlling capacities. Therefore, the most used wireless standards in the industry are WirelessHART and ISA100 Wireless [7]. WirelessHART network infrastructure is given in Figure 14.5.

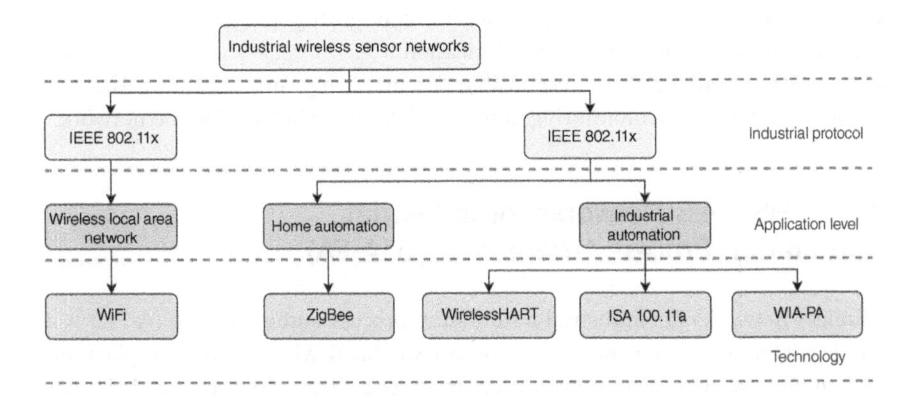

Figure 14.4 Classification of IWSNs.

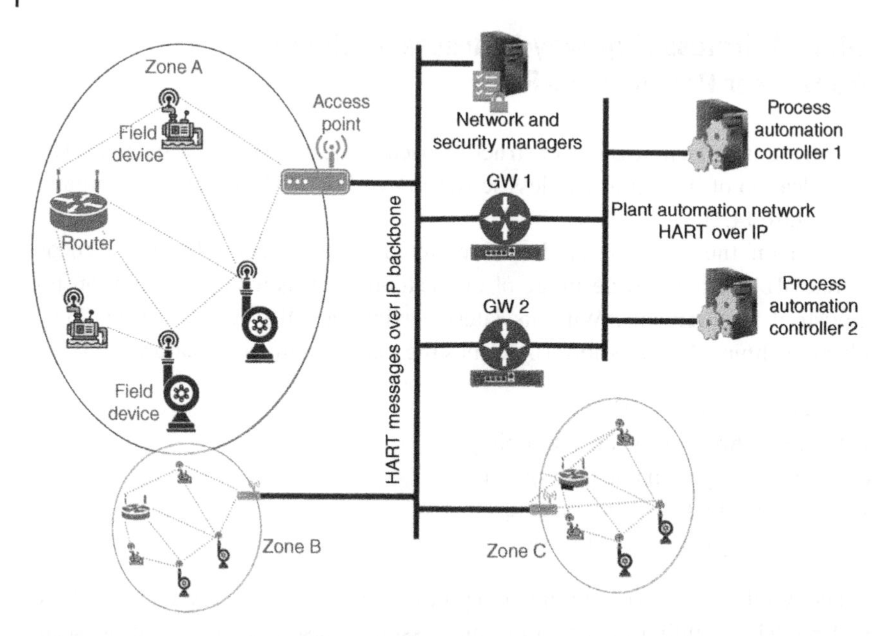

Figure 14.5 WirelessHART network structure (single backbone-multiple access points).

The essential elements of the WirelessHART network control system (WHNCS) are given as follows:

- **Field device:** It is a device connected to an industrial process plant.
- **Wireless handheld:** Used for diagnostics, device configuration, and calibration from a remote location.
- **Gateways:** Acts as a bridging device to connect host applications and field devices.
- **Network manager:** It is responsible for configuring, programming, routing the network, and managing communication.
- **Security manager:** It is responsible for managing and allocating security encryption keys and monitoring authorized devices connected to the network.

14.5 Wireless Networks for Industrial Automation-Process Automation (WIA-PA)

Wireless networks for industrial automation-process automation (WIA-PA) is a wireless communication network protocol validated with practical application and suitable for complex industrial environments. The advantages of WIA-PA, such as flexible TDMA, frequency (FHSS frequency hopping mechanism), and

Table 14.2 Comparison of industrial WSN standards.

	Zigbee	WirelessHART	ISA 100.11a	WIA-PA
Data security	*High*	*Very high*	*Very high*	*Very high*
Scalability	*Medium*	*High*	*High*	*High*
Power usage	*Low*	*Low*	*Low*	*Low*
Data rate	*20–250 kbps*			
Topology	*Star/Mesh/Tree*	*Star/Mesh*	*Hybrid*	
Data reliability	*Low*	*Very high*	*Very high*	*Very high*
Routing capability	*Limited*	*Full*	*Full/Limited*	*Limited*
Channel hoping	*No*	*Yes*	*Yes*	*Yes*
Frequency channels	*27 (all bands)*	*15 (2.4 GHz)*	*16 (2.4 GHz)*	*16 (2.4 GHz)*
Manager architecture	*Centralized/ Distributed*	*Centralized*	*Centralized*	*Centralized/ Distributed*

space (hybrid network topology consisting of mesh and star structure), make it a relatively simple and effective protocol. Thanks to this simplicity, it has a structure that does not require complex installation. Below is a comparison of IWSN standards mentioned in the book [8] (Table 14.2).

The WIA-PA standard is one of the three (WIA-PA, WirelessHART, or ISA 100.11a) that implements the beacon-enabled mode of the IEEE 802.15 standard. This structure allows WIA-PA to interoperate with other wireless sensor networks [9]. WirelessHART and ISA 100.11a adopt centralized resource management, which simplifies the implementation of global optimization but makes it difficult to manage large-scale networks. WIA-PA is based on a hierarchical framework and supports a hybrid (star and mesh infrastructure) centrally distributed management. Due to this high-scalability feature of WIA-PA compared to WirelessHART and ISA 100.11a, it is much more suitable for large-scale IWSNs. Therefore, this standard is used in steel mills and oil fields.

As can be seen from the Figure 14.6, the field is divided into clusters through the heads (routers) located in the cluster centers [10]. These clusters connect through field devices and handheld devices. Cluster heads are connected to a wireless mesh network structure. This structure connects to other networks (Internet, Intranet, etc.) with gateway devices at the top. The dynamic scalability feature is realized by adding/removing clusters from the system when necessary.

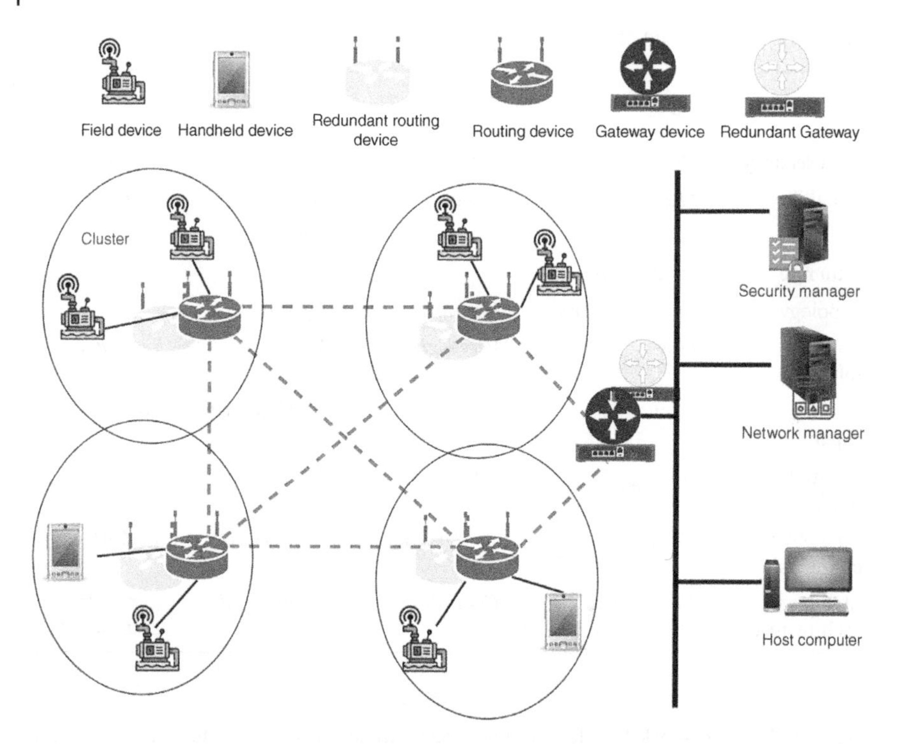

Figure 14.6 The standard topology of WIA-PA IWSNs.

The big picture showing the long-distance use of industrial wireless networks created with WLA-PA, WirelessHART, ISA 100.11a, and Zigbee standards and its place in the IoT system is given in the Figure 14.7 [11].

14.6 6LoWPAN

6LoWPAN is an acronym for IPv6 over low-power WPANs. The main reason for its development is to make IP networking available for powerful and space-constrained devices that do not require high bandwidth. This protocol can be used with 802.15.4 and WLAN communications such as Bluetooth, sub-1 GHz RF protocols, and a power line controller (PLC). The most basic feature of 6LoWPAN is that even the simplest sensors can be given IP addresses. These small sensors can operate as a network component over 3G/4G/LTE/WiFi/Ethernet routers. Another advantage is that it allows 2128 or 3.4 × 1038 unique addressing because it uses IPv6. Thus, it will be possible to connect billions of IoT devices to the network [12] (Figure 14.8).

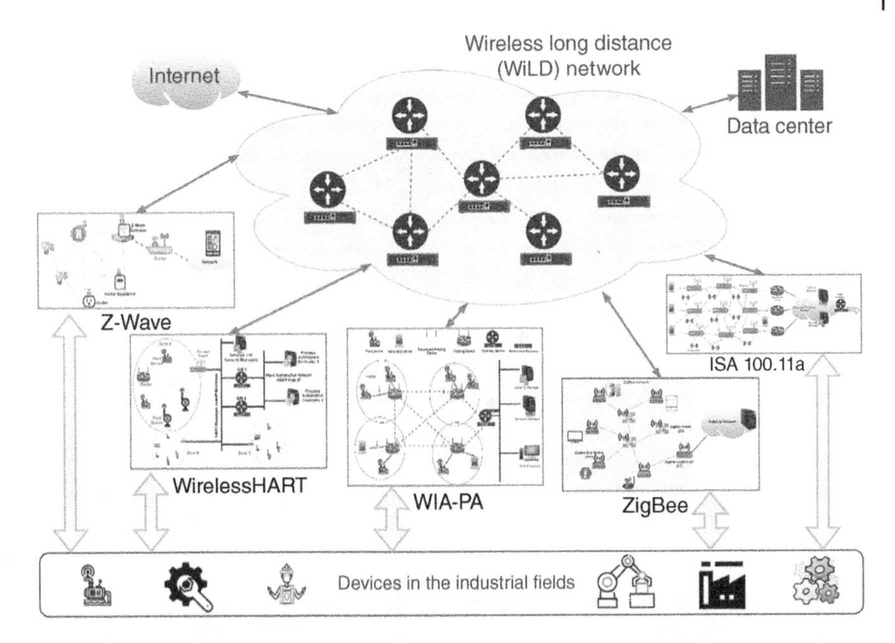

Figure 14.7 Integration of industrial WSN.

Because it works with IPv6, 6LoWPAN works in the OSI model's network layer (layer 3). A 6LoWPAN mesh network has three types of nodes:

- **Router nodes:** These nodes queue data from one 6LoWPAN network node to another. Routers can also communicate in WAN and Internet direction.
- **Host nodes:** Host nodes in a mesh network are endpoints that cannot route data in the network and only consume or generate data. Host nodes are allowed to be in a sleep state and wake up occasionally to generate data or retrieve data cached by their host router.
- **Edge routers:** As can be seen, these gateways and mesh controllers are usually located at the WAN edge. It is administered under a 6LoWPAN mesh edge router.

In cases where the wireless network does not need an Internet connection (ad-hoc 6LoWPAN), the network can be managed with router nodes without an edge router. When IPv6 addressing is not required, this practice is rarely carried out. In the ad-hoc WPAN application, the router performs the node's unique local unicast address generation and neighbor discovery (ND) registration functions. In Figure 14.9, the OSI layers and the protocols used by 6LoWPAN are given.

6LoWPAN advantages:

- Robust, scalable, and self-healing mesh network structure.
- Long-range communication capable of detecting signals below the noise level.

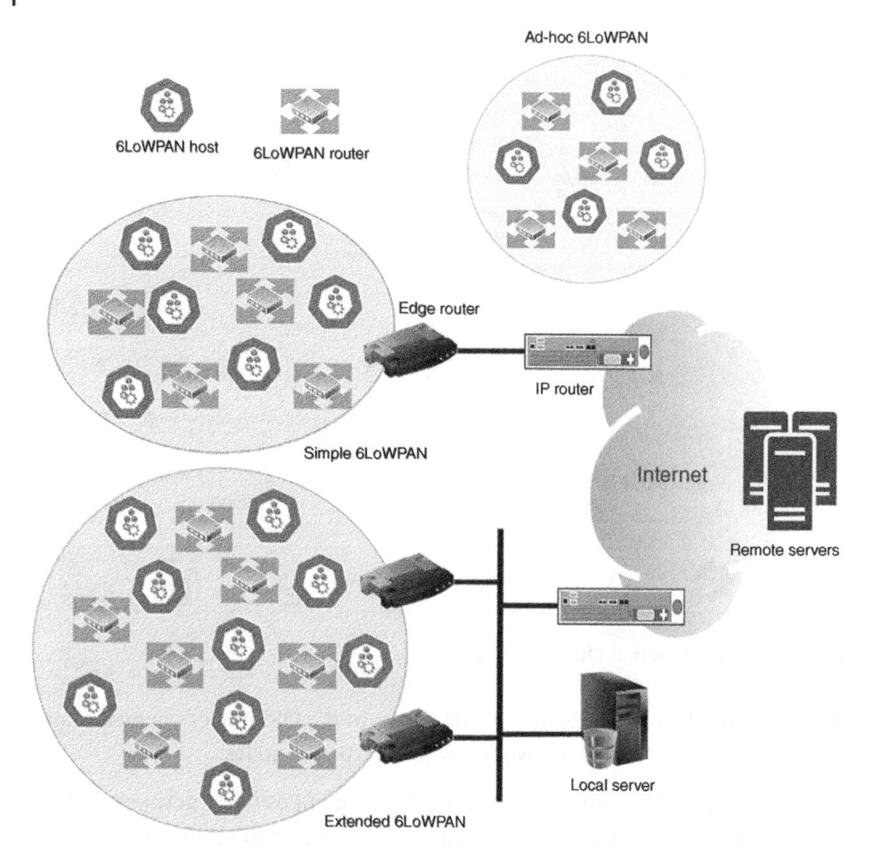

Figure 14.8 6LoWPAN architecture.

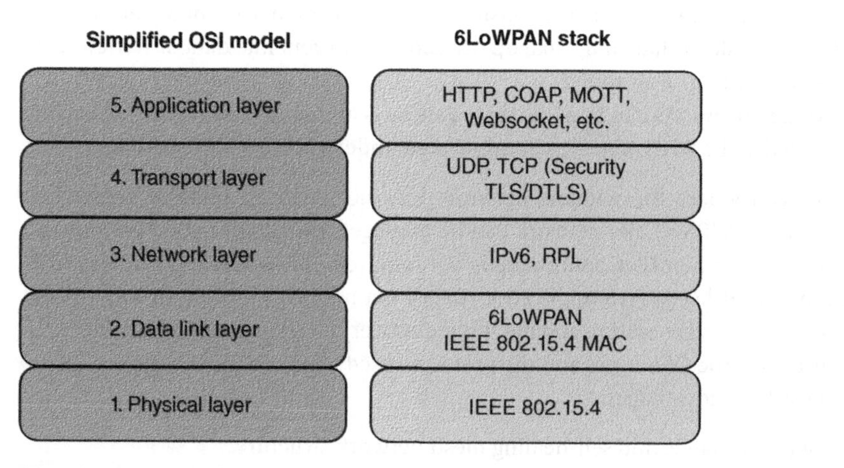

Figure 14.9 6LoWPAN stack.

- Very long battery life due to low power consumption and reduced transmission (short-time pulses).
- Wide network structures with the usage of millions of devices.
- Low-cost and secure communication for IoT.
- Transparent integration is possible due to the capacity of one-to-many and many-to-many routing.
- Direct access to cloud platforms as it uses IPv6.

In addition to all these advantages, it has disadvantages such as less secure communication compared to Zigbee, less immunity to interference compared to WiFi or Bluetooth devices, and short-distance support when mesh topology is not used. A comparison of narrowband IoT applications is given in Table 14.3.

Table 14.3 Comparison of narrow-band IoT standards.

	Sig Fox	LoRaWAN	6LoWPAN
Frequency	*902 MHz (US)* *868 MHz (EU)*	*902–928 MHz (NA)* *863–870 or 434 MHz (EU)* *779–787 MHz (CH)*	*2.4 GHz (Wordwide)* *902–929 MHz (NA)* *868–868.6 MHz (Eu)*
Channel	*360 with 40 reserved*	*80 (902–928 MHz)* *10 (779–787 and 863–870) MHz*	*16 (2.4 GHz)* *10 (915 MHz)* *1 (868.3 MHz)*
Bandwidth	*10 Hz to 1.2 kHz*	*125 and 500 kHz (in 915 MHz)* *125 and 250 kHz (in 868 and 780 MHz)*	*5 MHz (2.4 GHz)* *2 MHz (915 MHz)* *600 kHz (868.3 MHz)*
Range	*10–50 km*	*5–15 km*	*10–100 m*
Data rate	*980 bps to 21.9 kbps*	*100–600 bps*	*250/40/20 kbps*
Payload	*0–12 bytes*	*19–250 bytes*	*6 bytes (header)* *127 bytes (data unit)*
Channel coding	*Ultra narrow band coding*	*Chirp spread spectrum (CSS)*	*Direct sequence spread spectrum (DSSS)*
Security	*No encryption mechanism*	*NwkSKey (128 bits-for data integrity)* *AppSKey (128 bits-for data confidentiality)*	*Handled at link layer that includes secure and non-secure mode*

14.7 WPAN with IP Thread

Thread is a new networking protocol developed for IoT based on IPv6. Its primary use case is home connectivity and home automation. Based on the IEEE 802.15.4 protocol and 6LoWPAN, Thread has standard features with Zigbee and other 802.15.4 variants. But there is an essential difference between them: thread can be addressed with IP protocol. Thread networks operate in a low-power mesh networking topology.

Thread networks are straightforward to install, use, and automatically reuse when a device is added or removed. The network can heal itself in the event of a failure. Thread provides an infrastructure that connects hundreds or even thousands of products reliably and securely end-to-end (device to device, device to mobile, device to cloud).

Thread uses the proven, open standards of the Internet to build an IPv6-based network. As a result, Thread devices integrate seamlessly with more extensive IP networks and do not need dedicated gateways or translators. This reduces infrastructure investment and complexity, eliminates potential points of failure, and reduces maintenance burdens. Thread also securely connects devices to the cloud, making it easy to control IoT products and systems from personal or administrative devices such as mobile phones and tablets.

Thread's IP basis is independent of the application layer. It offers product manufacturers the flexibility to choose one (or more) application layer(s) for their use case to connect devices over multiple networks. With the growing demand for interoperability and coexistence, Thread's cross-industry membership reveals networking technologies underpin IP as a convergence point for the IoT industry. For example, a single application layer, such as a smart home standard developed by any manufacturer, can work with devices connected via both Thread and WiFi. This enables device manufacturers to choose the correct network technologies for their applications while creating a seamless network of interoperable products.

With Thread, developers can bring their apps, devices, systems, and services to market faster because they use the same rich set of tools available on the Internet. Moreover, the application layer and cloud services in Thread devices can change over time as Thread is independent of the application layer. Because Thread is IP-based, it can enable interoperability across various connected devices while providing a direct connection to manufacturers, their products, and their users. In addition, all communication can be performed securely using Advanced Encryption Standard (AES) as it is IPv6 compatible [13]. All nodes connected to the Thread network can be included in a fully encrypted and authenticated mesh network.

Based on the IEEE 802.15.4-2006 standard, Thread uses specifications that define MAC and physical addresses. It operates in the 2.4 GHz band with a speed of 250 kbps. Regarding topology, Thread communicates with other devices over a border router (a WiFi modem at home). The rest of the communication is based on 802.15.4, and a self-healing mesh is formed.

The devices in the topology are explained as follows:

- **Border router:** It works as a gateway. This device connects the thread network to the rest of your home network. It works like a kind of hub. All thread-based devices have a Bluetooth chipset as a backup. More than one border router can be used to ensure redundancy and failover in a network. It can act as a home network's border router for WiFi signals (Figure 14.10).
- **Lead device:** The lead device manages the registration of assigned router IDs. The lead also controls requests for Router Eligible End Devices (REED) to be promoted to the router. A lead device can act as a router and have device-end children. The protocol for assigning router addresses is called Constrained Application Protocol (CoAP). State information managed by a lead device can also be stored in other thread routers. This allows self-healing and failover in case of loss of lead device connection.
- **Thread routers:** These routers manage the routing services of the mesh network and do not go to a sleep state unless they are downgraded to REEDs (Figure 14.11).

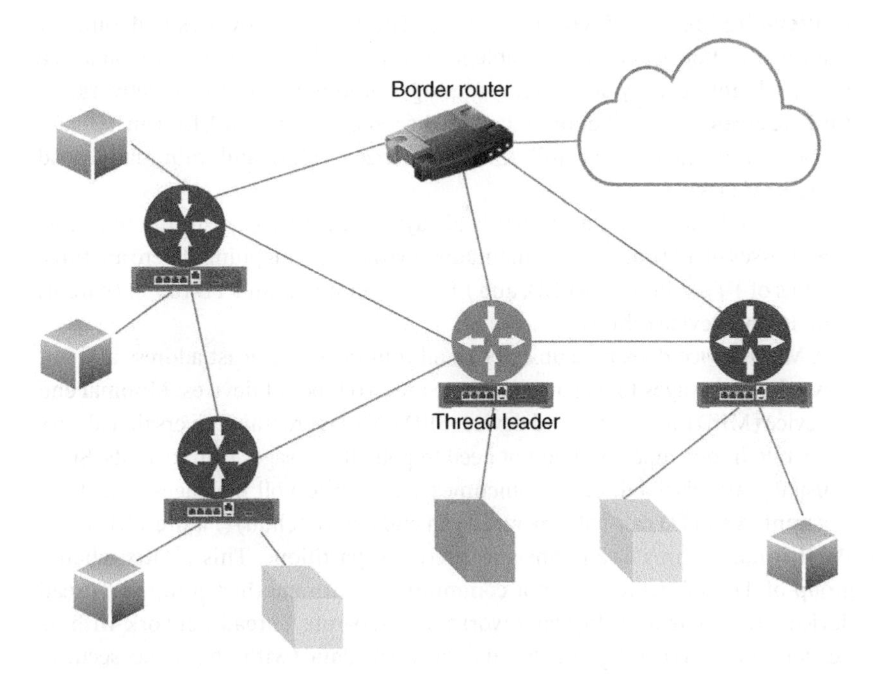

Figure 14.10 General architecture of IP thread.

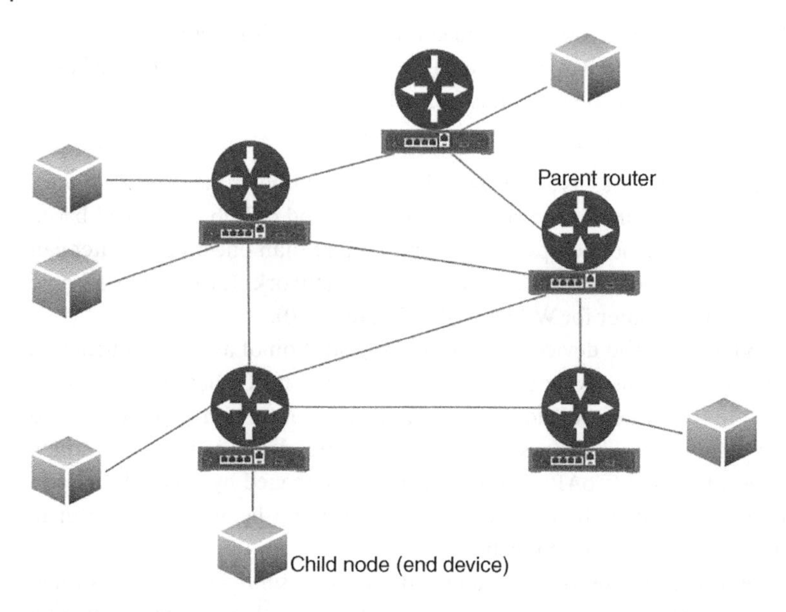

Figure 14.11 Parent-child structure.

- **Router-eligible end devices (REEDs):** These are a network's endpoints or leaf nodes. They are not responsible for routing unless they are promoted as a router. Therefore, they do not relay messages or join devices to the network.
- **End devices:** Some endpoints cannot become routers. REEDs can be subscribed to in two states: full thread devices (FTDs) and minimal thread devices (MTDs).
  - An FTD device is a device that is always open, subscribes to the multicast addresses of all routers, and maintains IPv6 address mapping. There are three types of FTDs: Router, REED, and FED. It operates as an FTD router (parent) or an end device (child).
  - A MED device does not subscribe to all routers to multicast address and forward all messages to its parents. There are two types of devices: Minimal end device (MED) and sleepy end device (SED). MEDs are transceivers that always operate in one mode and do not need to poll all messages from parents. SEDs usually are disabled. SEDs sometimes wake up to poll messages from their parent. An MTD can only operate as an end device (child) (Figure 14.12).
- **Partitions:** A thread can consist of network partitions. This occurs when a group of Thread devices cannot communicate with another group of Thread devices. Each partition logically works as a separate Thread network with its Leader, router ID assignments, and network data (with the same security credentials for all devices) (Figure 14.13).

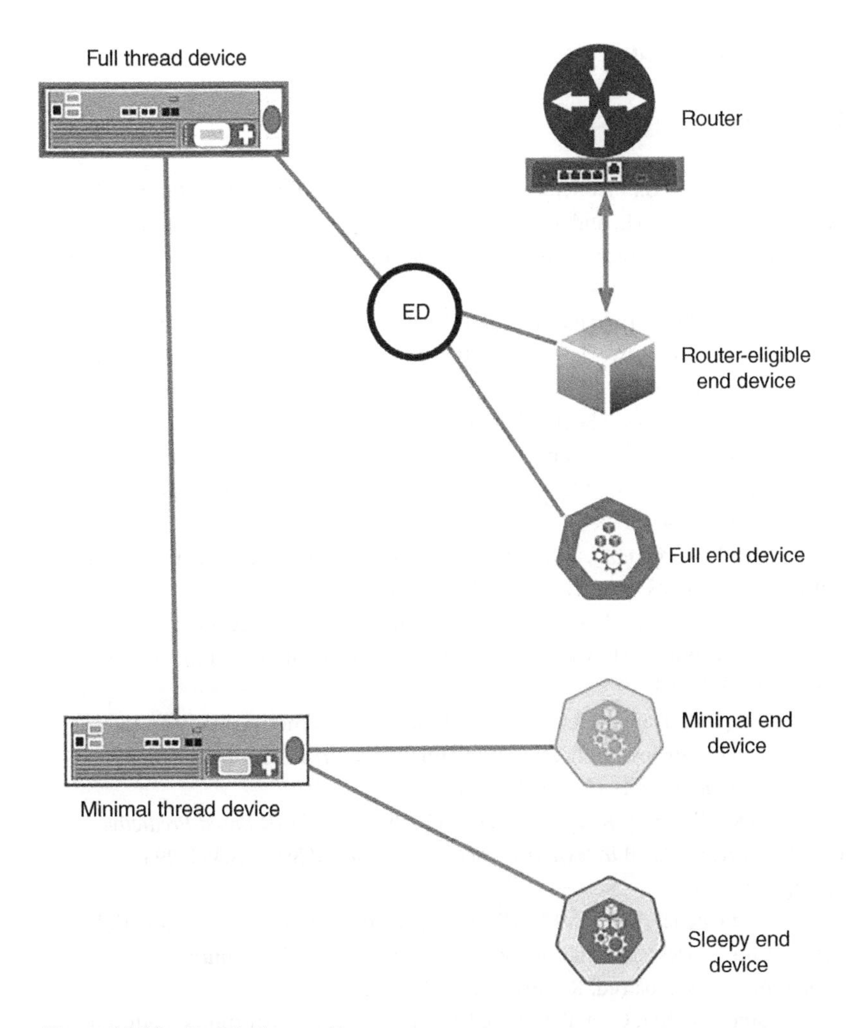

Figure 14.12 End device modes.

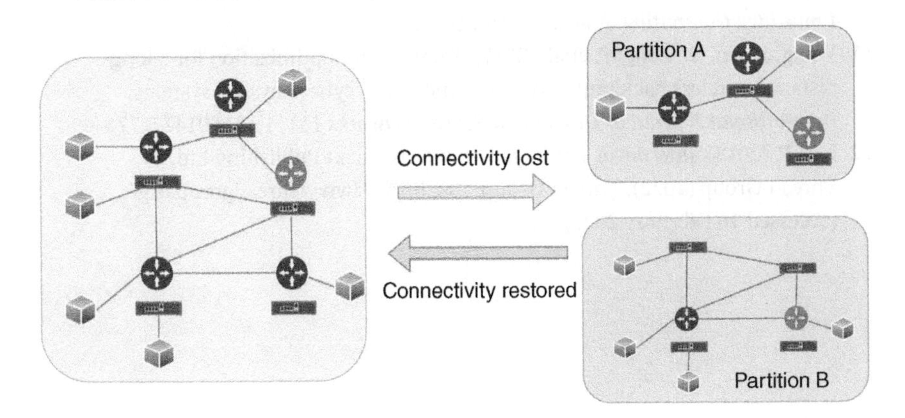

Figure 14.13 Partitioning.

References

1 Cirani, S., Ferrari, G., Picone, M., and Veltri, L. (2018). *Internet of Things*. Wiley.
2 Kim, K.-W., Han, Y.-H., and Min, S.-G. (2017). An authentication and key management mechanism for resource constrained devices in IEEE 802.11-based IoT access networks. *Sensors* 17 (10): 2170.
3 Hasegawa, T. (2017). International standardization and its applications in industrial wireless network to realize smart manufacturing. *2017 56th Annual Conference of the Society of Instrument and Control Engineers of Japan (SICE)*, Kanazawa, Japan (19–22 September 2017).
4 Parvez, I., Rahmati, A., Guvenc, I. et al. (2018). A survey on low latency towards 5G: RAN, core network and caching solutions. *IEEE Communications Surveys and Tutorials* 20 (4): 3098–3130.
5 Raptis, T.P., Passarella, A., and Conti, M. (2020). A survey on industrial Internet with ISA100 wireless. *IEEE Access* 8: 157177–157196.
6 Devan, P.A.M., Hussin, F.A., Ibrahim, R. et al. (2021). A survey on the application of WirelessHART for industrial process monitoring and control. *Sensors* 21 (15): 495.
7 Li, J.-Q., Yu, F.R., Deng, G. et al. (2017). Industrial Internet: a survey on the enabling technologies, applications, and challenges. *IEEE Communications Surveys and Tutorials* 19 (3): 1504–1526.
8 Hassan, S.M., Ibrahim, R., Saad, N. et al. (2020). *Hybrid PID Based Predictive Control Strategies for WirelessHART Networked Control Systems*, vol. 293. Springer Nature.
9 Silva, I. and Guedes, L.A. (2017). IEC 62601: wireless networks for industrial automation–process automation (WIA-PA). In: *Industrial Communication Technology Handbook* (ed. R. Zurawski). CRC Press.
10 Jin, X., Guan, N., Xia, C. et al. (2018). Packet aggregation real-time scheduling for large-scale WIA-PA industrial wireless sensor networks. *ACM Transactions on Embedded Computing Systems* 17 (5): 1–19.
11 Wang, J., Jin, X., Zeng, P. et al. (2017). Deployment optimization for a long-distance wireless backhaul network in industrial cyber physical systems. *International Journal of Distributed Sensor Networks* 13 (11): 155014771774499.
12 Lea, P. (2018). *Internet of Things for Architects*. Packt Publishing Ltd.
13 Thread Group (2022). Thread Group, Inc. https://www.threadgroup.org/ (accessed 16 February 2023).

15

Low-Power Wide-Area Networks

15.1 Introduction

A low-power wide-area network (LPWAN) is a wireless telecommunication wide-area network that allows low-bitrate long-range communication between connected objects such as battery-powered sensors.

At one end of the spectrum are cellular standards (3G, 4G), which offer a good range and reasonable data rates but consume a lot of power. On the other hand, short-range technologies such as Wi-Fi and Bluetooth consume less energy but are limited in range. LPWAN bridges the gap between requesting a more prolonged interval at lower power to send small amounts of data. The choice of which LPWAN technology should be used for applications is decided by looking at parameters such as reliability, security, network capacity, battery life, mobility support, public vs. private network, proprietary vs. standard, operating frequency, data rates, and variable payload size. The taxonomy of distance-based wireless communication technologies is given in Figure 15.1.

Four technologies account for over 96% of the globally installed base of LPWAN-enabled active devices: narrowband (NB)-IoT, long-range (LoRa), long-term evolution for machines (LTE-M), and Sigfox. NB-IoT leads with 47% of the globally installed base, followed by LoRa with 36%. In 2019, the share of these technologies was 94% of the market. In 2021, the share was increased by further two percentage points [1, 2].

15.2 General Architecture

LPWAN technologies used under the title of IoT can be examined in two categories: low throughput and cellular.

Evolution of Wireless Communication Ecosystems, First Edition. Suat Seçgin.

Published 2023 by John Wiley & Sons, Inc.

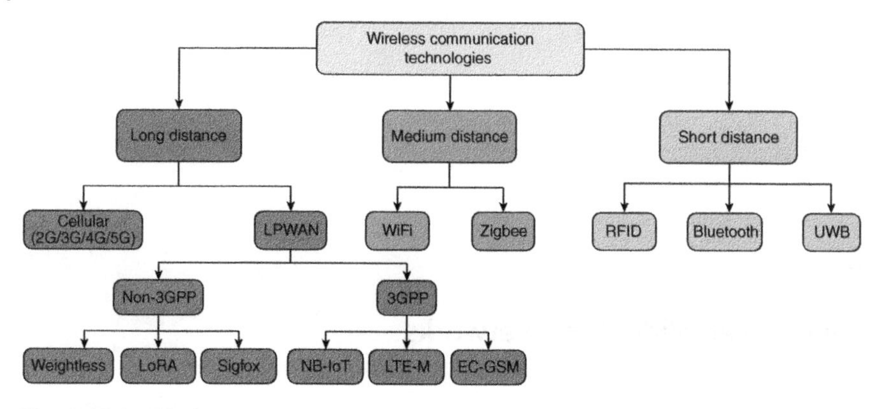

Figure 15.1 Wireless communication technologies (based on distance).

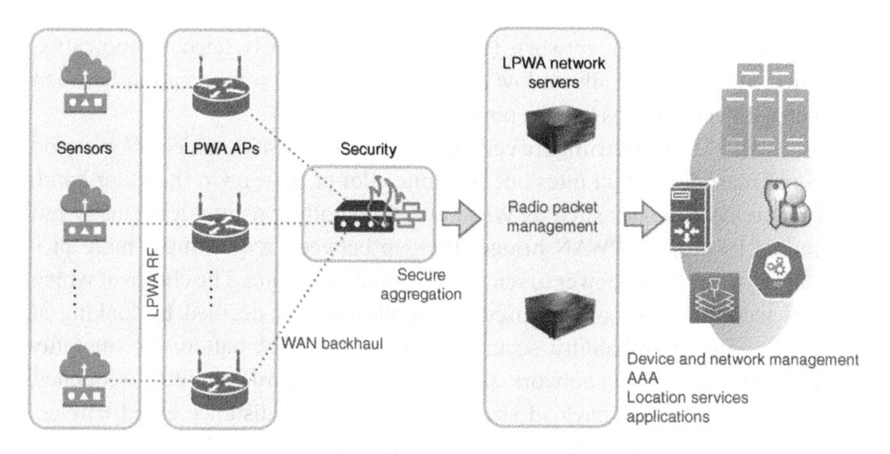

Figure 15.2 LPWAN general architecture.

The end-to-end LPWAN architecture is given in Figure 15.2 [3].
Comparisons between LPWAN technologies are given in Table 15.1 [4].

15.3 EC-GSM-IoT

Extended coverage GSM IoT (EC-GSM-IoT) is a low-power wide-area-based standard released with the 3GPP Release 13 specification. EC-GSM-IoT is a high-capacity, long-range, low-energy, and low-complexity cellular system based on eGPRS (enhanced General Packet Radio Service)-based technology designed for IoT communication. Optimizations made in EC-GSM-IoT, which should be made to existing GSM networks, can be made as a software upgrade, thus providing

Table 15.1 Comparison between LPWAN technologies.

	LoRa/LoRaWAN	Sigfox	NB-IoT	LTE-M	RPMA	Weightless-P	Symphony link
Origin	*France*	*France*	*USA (Global)*	*USA (Global)*	*USA*	*UK*	*US*
Proprietary or open	*P/O*	*Net:P, Devices:O*	*O*	*O*	*P*	*O*	*P*
Cellular	*No*	*No*	*Yes*	*Yes*	*No*	*No*	*No*
Spectrum	*Unlicensed*	*Unlicensed*	*Licensed*	*Licensed*	*Unlicensed*	*Unlicensed*	*Unlicensed*
Range (km)	*Urban: 2–5* *Rural: 15*	*U: 3–10* *R: 30–50*	*U: 1–5* *R: 10–15*	*U: 2–5*	*U: 1–3* *R: 25–50*	*U: 2*	*U: 2–5* *R: 15*
Uplink/downlink	*50 kbps/50 kbps*	*300 bps/—*	*250 kbps/250 kbps*	*1 Mbps/1 Mbps*	*634 kbps/156 kbps*	*100 kbps/100 kbps*	*100 kbps/100 kbps*
Power consumption	*High*	*Low*	*Low*	*High*	*Medium*	*Low*	*Medium*
Security	*Medium*	*Medium*	*High*	*High*	*High*	*High*	*High*
Availability of devices	*Medium*	*High*	*Medium*	*Low*	*Medium*	*Low*	*Medium*
Price	*Medium*	*Low*	*Medium*	*High*	*High*	*Low*	*Medium*
Application areas	*Precision farming manufacturing automation pipeline monitoring*	*Predictive maintenance capacity planning demand forecasting*	*Electric metering manufacturing automation retail PoS*	*Tracking objects wearables energy management utility metering, city infrastructure*	*Digital oilfield connected cities usage based insurance agriculture*	*Smart grid healthcare automotive smart cities asset tracking*	*Industrial control system lighting control alarm systems*
Supporting companies	*IBM, Semtech, Cisco, HP, Orange, Kerling, Actility*	*STMicroelectronic, Texas Instruments, Atmel, Silicon labs*	*Huawei, Ericsson, Qualcom, Vodafone*	*Verizon, AT&T, Nokia*	*Ingenu*	*Accenture, Sony, Uniik, ARM, Telensa*	*Link labs*

coverage and accelerating time to market. Up to 10 years of battery life can support various use cases.

EC-GSM-IoT network trials began with the first commercial launches planned for 2017. Supported by all major mobile equipment, chipset, and module manufacturers, EC-GSM-IoT networks will coexist with 2G, 3G, and 4G mobile networks. In addition, it will benefit from all security and privacy features such as user identity privacy, entity authentication, confidentiality, data integrity, and mobile equipment identification.

The features of EC-GSM technology are given as follows:

- Same frequency ranges as GSM network, 850–900 MHz, and 1800–1900 MHz
- Same architecture as GSM infrastructure
- TDMA/FDMA access type
- Gaussian Minimum Shift Keying (GMSK), 8PSK modulation types
- 200-kHz bandwidth per channel
- 70 kbps for GMSK, 240 kbps for 8PSK peak data rates (downlink/uplink)
- Coverage: 164 dB with 33 dBm power class, 154 dB with 23 dBm power class
- HD, FDD duplexing
- Power-saving mode PSM, ext. I-DRX (discontinuous reception) (Table 15.2)

15.4 Random Phase Multiple Access

RPMA is a built and designed wireless network technology for IoT and M2M communication. This LPWAN technology operates in the unlicensed, 2.4 GHz industrial, scientific, and medical (ISM) band. Since it works in the 2.4 GHz band, it can be applied anywhere in the world. The technology offers low power consumption, an independent broadcast channel for fast firmware updates, an excellent on-premises range, and AES 128-bit security encryption for various IoT applications. The technology was developed by the company Ingenu [4].

RPMA technology provides 31 kbps download, and 15.6 kbps upload speeds. Similar to NB-IoT, but with a maximum loss of connection (MCL) of 167 dB, its signals reach deep into buildings and underground. This technology is ideal for IoT/M2M applications because its low power requirements allow devices using this technology to operate for 10+ years on a single charge. This is ideal for devices that need to be placed in remote locations or hard-to-reach areas that are not connected to the mains. Ingenu is the creator of Machine Network™, a wireless network designed specifically for machine-to-machine (M2M) and Internet of Things (IoT) applications. This network uses Ingenu's proprietary LPWAN's random phase multiple access (RPMA) technology. RPMA technology can also be rolled out as a private network and is ideal for regions where the rollout of 3GPP LPWA

Table 15.2 Comparison of EC-GSM, Sigfox, and LoRa.

	EC-GSM	Sigfox	LoRa
Standard	*3GPP R.13*	*Private, ETSIGS LTN 001, 002, 003 (Low throughput networks)*	*Open, LoRa WAN Specification V1.O*
Spectrum	*Licensed like GSM 850-900 MHz 1800–1900 MHz*	*Unlicensed, 868 MHz (EU) 915 MHz (US)*	*Unlicensed 902–928 MHz (US) 863–870 MHz (EU) 779–787 MHz (CH)*
Channel bandwidth	*200 kHz*	*100 kHz*	*7.8–500 kHz*
System bandwidth	*1.4 MHz*	*100 kHz*	*125 kHz*
Peak data rate	*DL: 74 kbps UL: 74 kbps*	*DL: 600 bps UL: 100 bps*	*180 bps to 37.5 kbps*
Max. number of message per day	*Unlimited*	*Device: 140, BTS.50K*	*BTS.50K*
Device peak Tx power	*26 dBm*	*14 dBm*	*14 dBm*
Maximum coupling loss (MCL)	*164 dB*	*DL: 147 dB, UL: 156 dB*	*DL: 168 and 132 dB UL: 156 dB*
Device power consumption	*Low*	*Low*	*Low to medium*

technologies is delayed, cellular coverage is often poor, or users want complete control over their network deployments [5] (Figure 15.3).

15.5 DASH7

DASH7 stands for Developers Alliance for Standards Harmonization of ISO 18000-7. DASH7 Alliance (D7A) is an open-source active radio frequency identification (RFID) standard for the wireless sensor and actuator network (WSAN) protocol. The D7A complies with the ISO/IEC 18000-7 standard. ISO/IEC 18000-7 is a license-free 433, 868, and 915 MHz ISM band open source standard for wireless communication. For security, it uses AES 128 shared key encryption. The standard provides a longer propagation distance and better penetration. It provides a distance of up to 2 km, low latency and multiyear battery life, and connection to mobile objects.

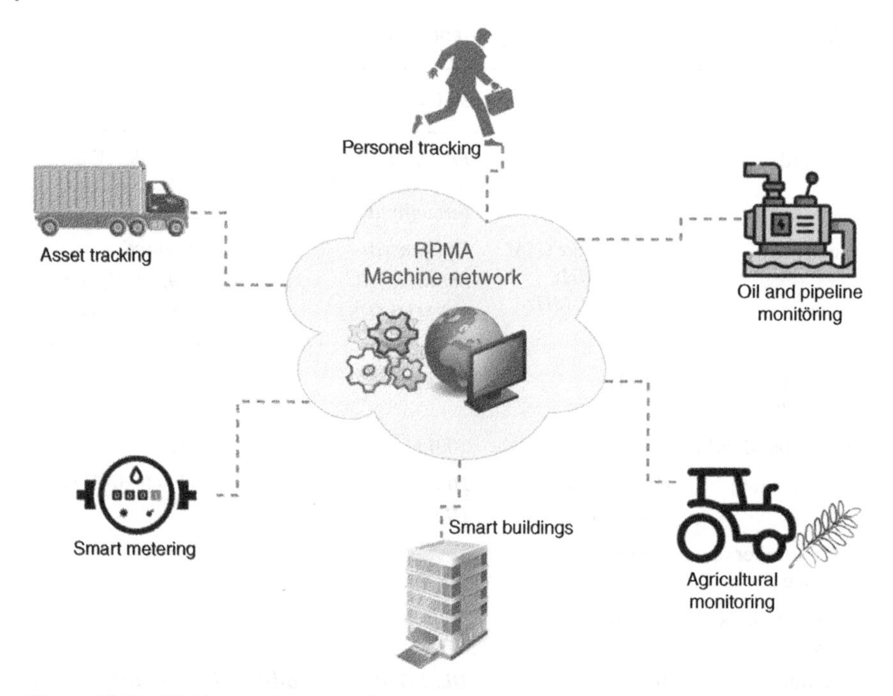

Figure 15.3 RPMA machine network.

D7A is a BLAST (bursty, light, asynchronous, stealth, and transitive) network technology [6]. These features are described as follows:

- Bursty: Short and irregular data transmission
- Light: Small packet size limited to 256 bytes
- Asynchronous: Command-based asynchronous communication. There is no periodic synchronization.
- Stealth: End device communication is done with a preapproved gateway. Periodic beacons are not required.
- Transitive: End devices can seamlessly move between different GW coverages.

D7AP architecture is similar to LoRa architecture. As shown in Figure 15.4, topology consists of end devices and gateways that allow connecting applications from the cloud. Here, end devices are simple devices consisting of sensors and actuators with a transceiver. End devices send the information they collect to gateways asynchronously. The devices have low power consumption, mostly in sleeping mode. These devices, which we say are simple, do not have all the D7AP features and will periodically wake up for possible packets and messages and switch to listening mode.

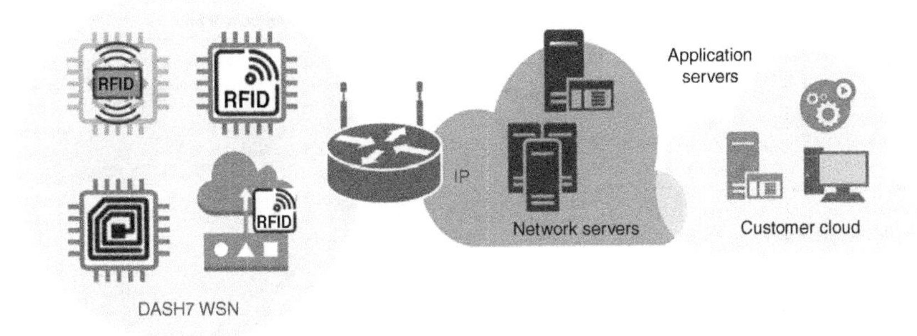

Figure 15.4 DASH7 topology.

A subcontroller device is similar to an end device and relays packets between an end device and the gateway. Unlike the end device, all D7AP features are defined on the subcontroller device. A gateway device also has all D7AP features and always operates in receiving mode unless it is in transmit mode. The gateway device is also called the DASH7 coordinator. It forwards the packets from the end devices to the IP network or another DASH7 network.

A network server installed on the cloud aggregates the incoming data, removes the duplicate data when necessary, and forwards it to the nearest gateway for downlink. The customer cloud is the structure where the end devices' applications are run, updated, or configured.

15.6 Long-Term Evolution for Machines

Long-term evolution (LTE) machine-type communication (MTC), briefly called LTE-M, is an LPWAN technology standard introduced by 3GPP in release 13. LTE Cat M1 has also enhanced machine-type communication (eMTC). LTE-M is a 5G technology that offers simplified device complexity, massive connection density, low power consumption, low latency, and extended coverage. It is mainly developed to fulfill cellular IoT device goals such as low device cost, in-depth coverage, longer battery life, higher cell capacity, etc. LTE-M can be defined as "in-band" within the standard LTE carrier or used as a standalone in a dedicated spectrum.

Despite the generational implementations to meet the increasing bandwidth need, most new IoT connections do not require the maximum speed and throughput of 4G or 5G. Based on this interpretation, LTE-M was developed. LTE-M is the preferred solution for alarm systems, asset trackers, digital city controllers, wearables, and industrial sensors.

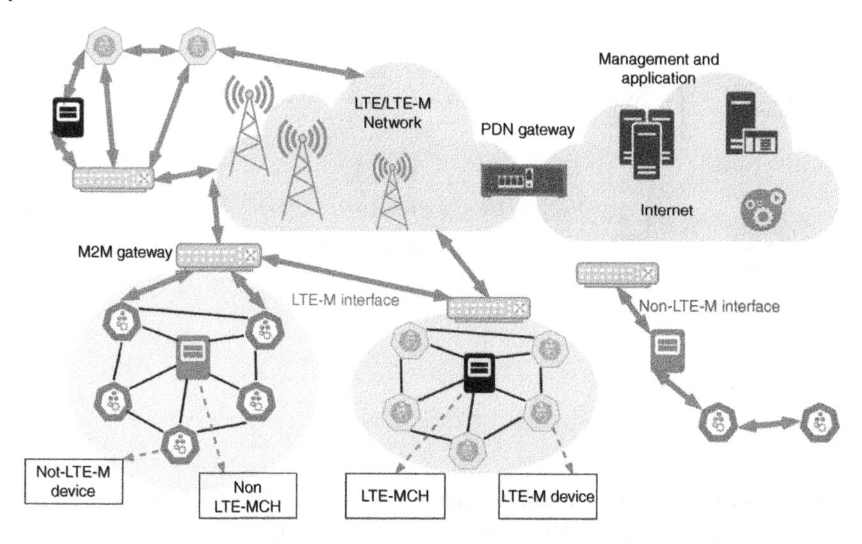

Figure 15.5 LTE-M architecture for M2M communication.

LTE has taken network efficiency and simplicity to a new level with Cat.M. In other words, LTE-M supports simple devices transmitting low data volumes over long periods. It also combines the advantages of low power consumption (long battery life), low latency (more use cases), and better penetration of radio waves (enhanced indoor coverage) (Figure 15.5).

15.7 Narrowband IoT

Similar to LTE-M, NB-IoT or LTE Cat. NB is a wireless communication standard for different IoT applications and is considered within LPWANs. The essential feature is that it provides small data, low bandwidth, and long battery life. NB-IoT is an NB radio technology used for M2M and IoT devices and applications that require low power usage for this extended battery life at a relatively low cost in an extended coverage area [7] (Table 15.3).

In line with the explanations above, we can say that CAT M (LTE-M) and NB-IoT (LTE Cat. NB) do not compete; on the contrary, both are different wireless technologies.

15.8 Massive IoT

We know that billions of devices will be connected with the spread of IoT technologies. The increased number of devices also causes high increases in data traffic. A natural consequence of this situation is the highly crowded spectrum. Due

Table 15.3 LTE Cat.M1 and LTE Cat. NB.

	LTE Cat. Ml (LTE CAT.M)	LTE Cat. NB (NB-IoT)
System bandwidth	*1.4 MHz CAT M1* *5 MHz CAT M2*	*180/200 kHz*
Peak UL/DL	*1 Mbps/1 Mbps CAT M1* *7 Mbps/4 Mbps CAT M2*	*63 kbps/27 kbps CAT NB1* *158 kbps/124 kbps CAT NB2*
Coverage/ Penetration	*20/23 dBm*	*20/23 dBm+ 14 dBm CAT NB2*
Latency	*10 ms to 4 s*	*1.4–10 s*
Mobility	*Connected mobility with some limitations (inter freq. handover)*	*Limited, changing cell without handover*
Voice	*Restricted voice for simple user case*	*No voice, data only*
Battery life	*Extended with PSM or eDRX*	*Extended with PSM or eDRX*
Antenna	*Single antenna*	*Single antenna*
Application	*FOTA capable*	*Incr. FOTA only*

to the billions of devices connected to the wireless network, there is more significant data loss in the network due to interference problems.

The limits seen in LPWAN technologies are given below [8];

- Low resistance to interference results in high packet error rates and information loss in noisy environments
- Coexistence problems with other radio networks cause the entire network infrastructure to become unstable, reducing the scope of applications and quality of service.
- High power consumption due to inefficient or repetitive data transmission causes low battery life.
- Due to long broadcast times, data packets are vulnerable to jamming attacks and can be tampered with.
- Limited availability for mobile apps reduces app spectrum.
- Proprietary solutions hinder IoT device interoperability and lock investment.

Mioty is a software-based LPWAN protocol developed to overcome current and future wireless connectivity limitations. With best-in-class reliability and scalability, mioty is designed for massive industrial and commercial IoT deployments.

The main breakthrough behind mioty technology is the Telegram Split Multiple Access (TSMA) method. As defined by the European Telecommunications Standards Institute (ETSI TS 103 357), Telegram Partition divides the data packets to be carried in the data stream into small subpackets at the sensor level (Figure 15.6).

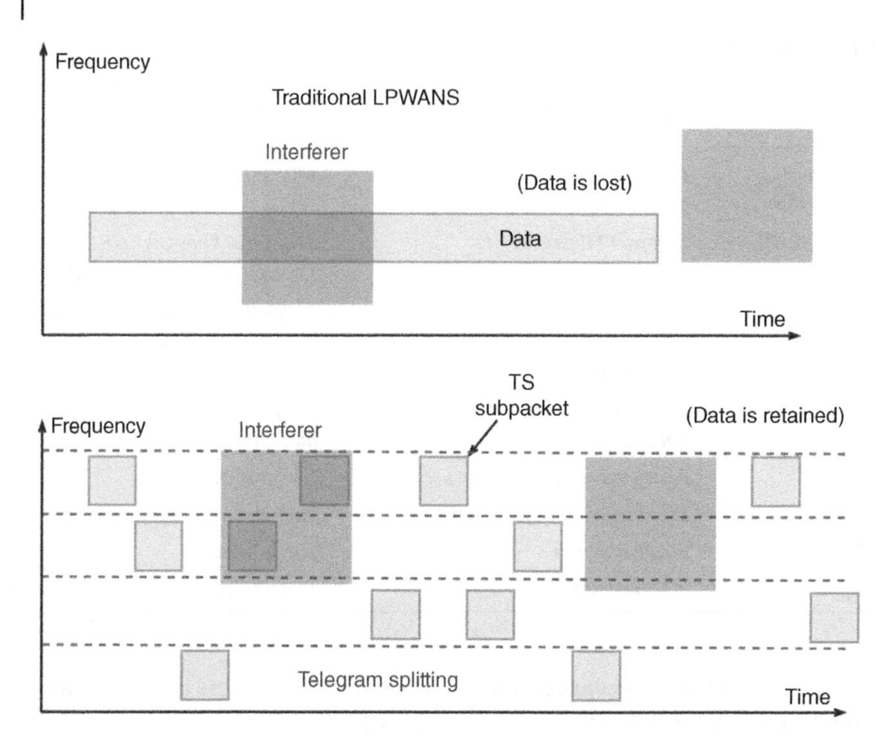

Figure 15.6 Telegram splitting.

These subpackets are then transmitted over different frequencies and times. An algorithm in the base station permanently scans the spectrum for mioty subpackets and reassembles them into a complete message. Due to advanced forward error correction (FEC), the receiver only needs 50% of the radio bursts to reconstruct the information. This reduces the effect of corrupted or lost bursts from collisions and increases resistance to interference.

The gains obtained with this approach are listed as follows:

- **Unmatched reliability:** Mioty's patented telegram splitting technology provides the most robust connectivity solution on the market. Even in a crowded spectrum, it reduces noise and the lowest packet error rates.
- **Maximum efficiency:** The protocol combined with battery-saving times enables the implementation of smaller batteries. With a power consumption of 17.8 μWh per message (endpoint, 868 MHz), Mioty makes a reality for 20+ years of battery life.
- **Unrivalled scalability:** Mioty works without interference via other networks or transmissions, enabling over one million devices per network and up to 1.5 million messages daily.

- **Best choice for mobility:** Unrivaled accuracy and reliability allow mioty nodes and base stations to operate at full performance at speeds up to 120 km/h.
- **Future proof:** Unlike today's current proprietary solutions, Mioty complies with the new ETSI Standard and is compatible with most hardware variants. This ensures that your network investments are not locked in and can scale and adjust to changing needs.
- **Flexible:** Software-based mioty networks have a low barrier to entry as the technology is compatible with most existing platforms. Each network can be private or public, fully customizable, and requires no frequency licensing.

Due to the features mentioned here, moiety can be an efficient solution for industrial IoT, smart cities/buildings, and agriculture applications.

15.9 IoTivity

The Open Connectivity Foundation (OCF) is an industry group and standards organization that aims to develop standards and certifications for IoT devices. IoTivity is an open-source reference implementation of the OCF standard specification but is not strictly limited to these requirements. It works as middleware and aims to provide seamless device-to-cloud and device-to-device connectivity. The software framework is licensed under the Apache license version 2.0. IoTivity is an open-source reference implementation of the OCF Secure IP Device Framework [9] (Figure 15.7).

It can run on multiple platforms, including IoTivity and connectivity technologies like Wi-Fi, Ethernet, BLE, NFC, and more. It works on various operating systems such as Linux, Android, Arduino, and Tizen [10]. The IoTivity framework runs as middleware across all operating systems and connectivity platforms and has some basic building blocks. Figure 15.8 shows that the discovery block supports multiple mechanisms to discover devices and resources near and far. The data transmission block supports information exchange and control based on a messaging and flow model. The data management block supports collecting, storing, and analyzing data from various sources. The CRUDN (create, read, update, delete, and notify) block supports a simple request/response mechanism with creating, reading, updating, deleting, and notifying commands.

The common resource model block defines real-world entities as data models and sources. The device management block supports device configuration, provisioning, and diagnosis. IoTivity's messaging block is based on the resource-based RESTful architecture model. So it presents everything (sensors or devices) as resources and uses the CRUDN model to manipulate resources using IETF CoAP. ID and addressing block supports OCF IDs and addressing for OCF entities such as devices, clients, servers, and resources [11].

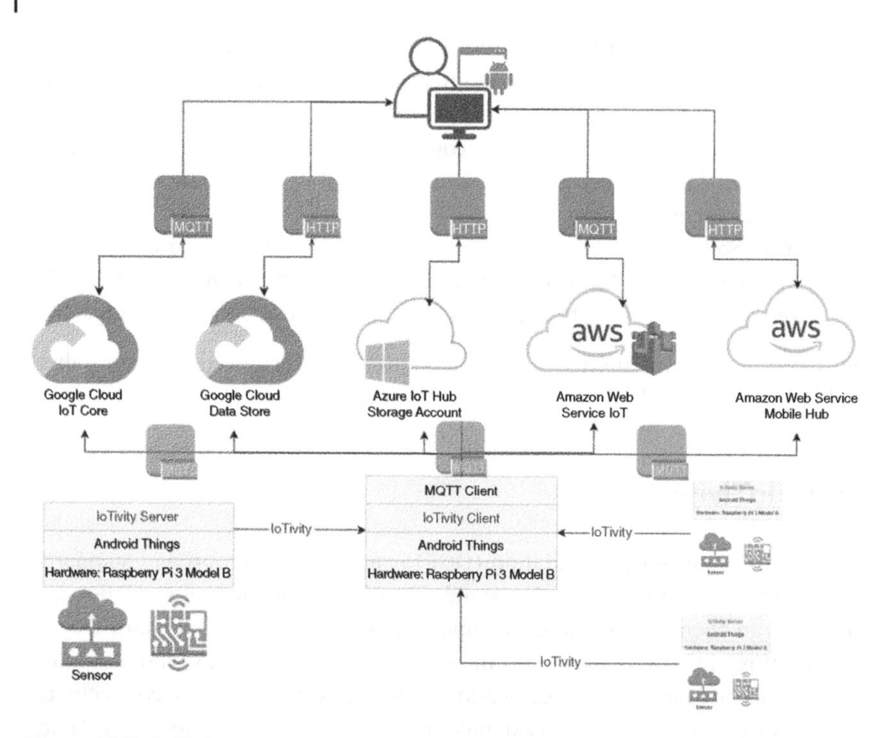

Figure 15.7 IoTivity structure.

15.10 LoRa and LoRaWAN

LoRa (long range) is an emerging technology for long-range communication operating in an unlicensed band below 1 GHz. LoRaWAN defines the communication protocol and system architecture, while LoRa defines the physical layer. LoRa Technology offers attractive features for IoT applications, including long-range, energy-efficient, and secure data transmission. The technology can be used by public, private, or hybrid networks and provides a more extended communication range than cellular networks. The bandwidth is built to provide data rates from 0.3 to 50 kbps, which is not much compared to IEEE 802.11 but sufficient for most applications in automation and data collection. It also maximizes the battery life for mobile or autonomous terminals. The concept is affordable for IoT applications due to reduced implementation costs, especially in long-range conditions.

To fully describe the LoRaWAN protocol and ensure interoperability between devices and networks, the LoRa Alliance develops and maintains documentation to explain the technical implementation, including MAC layer commands, frame content, classes, data rates, security, and flexible network frequency management [12] (Figure 15.9).

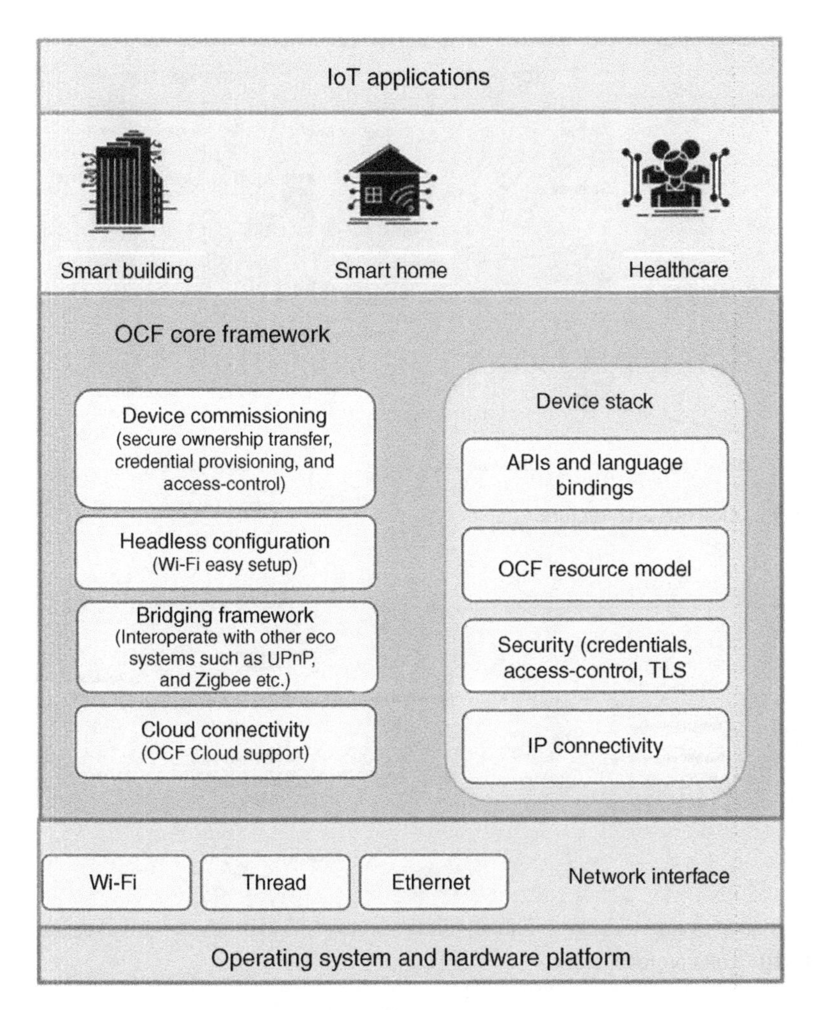

Figure 15.8 IoTivity device stack and modules.

The LoRaWAN network architecture is deployed in a star topology where gateways forward messages between edge devices and a central network server. Gateways connect to the network server via standard IP connections and act as a transparent bridge, simply converting RF packets to IP packets and vice versa. Wireless communication takes advantage of the long-range capabilities of the LoRa physical layer, allowing a single-hop connection between the end device and one or more gateways. All modes are capable of bidirectional communication. There is support for multicast-addressing groups for efficient spectrum use during tasks such as firmware over-the-air (FOTA) upgrades or other bulk delivery messages (Figure 15.10).

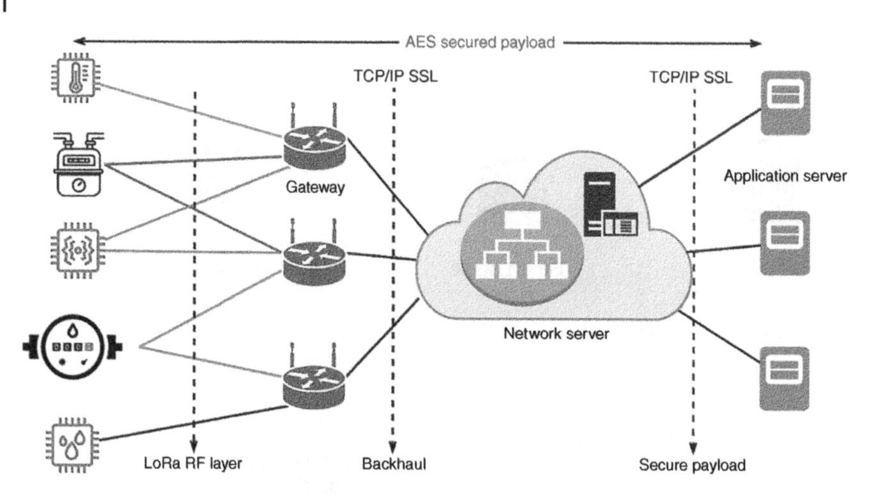

Figure 15.9 LoRaWAN architecture.

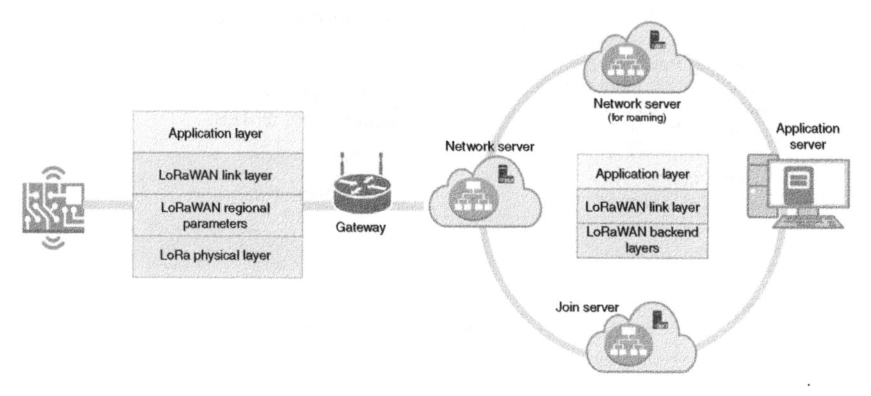

Figure 15.10 LoRa protocol stack.

Many legacy wireless systems use frequency shift keying (FSK) modulation at the physical layer because it is a very efficient modulation to achieve low power. LoRa is based on chirp spread spectrum modulation, which retains the same low power characteristics as FSK modulation, but significantly increases the communication range. Chirp spread spectrum has been used in military and space communications for decades due to achievable long communication samples and robustness against interference. LoRa was the first low-cost application for commercial use.

In the LoRaWAN™ network, nodes are not associated with a particular gateway. Instead, data transmitted by a node is typically received by multiple gateways. Each gateway will forward the received packet from the end node to the cloud-based network server via a backhaul (cellular, Ethernet, satellite, or Wi-Fi).

Gateways manage the network intelligence and complexity and filter out extra packets, perform security checks, schedule notifications through the optimum gateway, perform adaptive data rates, etc. No gateway-to-gateway transition is required if a node is mobile or moving; this is a critical feature for enabling asset-tracking applications.

LoRa has advantages in IoT applications. LoRa uses the ISM 868/915 MHz bands available worldwide. It has a broad coverage of about 5 km in urban areas and 15 km in suburbs. It consumes less power, and therefore, the batteries last longer. A single LoRa Gateway appliance is designed to take care of thousands of edge devices or nodes, easy to deploy thanks to its simple architecture. LoRa uses the adaptive data rate technique to change the output data rate of end devices. This helps to maximize battery life as well as the overall capacity of the LoRaWAN network.

However, LoRa also has its drawbacks. LoRaWAN network size is limited by a parameter called the duty cycle. The duty cycle is the percentage of time in which the channel can be busy. This parameter is restricted for regulatory reasons as the critical limiting factor for traffic in the LoRaWAN network. LoRaWAN supports limited-size data packets and has longer latency. Therefore, it is not an ideal candidate for real-time applications that require lower latency and limited jitter requirements.

In summary, the use of LoRa is strongly influenced by the characteristics of the environment in which the technology is applied. The type of IoT application should pay attention to these issues. Using LoRa in open areas such as rural areas has advantages in energy consumption, wide range, and flexibility in network development. However, for applications that require specific requirements, such as the amount of data exchanged in a given period, the network configuration must be considered primarily for indoor applications. LoRa network cannot transmit a large amount of data for a broader range of regions. LoRa technology is affected by obstacles such as tall buildings and trees, resulting in increased packet loss levels in the area. The use of GPS in the LoRa module has not been reliable, especially for real-time location-tracking software applications.

15.11 Sigfox

Sigfox was introduced by a French global network operator founded in 2009 and built wireless networks to connect low-power objects such as IoT sensors and smartwatches that need to be constantly on and emit tiny amounts of data. Sigfox wireless technology is based on low throughput network (LTN) and ultra-narrow band (UNB) concepts. In other words, it is an LPWAN technology that supports low data rate communication over greater distances [13] (Figure 15.11).

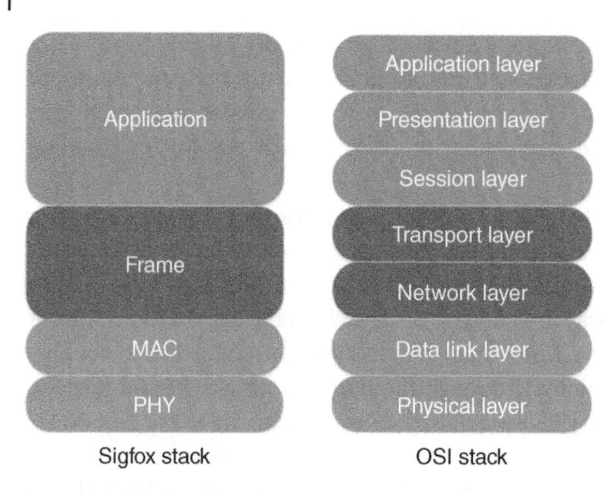

Figure 15.11 Sigfox protocol stack.

All data on Sigfox is managed on the Sigfox backend network. All messages from Sigfox base stations reach the backend server via IP connection. In other words, a Sigfox backend cloud service is the final destination for a package. After verifying a client node's authenticity and duplication, the backend stores the message and sends it to the client. If data needs to be forwarded to an endpoint node, the backend server selects a gateway with the best connection to the endpoint and forwards the message on the downlink channel. Since the packet ID already identifies the backend device, this preprovisioning will force the data to be sent to a destination.

The Sigfox cloud routes the data to the selected destination. The cloud service provides APIs through a pull model to integrate Sigfox cloud functions with a third-party platform. Devices can be registered via another cloud service. Sigfox also offers callbacks to other cloud services. This is the preferred method to retrieve data. The Sigfox protocol uses the proprietary non-IP protocol and converges data as IP data to a Sigfox network backend.

Multiple gateways can receive messages from one node to ensure the data integrity of the burn-and-forget communication model; all subsequent messages will be forwarded to the Sigfox backend, and duplicates will be removed. This adds a level of redundancy to data retrieval. Thus, a Sigfox endpoint node is designed to make it easy to install. No extra pairing or signaling is required (Figure 15.12).

Sigfox is a protocol with good long-range performance. However, sending a single packet takes a long time (6–12 seconds). Also, the 868 MHz band in Europe has a limit of 140 packets per day with a 12-byte payload due to ETSI regulations. Applications that are not recommended for using Sigfox are given as follows [14]:

- Sigfox is not recommended for regular duty-cycle projects that require a frame sent every few minutes. Exceeding the 140-package-limit per day may result in additional charges or license revocation by Sigfox.

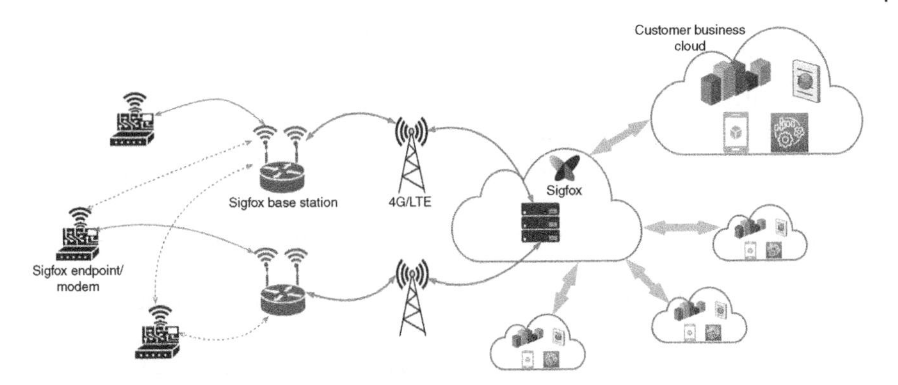

Figure 15.12 Sigfox architecture.

- Sigfox is recommended for long-range device communication in cities where base stations are deployed.
- Sigfox is not recommended for bidirectional communication. Although the package provides acknowledgments and callbacks, there is no actual data downlink.
- Sigfox is not recommended for streaming. Transmission is not done in real-time as there is a minimal delay for packet arrival.
- It is not recommended to use Sigfox for the transmission of large amounts of data. The maximum payload is 12 bytes. It also has a capacity of 100 bps uplink and 600 bps downlink throughput.

Besides the disadvantages described here, Sigfox has many advantages:

- Easy to install thanks to devices that do not require complex setup and wireless technology
- Sigfox's radio modules generally cost less than other LPWAN technologies.
- Sigfox works best for simple devices that rarely transmit small amounts of data slowly, which has the advantage of extending the battery life of the sensors.
- Sigfox has a data transmission capacity range of approximately 10 km in urban and 40 km in rural areas.
- Supports up to one million nodes per Sigfox base station.

Essentially, Sigfox significantly reduces the energy consumption and costs associated with connected devices. The platform is ideal for low-power applications that send only small amounts of data infrequently over large distances.

References

1 BehrTech (2019). LPWAN Technologies Comparison. https://behrtech.com/blog/tag/lpwan-comparisons/ (accessed 16 February 2023).

2 Devopedia (2020). Low-power wide-area network, Version 68, July 31. https://devopedia.org/low-power-wide-area-network (accessed 16 February 2023).
3 Cisco (2023). Networking, cloud, and cybersecurity solutions. https://www.cisco.com/ (accessed 16 February 2023).
4 Flespi (2023). Universal telematics gateway, parser & API. https://flespi.com/ (accessed 16 February 2023).
5 Ingenu (2023). Technology. https://www.ingenu.com/technology/ (accessed 16 February 2023).
6 Ayoub, W., Samhat, A.E., Nouvel, F. et al. (2019). Internet of Mobile Things: overview of LoRaWAN, DASH7, and NB-IoT in LPWANs standards and supported mobility. *IEEE Communications Surveys and Tutorials* 21 (2): 1561–1568.
7 Thales Group (2023). Thales – Building a future we can all trust. https://www.thalesgroup.com/en (accessed 16 February 2023).
8 Mioty Alliance (2023). Mioty Allience – mioty alliance. https://mioty-alliance.com/ (accessed 16 February 2023).
9 IoTivity (2023). http://iotivity.org/ (accessed 16 February 2023).
10 Xuan, S. and Kim, D. (2020). Development of cloud of things based on proxy using OCF IoTivity and MQTT for P2P internetworking. *Peer-to-Peer Networking and Applications* 13 (3): 729–741.
11 Yang, Y., Chen, X., Tan, R., and Xiao, Y. (2021). *Intelligent IoT for the Digital World*. Wiley.
12 LoRa Alliance (2023). Homepage – LoRa Alliance®. https://lora-alliance.org/ (accessed 16 February 2023).
13 Sigfox (2023). https://www.sigfox.com/en/use-cases (accessed 16 February 2023).
14 Sigfox (2021). When is Sigfox recommended? Sigfox networking guide. https://development.libelium.com/sigfox_networking_guide/when-is-sigfox-recommended (accessed 16 February 2023).

16

IoT Edge to Cloud Protocols

16.1 Introduction

IoT protocols are a crucial part of the IoT technology stack. Without them, the hardware becomes unusable as Internet of Things (IoT) protocols allow it to exchange data in a structured and meaningful way. We always think of communication when we talk about the IoT. The interaction between sensors, devices, gateways, servers, and user applications is the key feature of an IoT system. But what allows all these smart things to talk and interact are IoT protocols, which can be seen as the languages that IoT hardware uses to communicate [1] (Figure 16.1).

The natural question is why there is any protocol other than HTTP to move data over WAN? HTTP has provided essential services and capabilities for the Internet for over 20 years but is designed and configured for general-purpose computing in client/server models. IoT devices can be very constrained, remote, and bandwidth limited. Therefore, more efficient, secure, and scalable protocols are required to manage large numbers of devices in various network topologies such as mesh networks.

While many of these protocols are message-oriented middleware (MOM) implementations, there are also protocols implemented with the concept called representational state transfer (REST), which is an alternative to MOM implementation.

The main idea behind MOM is that communication between two devices is done via distributed message queues. A MOM delivers messages from one user-space application to another. Some devices generate data to be added to the queue, while others consume data stored in the queue. Some implementations require a broker or an intermediary on the central service. In such a case, producers and consumers have a publish- and subscribe-type relationship with the broker.

Evolution of Wireless Communication Ecosystems, First Edition. Suat Seçgin.

Published 2023 by John Wiley & Sons, Inc.

Business applications	Device management		
	Business procesess		
	Analytics		
Business model	Cloud	Open	Closed
	Integrated	Cassandra	On premise
	Indirect	Platform	Direct
Data retreival and storage	HBase	MongoDB	
	Hadoop	Cassandra	
Data aggregation and processing	RapidMQ	Fluentd	
	Scribe	Flume	Strom
	Kafka	Luxun	
Session and communication	XMPP	HTTP	TCP/UDP
	Telnet	SSH	FTP
	CoAP	MQTT	DDS
	OPC-UA	AMQP	
Transport layer	IPv4	IPv6	
	6LowPAN	RPL	
Link layer	BLE	RFID	GSM
	Dash7	CDMA	ZigBee
	Bluetooth	Ethernet 802.3	802.15.4e
	WiFi 802.11/a/b/g/n	Bluetooth	
Connectivity/PHY layer	PLC	RS-232	Wireless
	OBD2	Modbus	USB
	RJ45	SPI	
Devices and smart gateways sensors			

Figure 16.1 IoT protocols and applications.

AMQP, MQTT, and STOP are implementations of MOM. Others include CORBA and Java messaging services. A MOM application that uses queues can use them for flexibility in design. Data may remain in queues even if the server fails (Figure 16.2).

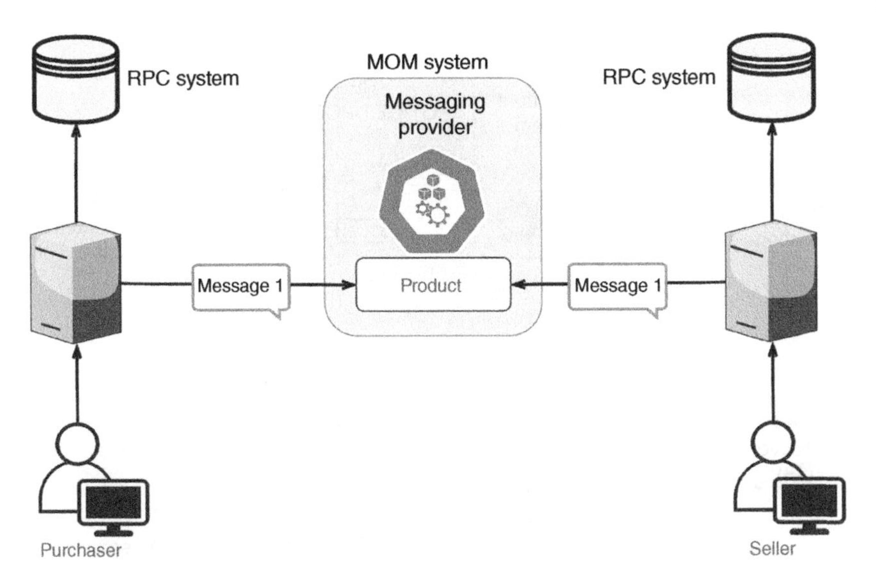

Figure 16.2 MOM system architecture.

Alternatives to MOM implementations are RESTful implementations. In the RESTful model, a server has the state of a resource, but the state is not transmitted in a message from the client to the server. RESTful designs use HTTP methods such as GET, PUT, POST, and DELETE to place requests upon a resource's universal resource identifier (URI). In this architecture, no broker or middle agent is required. Because it is based on the HTTP stack, it takes advantage of most of the services, such as HTTPS security. RESTful architectures are typical client-server architectures. Clients initiate access to resources with synchronous request-response patterns. RESTful design is an architecture built on HTTP and uses HTTP paradigms for client-to-server communication. In addition, clients are responsible for errors even if the server fails (Figure 16.3).

From another perspective, IoT protocols and standards can be evaluated in two categories: IoT data protocols (presentation/application layers) and network protocols for IoT (datalink/physical layers).

IoT data protocols are used to connect low-power IoT devices. They provide communication with the hardware on the user side without needing an Internet connection. Connections in IoT data protocols and standards are carried out over a wired or cellular network. Protocols such as MQTT, CoAP, AMQP, DDS, and WebSocket are IoT data protocols.

IoT network protocols are used to connect devices to the network. This set of protocols is typically used over the Internet. Protocols such as WiFi, Bluetooth, ZigBee, Z-Wave, and LoRaWan that we describe in the book are IoT network protocols. In this section, IoT data protocols will be explained.

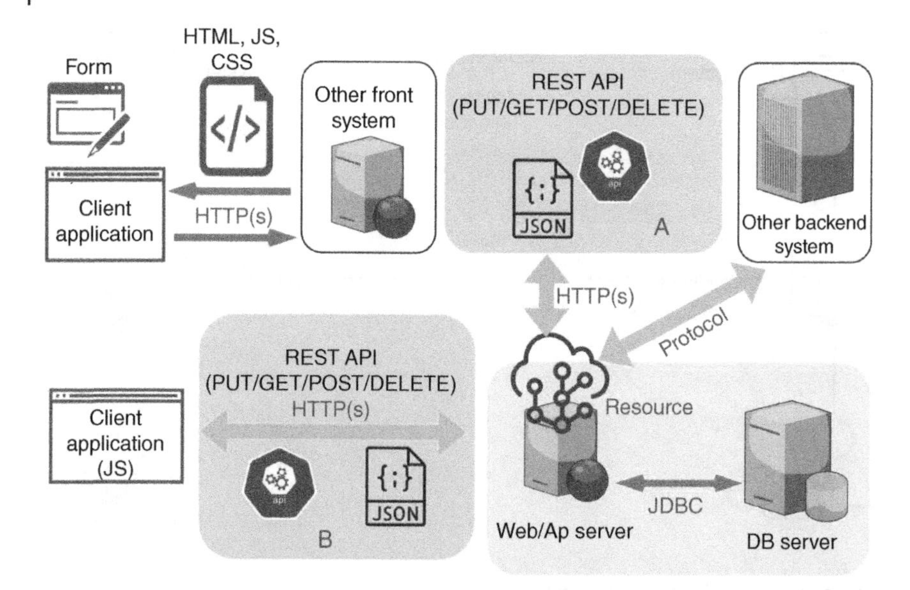

Figure 16.3 RESTful architecture.

16.2 Message Queue Telemetry Transport Protocol

MQTT stands for Message Queuing Telemetry Transport. It is an extremely simple, lightweight, publish/subscribe messaging protocol for restricted devices and low-bandwidth, high-latency, or unreliable networks. The design principles seek to ensure reliability and some degree of delivery assurance while minimizing network bandwidth and device resource requirements. These principles also make the protocol ideal for the emerging "machine-to-machine" (M2M) or "Internet of Things" world of connected devices and mobile applications where bandwidth and battery power are high [2].

For many years, client–server architecture has been the mainstay for data center services, while publish-subscribe models have emerged as an alternative for IoT use.

It has a working principle similar to TV broadcasting. A broadcaster broadcasts on a particular channel. The audience has to tune in to this channel to watch this broadcast. There is no direct connection between the broadcaster and the audience. The MQTT protocol also works on this logic. In MQTT, a publisher publishes a message on a topic. A subscriber has to subscribe to this topic to view this message. MQTT uses a central broker for this transaction [3] (Figure 16.4).

Here are the key points;

- Clients do not have an address as in the email system (the subscriber in the TV example).

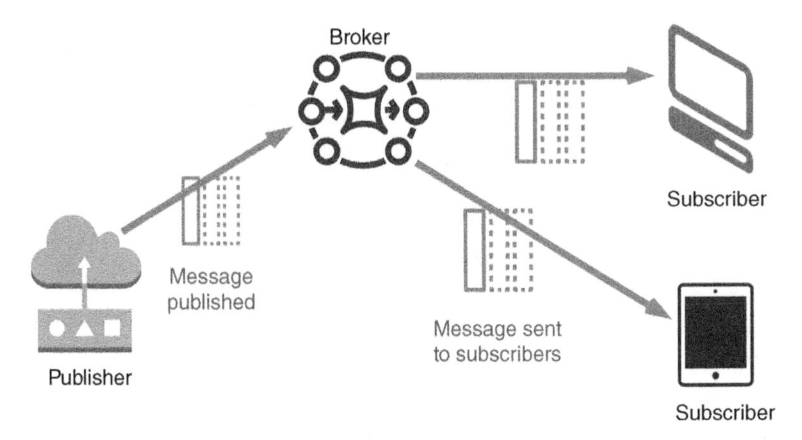

Figure 16.4 Publish–subscriber model.

- Messages are sent to a broker (TV device in the TV example) on a topic (TV channels).
- The responsibility of the MQTT broker is to filter the messages (separate channels) and distribute them to the end user.
- A client receives these messages by subscribing to the channel on the same broker (selects the channel with the remote control).
- There is no direct connection between the publisher and the subscriber.
- All clients can publish (broadcast) or subscribe (receive).
- MQTT brokers generally do not store messages (TVs do not record the broadcast).

The advantages of MQTT are given as follows:

- **Lightweight and efficient:** MQTT clients are tiny and require minimal resources. So they can be used on small microcontrollers. MQTT message headers are small to optimize network bandwidth.
- **Bidirectional communications:** MQTT allows device-to-cloud and cloud-to-device messaging. This will enable messages to be quickly passed to a group of objects.
- **Scale to millions of things:** MQTT can scale up to millions of IoT device connections (Figure 16.5).
- **Reliable message delivery:** MQTT offers three levels of service quality: 0–At most once, 1–at least once, 2–exactly once.
- **Support for unreliable networks:** Many IoT devices connect over unreliable cellular networks. MQTT's support for persistent sessions reduces the time it takes to reconnect the client to the agent.
- **Security enabled:** MQTT makes it easy to encrypt messages using TLS and authenticate clients using modern authentication protocols like OAuth.

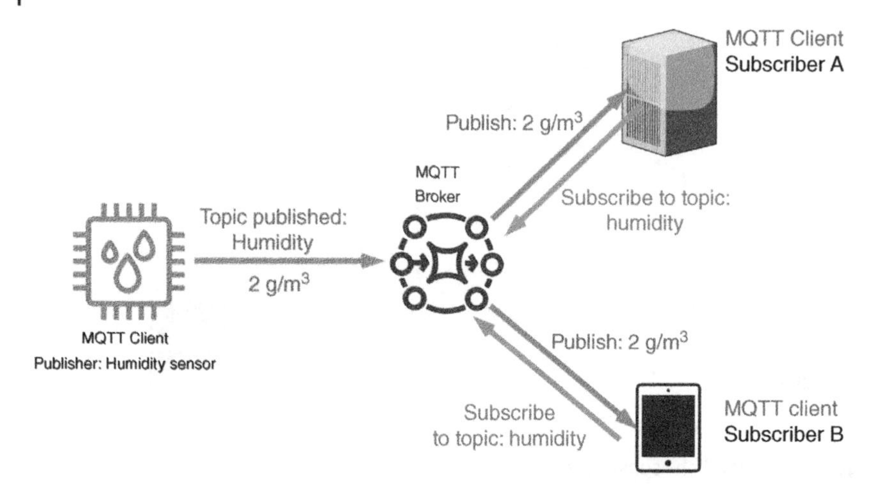

Figure 16.5 MQTT example.

MQTT successfully separates publishers from consumers. As the broker is the governing body between publishers and consumers, there is no need to directly identify a publisher and a consumer based on physical aspects (such as the IP address). This is useful in IoT deployments where the physical identity is unknown or ubiquitous. MQTT and other pub/submodels do not change over time either. This means that a message published by a client can be read or reacted to by the subscriber at any time. A subscriber may be in a deficient power-/bandwidth-limited state (for example, in a Sigfox architecture) and respond to the message within minutes or hours. Due to the lack of physical and time relations, pub-/submodels scale very well to extreme capacity. Cloud-managed MQTT agents can typically receive millions of messages per hour and support tens of thousands of publishers.

MQTT is data format-independent. The payload can contain any data type. So publishers and subscribers must understand and accept the data format. It can transmit text messages, image data, audio data, encrypted data, binary data, JSON objects, or virtually any other structure in the payload. However, JSON text and binary data are the most common data payloads. The maximum packet size allowed in MQTT is 256 MB, which allows for an extremely large payload.

MQTT architecture details are given as follows:

- There are no message queues in MQTT. Queuing messages is possible but not necessary and often not performed. MQTT is based on TCP. Reliable transfer of packets is guaranteed.
- HTTP is symmetric (just like Telnet protocol) while MQTT is an asymmetric (just like FTP) protocol.

- MQTT can store a message in an agent indefinitely. A flag in standard message transmission controls this mode of operation.
- MQTT defines an optional facility called last will and testament (LWT). LWT is a message that the client specifies during the connection phase.
- Although MQTT is based on TCP, connections can still be lost, especially in the wireless sensor environment. A device may lose power, signal strength, or crash in the field, and a session goes into a half-open state. Here the server will believe the connection is still reliable and wait for data. To remedy this half-open situation, MQTT uses a keep-alive system. Using this system, the MQTT broker and client have the assurance that the connection is still valid even if there has been no transmission for a while.

Connections are confirmed by the agent using a connection acknowledgment message. MQTT is a command-response protocol. Each command is confirmed. You cannot be public or subscribe unless you have connected.

16.3 MQTT over WebSockets

We explained that MQTT is a lightweight publish/subscribe model and is very suitable for low-power sensors. WebSocket is a transport layer protocol. Because it works at this layer, WebSocket provides full-duplex communication channels over a single TCP connection between the client and server. The combination of MQTT and TCP protocols offers many possibilities in the IoT world. Displaying live sensor data via an HTML page without refreshing or polling the Web Server is just one of them (Figure 16.6).

Most of the MQTT brokers currently support WebSocket communication. All needed here is a client-side code to be written on the client side. As can be seen in Figure 16.6, the data can be viewed and processed by connecting to a Web Server that receives MQTT data.

16.4 MQTT for Sensor Networks

Due to the many sensors in IoT sensor networks, these sensors work wirelessly. As we mentioned in the previous sections, these networks are designed with low-power battery sensors working with limited processing power and storage, limited payload size, and not always on (sleeping) principles. MQTT-SN can be considered a version of MQTT and is based on publish/subscribe model like MQTT (Figure 16.7).

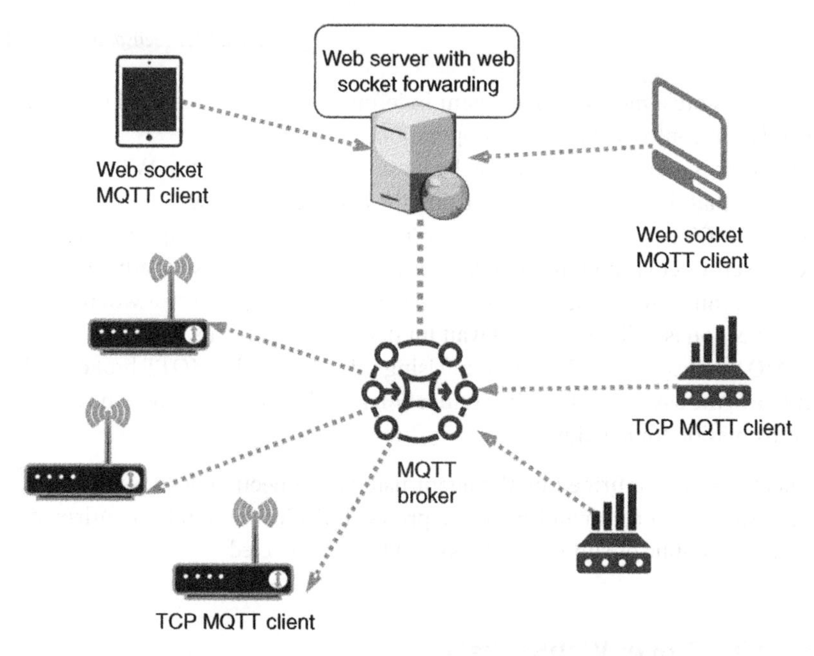

Figure 16.6 MQTT over WebSockets.

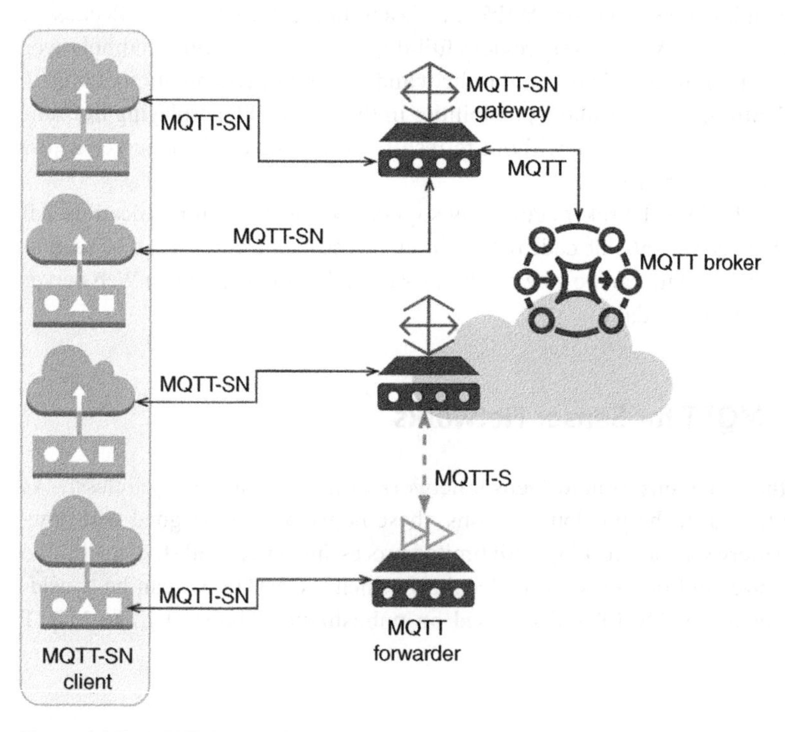

Figure 16.7 MQTT-SN architecture.

There are two main differences between MQTT-SN and MQTT: reduced message payload and eliminating the need for persistent connection by using the UDP protocol.

MQTT-SN specifications are listed as follows:

- The link message is split into three messages. Two of these are optional and used as request messages.
- Short topic names are used instead of Topic IDs.
- There are pre-defined topics.
- Discovery process for clients to discover the Gateway
- The request topic and messages can be changed during the session.
- Offline keep alive procedure for sleeping clients

The MQTT-SN architecture has three components: MATT-SN client, MQTT-SN gateway, and MQTT-SN forwarder. There are two types of gateways: transparent and aggregating gateways.

The first message sent from an MQTT-SN client to a broker/gateway is the connection message, and this message is acknowledged with the CONNACK message. Once the connection is established, the client can subscribe and publish the message. Like in MQTT, a connection can be persistent or nonpersistent, as indicated in the clean session flag.

16.5 Constrained Application Protocol

Constrained Application Protocol (CoAP) is a protocol that makes it possible to provide RESTful Web Service functions to resource-constrained devices and, consequently, to integrate WSNs and smart objects with the Web. Using Web Services over IP-based wireless sensor networks (WSN) simplifies software reusability and reduces the complexity of application development. One of the most significant benefits of IP-based networking in Low-power and Lossy Networks (LLN) is that it allows using standard Web Service architectures without application gateways. As a result, the Web of Things (WoT) concept has emerged, integrating smart objects with the Internet and the Web. The advantage of WoT is that smart object applications can be built on the best REST architectures [4].

As stated in the chapter introduction, REST architectures allow applications to rely on loosely coupled services that can be shared and reused. In a REST architecture, a resource is an abstraction controlled by the server and defined by a universal resource locator (URL). Services parse resources. Therefore, resources can be represented arbitrarily via various formats such as XML or JSON. Resources are accessed and modified by an application protocol based on client/server

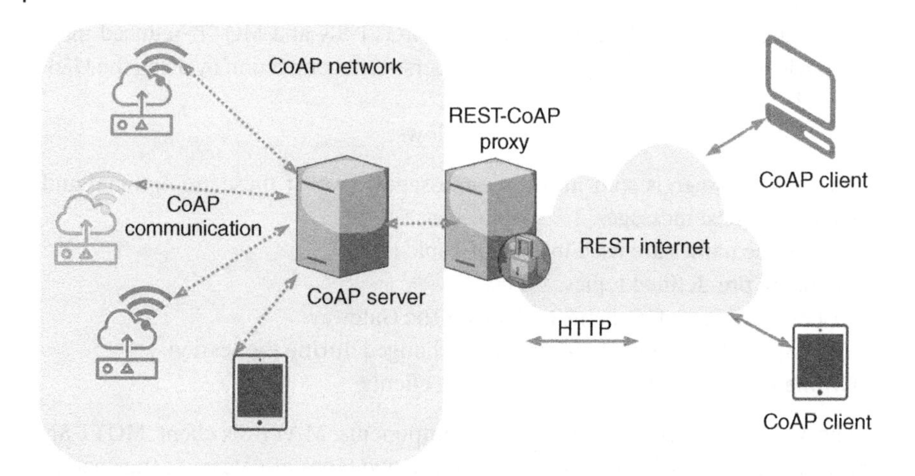

Figure 16.8 CoAP architecture.

requests/responses. REST is not tied to any particular application protocol. However, most REST architectures today use the Hypertext Transfer Protocol (HTTP). HTTP manages resources via GET, POST, PUT, etc. [5] (Figure 16.8).

The application layer usually uses HTTP to provide Web Service, but HTTP has high computational complexity, low data rate, and high energy consumption. For this reason, IETF has developed several lightweight protocols such as CoAP, embedded binary HTTP (EBHTTP), and Lean Transport Protocol (LTP). Constrained Application Protocol (CoAP) is a proprietary Web transport protocol for use with constrained nodes and constrained (for example, low-power and lossy) networks. CoAP provides a request/response interaction model between application endpoints, supports built-in discovery of services and resources, and includes core Web concepts such as URIs and Internet media types. CoAP is designed to easily interface with HTTP for integration with the Web while meeting specific requirements such as multicast support, low overhead, and simplicity for constrained environments [6].

CoAP specifications are given as follows [7]:

- Constrained Web protocol fulfilling M2M requirements
- Security binding to Datagram Transport Layer Security (DTLS)
- Asynchronous message exchanges
- Low header overhead and parsing complexity
- URL and content-type support
- Simple proxy and caching capabilities
- UDP banding with optional reliability supporting unicast and multicast requests

- A stateless HTTP mapping allows proxies to be built and provides access to CoAP resources via HTTPS. It also provides a uniform way for HTTP simple interfaces to be realized alternatively over CoAP.

It can be said that CoAP is a subset of HTTP functions designed considering the low processing power and energy constraints of small embedded devices such as sensors. In addition, various mechanisms and some newly modified operations have been added to make the protocol suitable for IoT and M2M applications. The first significant difference between CoAP and HTTP is the transport layer. It is based on the HTTP Transmission Control Protocol. However, TCP's flow control mechanism is unsuitable for LLNs, and its overhead is too high for shortlived transactions. In addition, TCP does not support multicast and is very sensitive to mobility.

On the other hand, CoAP is built on User Datagram Protocol (UDP). Therefore, it has both multicast support and low overhead. CoAP is designed as two layers over UDP, request/response and transaction. HTTP has 1451 bytes per transaction, 1333 mW and a lifetime of 84 days compared to 154, 07.44, and 151 in CoAP, respectively.

A CoAP system consists of seven key actors [8]:

- **Endpoints:** These are the sources and destinations of CoAP messages. The specific destination of an endpoint depends on the transport being used.
- **Proxies:** Proxies are CoAP endpoints commissioned by CoAP clients to perform requests on their behalf. Reducing the network load, accessing sleeping nodes, and providing a layer of security are some of the roles of the proxy. Proxies can be explicitly selected by a client (forward proxying) or used as in situ servers (reverse proxying). Alternatively, a proxy can map from one CoAP request to another CoAP request or even cross-proxying. In common use, it is an edge router that proxies HTTP services from a CoAP network for cloud-based Internet connections. It is a popular application for edge routers proxying from the CoAP network to HTTP services for cloud-based Internet connections. The proxy architecture is given in Figure 16.9.
- **Client:** It is the starting point of a request and the destination of a response.
- **Server:** The destination endpoint of the request and the starting node of the response.
- **Intermediary:** A client that acts as both a server and a client against a source server. Proxy is a broker.
- **Origin servers:** The server where a particular resource resides.
- **Observers:** An observing client can register using a modified GET message. The observer then connects to a resource, and the server sends a notification to the observer if that resource's state changes. Observers are unique in CoAP and allow a device to monitor changes in a particular resource. It is similar to the MQTT subscriber model, where a node will subscribe to an event.

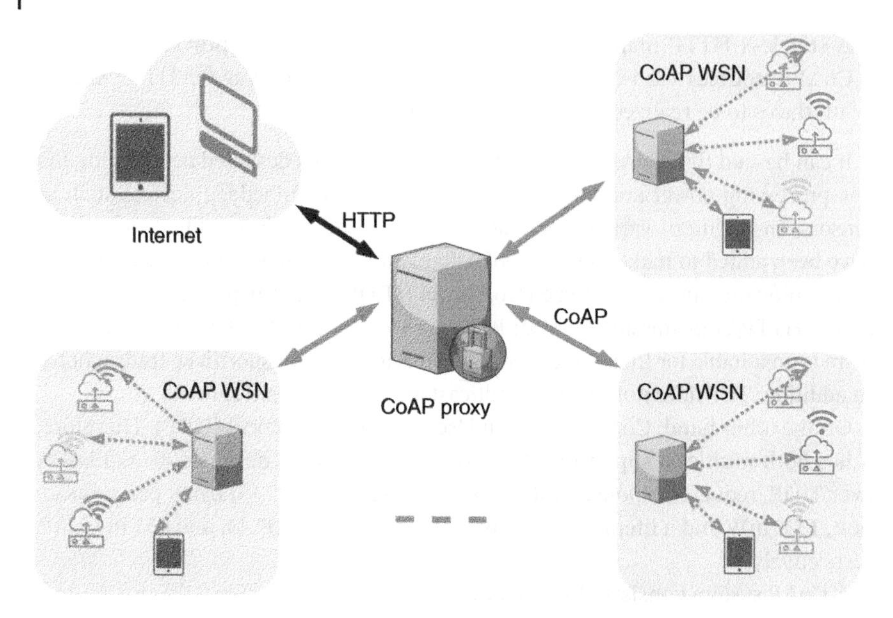

Figure 16.9 CoAP proxy.

16.6 Embedded Binary HTTP

Embedded Binary HTTP (EBHTTP) is a space-efficient binary format and stateless encoding of the standard HTTP/1.1 protocol. EBHTTP is designed for transporting small named data items such as sensor readings between resource-constrained nodes. EBHTTP is a RESTful protocol running on the OSI application layer.

The implementation features of the EBHTTP protocol are given as follows [9]:

- EBHTTP hosts must support UDP as a protocol for transporting EBHTTP messages.
- EBHTTP requests are forwarded from clients to servers over UDP, and HTTP responses are forwarded back to clients using the client's UDP source port.
- Multiple EBHTTP messages can be packed into a single UDP datagram, and EBHTTP servers must unpack these messages and treat them as separate requests. All lengths are specified inside the message, which makes this possible. EBHTTP message packaging allows more efficient network use by reducing the number of redundant UDP and IP headers used on constrained networks.
- Multiple EBHTTP responses can also be packed into a single UDP message. EBHTTP clients should unpack these responses and treat them like those found in individual datagrams.

- A single EBHTTP message should not span multiple UDP datagrams. EBHTTP does not support fragmentation. TCP should be used if EBHTTP needs to carry a message larger than a single datagram.
- EBHTTP hosts may run on EBHTTP over TCP.
- EBHTTP requests and responses must be packaged back-to-back (with no padding) into a TCP byte stream. Any requests and responses can be sent over a single TCP connection.

Publish/subscribe functionality can be implemented on EBHTTP using application-defined mesh formats. In addition, EBHTTP supports the HTTP-level publish/subscription mechanism defined in the General Event Notification Architecture (GENA) Internet draft. It supports the HTTP proxy as a proxy.

16.7 Lean Transport Protocol

The Lean Transport Protocol (LTP) allows transparent message exchange between resource-constrained devices and server or PC-type systems. It consumes only very little memory and CPU power and has small message sizes. To understand LTP, it is helpful to mention the concepts of the simple object access protocol (SOAP), Web Service description language (WSDL), and Web Service Addressing (WS-Addressing).

- **SOAP:** It describes the message format of Web Service communications. Messages are usually serialized using XML and exchanged via HTTP. The SOAP standard explicitly allows different serialization technologies and transport protocols. It has SMTP, JMS (Java message service), TCP, and UDP specifications.
- **WSDL:** It is an XML-based language that describes a Web Service's method signatures, transport bindings, and endpoint addresses. Transport binding describes the signature sequence and layout of the message while specifying how the message is serialized and via which transport protocol the message will be exchanged. For the specification of message flow between the consumer and the provider of a Web Service, WSDL defines four abstract mesh exchange patterns: "one-way," "request/response," "notification," and "solicit/response."
- **WS-addressing:** WS-addressing specifies addressing mechanisms for Web Service messages beyond simple URLs. Using WS-addressing, the addressing information is encoded within the SOAP message rather than part of the URL of the transport protocol. Thus, it provides a mechanism independent of a particular transport binding (such as HTTP) and additional features (such as the use of gateways). WS-addressing is mandatory if a transport protocol cannot address the Web Service endpoints (for example, TCP).

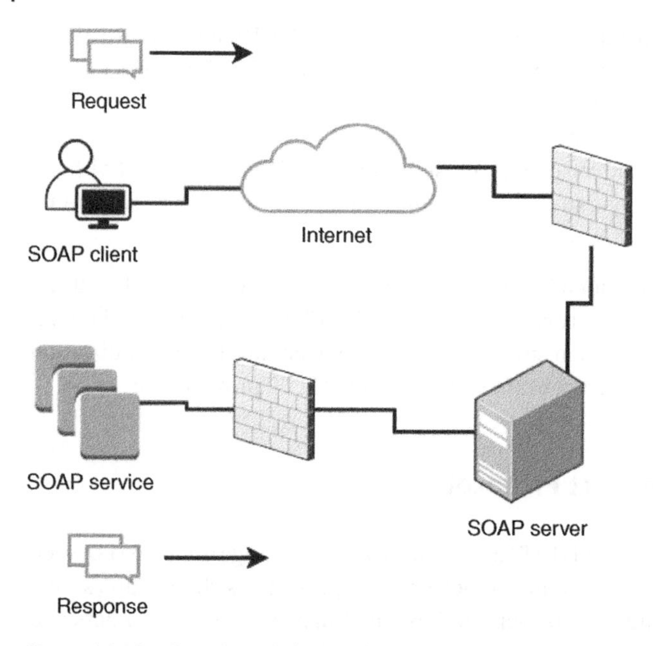

Figure 16.10 Overview of simple object access protocol.

LTP is not limited to exchanging messages between different types of networks, such as the Internet and RCNs (Network of Resource Constrained Devices) but can also be used within a single network. LTP provides transparent end-to-end message communication between Web Service endpoints in different types of networks. This is realized with "Platform-Independent Messaging" using "Platform-Independent Addressing Schema." So-called connectors on gateways adapt LTP to a particular underlying network. These gateways create self-contained subnets that perform internal end-to-end communication. The Platform-Independent Messaging layer defines a transparent end-to-end message exchange between endpoints in random subnets [10] (Figure 16.10).

16.8 Advanced Message Queuing Protocol

AMQP is a protocol designed as an open replacement for existing proprietary messaging middleware. The most significant benefit of AMQP is that it is reliable and interoperable. As the name suggests, AMQP provides messaging, reliable queuing, topic-based publish and subscribe messaging, flexible routing, transactions, and security. AMQP exchanges routes based on topic and headers.

The AMQP network protocol defines [11]:

- A peer-to-peer protocol; commonly, in AMQP, one peer plays the role of a client application, and the other performs the role of trusted message routing and delivery service or agent.
- Defines methods of forwarding failing connections to alternative services.
- Enables peers to discover each other's capabilities.
- Provides comprehensive security mechanisms for seamless end-to-end confidentiality, including SSL and Kerberos.
- It allows multiplexing of TCP/IP connections to enable multiple conversations over a TCP/IP connection. This mainly simplifies firewall management a lot.
- Specifies how to address the message source with network peers and which messages are interesting.
- Lifecycle of a message through fetching, processing, and acknowledgment: AMQP makes it very clear when responsibility for a message is transferred from one peer to another, thus increasing reliability.
- If desired, it enhances the performance by prefetching the messages ready for the client to process across the network.
- Defines the method of processing batches of messages within a transaction.
- Lightweight applications have a mechanism that allows complete message transfer from logging in to logging out in a single network packet.
- It provides skillful flow control that enables message consumers to slow down producers to a manageable pace. Thus, it allows different workloads to run parallel to varying speeds over a single connection.
- It provides mechanisms to resume message transfers when connections are lost and re-established. This mechanism is handy in case of service failover or intermittent connection.
- Security is held with transport layer security (TLS) and simple authentication and security layer (SASL) protocols over TCP (Figure 16.11).

AMQP is a binary wire protocol designed for interoperability between different vendors. The adoption of AMQP has been strong whereas other protocols have failed. Companies like JP Morgan use it to process a billion messages a day. NASA uses Nebula for cloud computing. Google uses this for complex event handling. Here are a few additional AMQP examples:

- It is used in India's Aadhaar project, one of the world's largest biometric databases hosting 1.2 billion identities.
- Used in the ocean observatories initiative, an architecture that collects 8 TB of data per day.

AMQP is all about queues. Sends transactional messages between servers. As a message-centric middleware, it can handle thousands of reliable queued

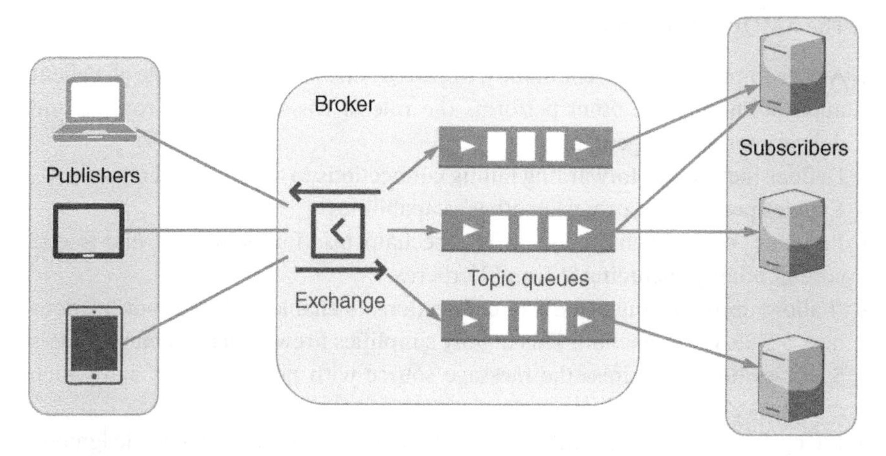

Figure 16.11 AMQP architecture.

transactions. AMQP is focused on not losing messages. All communications from publishers to exchanges and queues to subscribers use TCP, which ensures reliable P2P connectivity. Also, endpoints must acknowledge the acceptance of each message. The standard also describes an optional transaction mode with a proper multiphase commit sequence [12].

AMQP performs asynchronous transfer of messages regardless of the operating system, hardware, or programming language. It is advanced and optimized on a data frame with a buffering approach, improving server performance.

One or more virtual hosts with their namespaces, exchanges, and message queues reside on central servers. Producers and consumers subscribe to the Exchange service. The Exchange service receives messages from the publisher and forwards the data to an associated queue. This relationship is called "binding." It can be bonded directly to one queue or multidirected multiple queues as in broadcast. Alternatively, binding can associate an exchange with a queue using a routing key. This situation is formally called a direct exchange. Another Exchange type is topic Exchange [18].

16.9 Data Distribution Service

Data distribution service (DDS) is a middleware protocol and API standard defined by the Object Management Group (OMG). DDS directly targets devices that use device data and distributes the data to other devices. Although interface support with IT infrastructure is supported, the primary purpose of DDS is to connect devices to other devices. OMG Data distributor service (DDS) for real-time systems is the first open international middleware standard that handles publish-subscribe communication for real-time and embedded systems. DDS provides a virtual global data space where applications can only share information by reading and writing data objects addressed via an application-defined name (Topic) and a key. DDS features fine and extensive control of QoS parameters, including reliability, bandwidth, delivery dates, and resource limits. DDS also supports the creation of native object models on top of the Public Data Field.

It includes many real systems devices, servers, mobile nodes, and more. There are different communication needs, but it is better and easier to use a single communication paradigm whenever possible. System designers must determine which protocols meet the primary challenges of their intended application. Then, if possible, extending this primary choice to all aspects of the system is necessary. For example, cross-device data usage is a different use case than device data collection. The requirements for turning a light switch (CoAP is the best choice in this case) differ from the protocol required to manage this power generation (best with DDS) and the necessary protocol to monitor transmission lines (best with MQTT). Or, the protocol to be chosen for communication power usage within the data center (the best choice in this case, AMQP) is different.

Overall, DDS is the most versatile of these protocols. DDS can manage small devices, connect large, high-performance sensor networks, and close time-critical control loops. It can also present and receive data from the cloud. DDS communication is peer-to-peer. Eliminating message brokers and servers simplifies setup, minimizes latency, maximizes scalability, increases reliability, and reduces cost and complexity. Using DDS requires building a data model and understanding data-centric principles. It is ideal for IoT applications that require a durable, reliable, and high-performance architecture [12].

High-level data-centric interfaces have replaced message-centric programming in DDS. The main goal of DDS is to share the correct data in the right place and at the right time, even between the time-decoupled publisher and the consumer. DDS creates a global data space by carefully replicating the relevant parts of the logically shared data space [13].

- **The correct data:** Since not all data needs to be ubiquitous, middleware delivers only the data consumers need. This is accomplished by applying an interest-based filter to content and data rates. Thus, DDS saves bandwidth and processing power while minimizing overall application complexity. As a data-centric solution, DDS can understand the schema of shared data. This allows filtering only needed applications on content, age, and lifecycles. For example, you can only send boiler temperatures above 300 (content filter) at a maximum of ten updates per second (rate). This effective approach can save up to 90% of data transmission overhead in many systems (Figure 16.12).

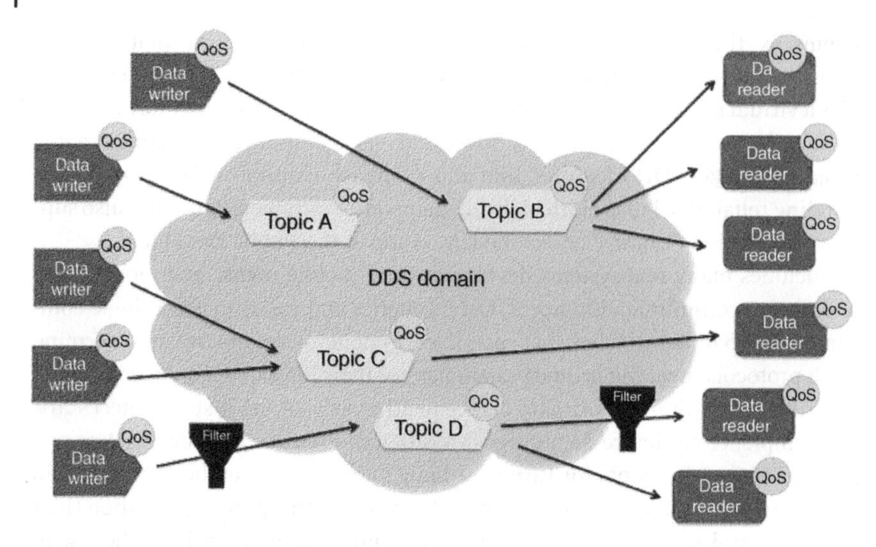

Figure 16.12 Data distribution service.

- **The right place:** DDS distributes and maintains data readily available. DDS publishers and subscribers dynamically discover the data types they want to share and related QoS (Quality of Service). After a successful match, DDS enforces the timely distribution according to QoS. It creates QoS-enforced logical channels for each data stream between each publisher-subscriber pair. A DDS subscriber can be assured that its peer publisher is alive and that any data produced will be delivered. This greatly simplifies application development and error handling.
- **The right time:** Real-time systems interact with the real world. Data must be available in time (a late right data is a failure). Data may differ in priority, reliability, timing, and other nonfunctional properties. DDS balances the use of scarce resources to distribute data at the right time. DDS middleware uses logical QoS principles set by applications at runtime to balance efficiency and determinism. The QoS contracts ensure these timing relationships. For example, if a subscriber needs to update every 10 ms and the matching publisher does not deliver, the system reports an error and activates corrective action. QoS policies cover many aspects, including urgency, importance, reliability, persistence, and vitality.

The primary technical concepts are also given as follows:

- **Relational data modeling:** DDS addresses data like relational databases. It can manage data by using key fields and allow ad-hoc queries and filters on content and time. So applications can extract specific data as needed.

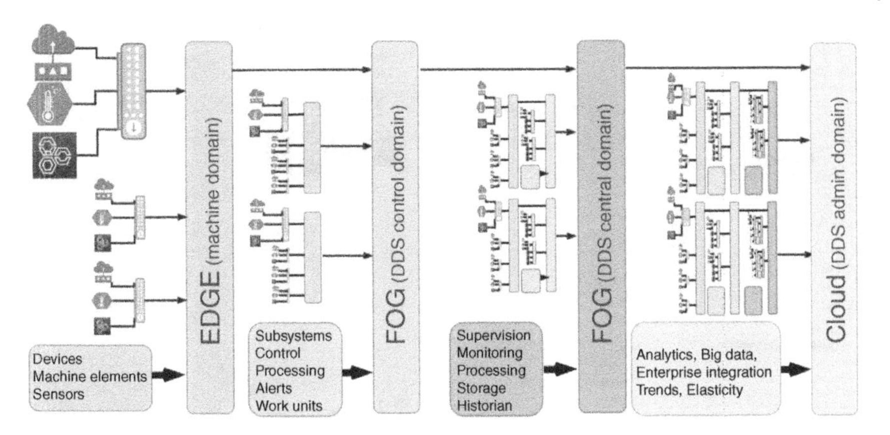

Figure 16.13 DDS architecture.

- **Pub-sub messaging:** DDS uses the publish/subscribe paradigm for dynamic discovery and primary management of data flows between related DDS entities, including publishers, subscribers, durability services, recording and playback services, and linked databases. Request-replay and other patterns are built on this powerful substrate.
- **Reliable multicast:** The DDS standard wire protocol implements reliable multicast over plain UDP sockets, enabling systems to efficiently take advantage of modern network infrastructures.
- **Lifecycle awareness:** Unlike message-oriented products, DDS offers open application support for information lifecycle awareness. For example, it detects, communicates, and informs applications about the first and last views of data (topic instance) updates. This makes it easy to respond to new and outdated information promptly.
- **Trigger patterns:** DDS offers a variety of trigger patterns that keep track of updates on subscribed information. Examples include polling, callbacks (typical for GUIs), and WaitSets ("select" similar to Unix to provide complete application control for prioritized handling of selective trigger events) (Figure 16.13).

16.10 Simple Text-Oriented Messaging Protocol

STOMP is a text-based, interoperable wire format protocol with MOM. This format allows clients to communicate with the message broker, enabling easy and pervasive messaging interoperability. This communication is independent of the languages, platforms, and agents used in the architecture. STOMP protocol provides a message header with a frame body and features like AMQP. It offers a

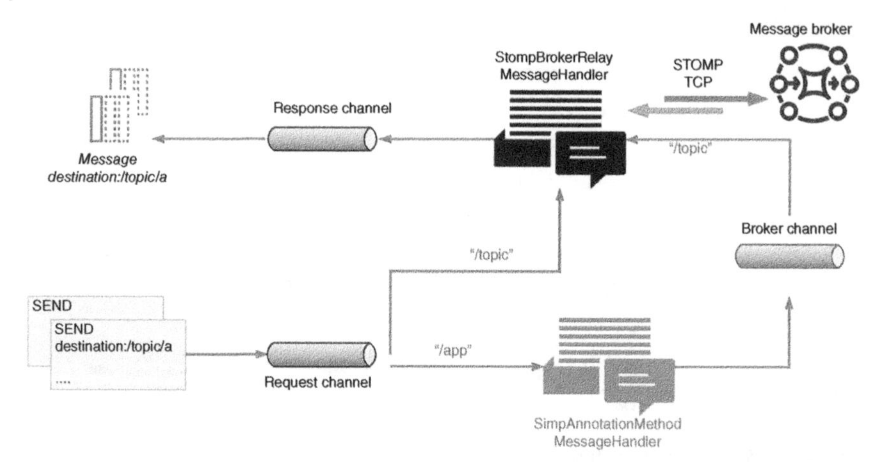

Figure 16.14 STOMP protocol architecture.

simple and lightweight connection with a wide variety of language bindings. Some transactional semantics are also provided by it. STOMP is a bidirectional protocol that additionally uses textual headers similar to HTTP.

Therefore, care should be taken when moving code between different brokers, as it is not easy. Like other broker-based protocols, STOMP does specify the publish-subscribe mechanism. Here, clients subscribe to topics, and the broker notifies the client when a message about that topic is sent. STOMP does not require the use of queues and topics. Instead, it uses "SEND" semantically with the string "destination." The broker internally maps the "destination" string to the topic, queue, or exchange. Consumers also subscribe to these destinations. Different brokers may support various destinations as it is not mandatory in the specifications. So care should be taken when moving code between other brokers as it is not easy [14] (Figure 16.14).

STOMP has clients written in python (Stomp.py), TCL (tStomp), and Erlang (stomp.erl). A few servers, such as RabbitMQ (via a plugin), have native STOMP support, and some servers are designed in specific languages (Ruby, Perl, or OCaml).

16.11 Extensible Messaging and Presence Protocol

XMPP is a client/server architecture in which XMPP clients communicate with the XMPP server using TCP sockets. It can also work over HHTP using WebSocket implementation. It is an open protocol that streams XML elements for exchanging messages in real-time. In other words, it is an XML-based asynchronous protocol.

Because of its scalability, it is applied to long-distance messaging and includes human intervention. XMPP originated from XML.

The protocol is specified with different RFCs. For example, RFC2778 and RFC2770 define the presence and instant messaging model. RFC3920 defines XMPP core while RFC3921 defines XMPP instant messaging. XMPP addressing, scalability, federation, and security features make it ideal for IoT applications. It also uses the publish/subscribe mechanism for data sharing, similar to the MQTT protocol. The specifications for communication between XMPP client and server are given as follows [15]:

- XMPP client to server (C2S) uses TCP port 5222 for communication.
- Server to server (S2S) uses TCP port 5269 for communication.
- Discovery and XML streams are used for S2S and C2S.
- XMPP uses Transport Layer Security (TLS) and Simple Authentication and Security Layer (SASL) for security mechanisms.
- Unlike the email mechanism, there are no intermediate servers for federation (Figure 16.15).

The federation specified here means the feature that two business domain users can talk to each other. As an example, we can give a connection from a thermostat to a Web server. This server can be easily accessed with a smartphone in the future.

XMPP can be used in instant messaging applications (Google Talk and WhatsApp), presence status, message delivery, conferencing, roster management, voice and video calls, online gaming, news websites, and VoIP applications.

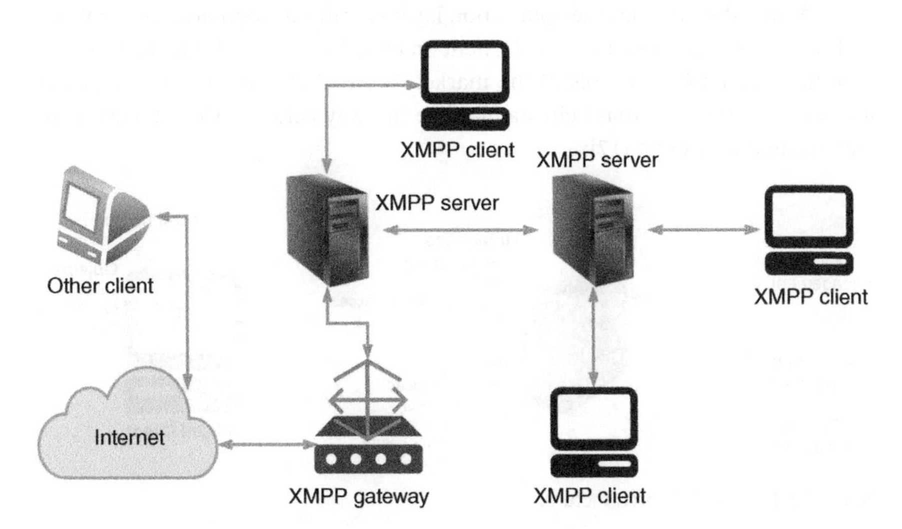

Figure 16.15 XMPP architecture.

Advantages of the XMPP Protocol:

- **Extensible:** It can be customized according to individual user requirements.
- **Messaging:** Short messages are used for fast communication between the user and the server.
- **Presence:** It is reactive to the user's presence and state.
- **Protocol:** It is an open platform that is constantly evolving.
- **Secured:** Uses TLS and SASL mechanisms for reliable end-to-end connectivity.
- It provides a permanent connection.
- It is decentralized in nature as there is no need for a central XMPP server.
- It allows servers of different architectures to communicate.

Disadvantages of the XMPP Protocol:

- It does not have the QoS mechanism used by the MQTT protocol.
- Streaming XML has overhead due to text-based communication compared to binary-based communication.
- XML content is moved asynchronously.
- The server can be overloaded with presence and instant messaging.

16.12 Lightweight M2M

Lightweight M2M is an Open Mobile Alliance (OMA) protocol developed for machine-to-machine or IoT device management and service enablement. The LwM2M standard defines the application layer communication protocol between an LwM2M Server and an LwM2M Client on an IoT device [16] (Figure 16.16).

Unlike other IoT protocols in the market, the LwM2M architecture supports four logical interfaces that help standardize the way actual device management and telemetry are done [17]:

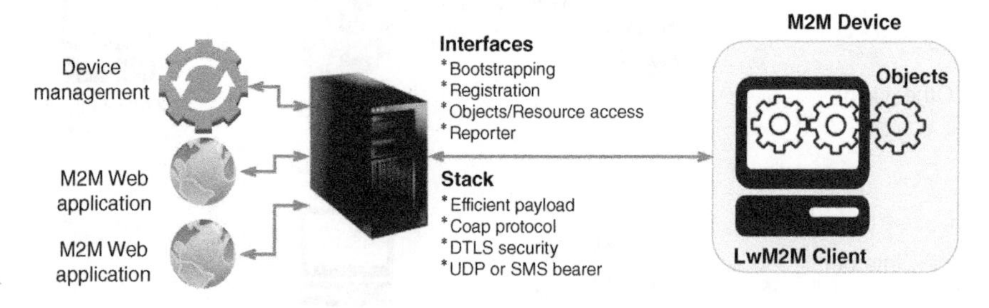

Figure 16.16 LwM2M architecture.

- **Bootstrapping interface:** This interface provides headless device management. This means it is possible to configure a device to provide the correct service without needing preconfiguration at the factory. This significantly reduces cost and optimizes time-to-market for the product or service.
- **Client registration interface:** This interface informs the server about the 'presence' of the client and the supported functionality. It also allows over-the-air-firmware and software updates.
- **Device management and service enablement interface:** LwM2M allows the provider to access object instances and resources to change device settings and parameters.
- **Information reporting interface:** Thanks to this publish/subscribe interaction of this interface, the user can receive error reports from devices and send queries about device status when the service is not working correctly.

Apart from being a simple and efficient protocol for managing low-power devices, LwM2M also has several features that put it ahead of the protocol competition. Unlike traditional M2M solutions, where a device often needs to maintain multiple technologies, protocols, and security services, the lightweight M2M model allows users to have a single technology stack at the device and application levels. Moreover, LwM2M offers cross-vendor and cross-platform interoperability, making it ideal for service providers who want to avoid vendor lock-in. It combines Datagram Transport Layer Security (DTLS), CoAP, Block, Observe, SenML LwM2M, and Resource Directory to create a device-server interface with a defined object structure. Combining all the above advantages, Lightweight M2M can provide an excellent time to market as it is available for instant distribution.

LwM2M has a well-defined data model, unlike many industry-proven protocols (like MQTT). The idea is straightforward: a tree with a maximum depth of 4, consisting of Objects (e.g. temperature sensor), Object Instances (sensor1, sensor2, etc.), Sources (e.g. current temperature), and Source Instances (e.g. 10° and 15°), respectively.

Finally, the Lightweight M2M protocol offers a well-defined device and data management model structure to enable several vendor-neutral features such as secure device boot, object or resource access, and device reporting.

To summarize, the Lightweight M2M protocol offers flexible, scalable, and vendor-independent device management with improved time to market, making it particularly suitable for low-power devices with limited processing and storage capabilities. With all this in mind, LwM2M is the best solution to consider for large, complex, and long-lived deployments involving cross-platform and cross-standard IoT services.

16.13 Health Device Profile Protocol (Continua HDP)

Bluetooth is used in many medical applications as a secure and reliable connection method. Typical implementations are based on the Bluetooth Serial Port Profile (SPP) and manufacturer-specific proprietary applications and protocols. Due to these manufacturer-specific applications, the functional performance of systems from different manufacturers is poor.

For these reasons, the Bluetooth Special Interest Group (SIG) established the Medical Device Working Group (MED WG). The primary purpose of this group is to create a profile that will ensure the interoperability of medical and collection devices from different manufacturers. As a result of these efforts, Multichannel Adaptation Protocol (MCAP) and Bluetooth Health Device Profile (HDP) were adopted in 2008. Application-level interoperability is provided by ISO/IEEE 11073-20601 Personal Data Exchange Protocol and IEEE 11073-104xx device specifications.

HDP mainly aims to support various domestic or in-hospital applications. The most typical use cases are different portable sensors such as EKG transmitters, blood glucose meters, or blood pressure meters that transmit measurements in the hospital to a monitoring computer. In an in-home application, sensor measurements can be sent to a gateway device that sends the information to remote servers for further processing [18] (Figure 16.17).

HDP advantages are listed as follows:

- **Medical, healthcare, and fitness applicability:** HDP is a custom profile designed to allow interoperability between medical, health, and fitness apps from different vendors. This gives HDP a significant advantage over more general profiles such as the SPP or others that only provide a base layer for specific protocols and data formats.
- **Wireless service discovery:** HDP provides a standard wireless discovery method by which a device's device type and supported application data type are

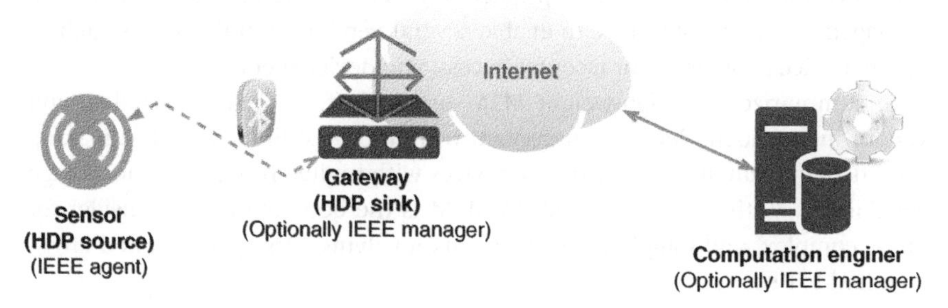

Figure 16.17 HDP structure.

determined. This discovery is accomplished using the generic access profile (GAP) discovery procedures and the service discovery protocol (SDP).

- **Reliable connection-oriented behavior:** HDP uses the connection-oriented property of Bluetooth sublayers to provide more reliable behavior when a source is out of range or disconnected (accidentally or intentionally). This will allow both Source and Sink to understand that the connection has been broken and take appropriate actions. In addition, the reliable data channel detects and retransmits packets corrupted by radio link interference. The optional frame check sequence (FCS) for L2CAP is critical when working with high noise levels, such as near Wi-Fi or other ISM band devices.
- **Reliable control channel:** HDP control channel uses enhanced retransmission mode for enhanced reliability desired for signal commands.
- **Support for flexible data channel configurations:** HDP data channels allow independent design, providing flexibility to applications. Data channels configured as "reliable" use enhanced retransmission mode, while data channels configured as "streaming" use streaming mode. The use of FCS is optional for the data channels and mandatory for the control channel.
- **Application level interoperability:** HDP, together with the ISO/IEEE 11073-20601 optimized Exchange protocol, provides a structured approach to establishing control and data channels to exchange information between communicating health devices. As device specializations are added to ISO/IEEE 11073-104XX and adopted by the MED WG, the Bluetooth Assigned Numbers document allows the addition of adopted specializations without updating HDP specifications.
- **Efficient reconnection mechanism:** HDP allows devices to maintain the system state and eliminates unnecessary configuration steps after reconnection. This process will enable devices to disconnect when data is not being received or transmitted and reconnect as data becomes available. This method reduces the average power consumption.
- **High-resolution clock synchronization:** HDP also defines an optional clock synchronization protocol (CSP) that allows precise timing synchronization (note: theoretically in the microsecond range) between health devices. This feature is for healthcare devices such as high-speed sensors that require close synchronization.
- **Optimized for devices with low resources:** HDP has a small set of simple control commands, making it relatively inexpensive to implement. Depending on the device role and individual application requirements, it is also possible for devices to support an even smaller subset of available commands. This is useful for product requirements that define limited code and memory space.

16.14 Devices Profile for Web Services

Devices Profile for Web Services (DPWS) was developed to enable secure Web Service features on resource-constrained devices. DPWS was developed mainly by Microsoft and some printer device manufacturers. DPWS allows sending secure messages to and from Web Services, dynamically discovering a Web Service, defining a Web Service, subscribing to a Web Service, and receiving events from a Web Service.

The Web Service specifications on which DPWS is based are given as follows [19]:

- WS-Addressing for advanced endpoint and message addressing
- WS-Policy for policy exchange
- WS-Security for managing security
- WS-Discovery and SOAP-over-UDP for device discovery
- WS-Transfer/WS-Metadata exchange for device and service description
- WS-Eventing for managing subscriptions for event channels

In addition to these specifications, DPWS SOAP, WSDL, and XML-Schema W3Cs Web Services architecture are partially based on W3Cs Web Services [20].

The original architecture suggested by the standard is given in Figure 16.18. Proxy-based structure confirmed for more flexible connections is shown in Figure 16.19.

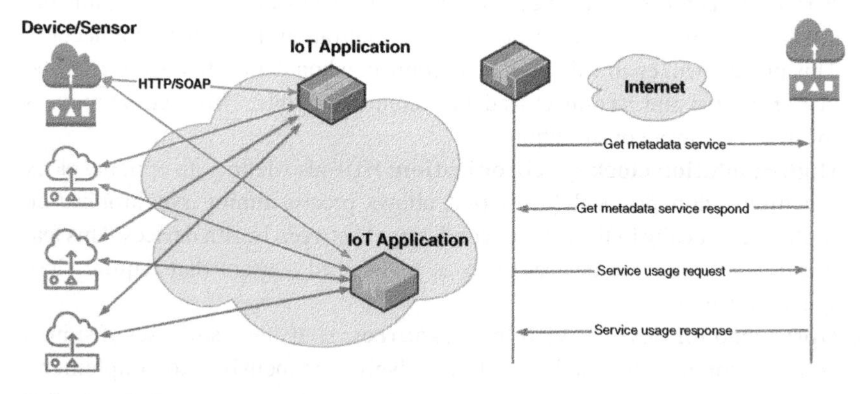

Figure 16.18 Original DPWS.

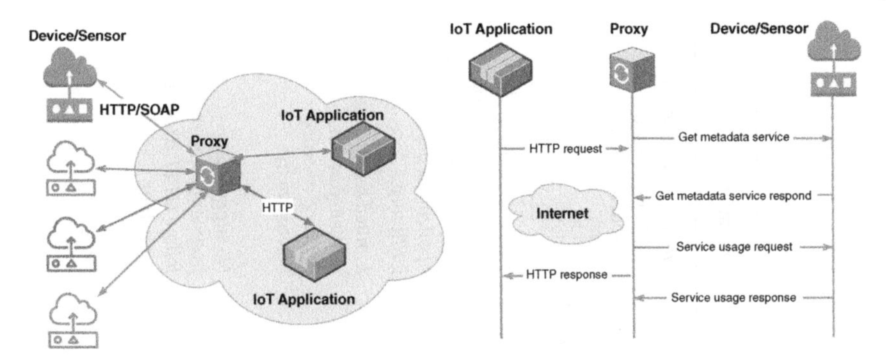

Figure 16.19 Proxy-extended DPWS.

16.15 Protocol Comparisons

Until this section, protocols used in IoT and M2M systems have been explained. An appropriate protocol should be selected considering the cost/performance criteria for the application. It is important to connect sensors and objects to identify the right applications and proper protocols for possible scenarios. Aspects of the application such as consumer/industrial, Web Service, IoT service, publish/subscribe, request/response, and real-time/nonrealtime should be considered while making these choices. Comparisons of some of the protocols described in Table 16.1 are given [21].

Table 16.1 Comparison of IoT protocols.

Protocol	Architecture	Transport	QoS	Security	Application areas	Advantages	Limitations
MQTT	*Asynchronous*	*TCP*	*Yes*	*SSL*	*Healthcare, energy and utilities, industry and irrigation, social networking, IoT based applications*	*Low overhead, delay and power consumption, high latency, better than CoAP in traffic management, higher throughput, optimal memory and CPU usage*	*Moderate bandwidth and battery usage compared to CoAP*
CoAP	*Synchronous*	*UDP*	*Yes*	*DTLS*	*Live data communication, sensor networks, IoT-based applications*	*URI and content-type support, enhanced reliability, reduced latency, single parsing, multicasting, reduced bandwidth usage, good PDR*	*Packet loss due to TCP retransmissions, high cost, network robustness, application deployment gullibility*
DDS	*Asynchronous*	*TCP/ UDP*	*Yes*	*SSL/ DTSL*	*M2M and IoT-based applications, air traffic and vehicle control* systems, *industrial automation systems*	*Excellent QoS control, configurable reliability, pervasive redundancy, multicasting*	*Limited scalability, resiliency in data delivery, network heterogeneity*
EBHTTP	*Asynchronous*	*UDP*	*No*	*SSL*	*Applications involving transfer of smaller messages in constrained hypermedia information systems*	*Resource discovery due to RESTful design, extensibility of HTTP to suit highly constrained networks*	*No support for fragmentation, must follow HTTP caching behaviour*

LTP	*Synchronous*	*TCP/ UDP*	*Yes*	*SSL*	*Web Service message exchanges*	*Standard-compliant with Web Services, combines with microfiber to give SOAP messages, header compression, message fragmentation*	*High implementation and maintenance cost*
XMPP	*Syn/Async*	*TCP*	*No*	*SSL*	*Voice and video calls, chatting and message exchange applications*	*Good to use if application is already built and running with XML*	*High power consumption due to complex computations, additional overhead, no QoS and suitable for M2M*
AMQP	*Asynchronous*	*TCP*	*Yes*	*SSL*	*Application based on the control plane and server-based analysis functions*	*Can connect across technologies, organizations and time domains, store-and-forward strategy for good reliability*	*Not suitable for constrained real-time applications, no support for automation discovery*

References

1 Timčenko, V., Zogović, N., and Đorđević, B. (2018). Interoperability for the sustainability assessment framework in IoT-like environments. *8th International Conference on Information Society and Technology*, Kopaonik, Serbia (11–14 March 2018), p. 21.

2 MQTT (2022). The standard for IoT messaging. *MQTT*. https://mqtt.org/ (accessed 16 February 2023).

3 Steve's Internet Guide (2018). Beginners guide to the MQTT protocol. http://www.steves-internet-guide.com/mqtt/ (accessed 16 February 2023).

4 Colitti, W., Steenhaut, K., De Caro, N. et al. (2011). REST enabled wireless sensor networks for seamless integration with web applications. *2011 IEEE Eighth International Conference on Mobile Ad-Hoc and Sensor Systems*, Valencia, Spain (17–22 October 2011).

5 Shelby, Z. (2010). Embedded web services. *IEEE Wireless Communications* 17 (6): 52–57.

6 Shelby, Z., Hartke, K., and Bormann, C. (2014). The Constrained Application Protocol (CoAP) (No. rfc7252). https://www.rfc-editor.org/rfc/rfc7252 (accessed 16 February 2023).

7 Andreev, S. and Koucheryavy, Y. (2012). *Internet of Things, Smart Spaces, and Next-Generation Networking*, LNCS, vol. 7469, 464. Springer.

8 Lea, P. (2018). *Internet of Things for Architects*. Packt Publishing Ltd.

9 IETF (2022). Internet of things. *IETF*. https://www.ietf.org/topics/iot/ (accessed 16 February 2023).

10 Glombitza, N., Pfisterer, D., and Fischer, S. (2010). LTP: an efficient web service transport protocol for resource constrained devices. *2010 7th Annual IEEE Communications Society Conference on Sensor, Mesh and Ad Hoc Communications and Networks (SECON)*, Boston, MA (21–25 June 2010), pp. 1–9.

11 AMQP (2023). Architecture. *AMQP*. https://www.amqp.org/product/architecture (accessed 16 February 2023).

12 Raj, P. and Raman, A.C. (2017). *The Internet of Things: Enabling Technologies, Platforms, and Use Cases*. CRC Press.

13 DDS-Foundation (2023). What is DDS? https://www.dds-foundation.org/what-is-dds-3/ (accessed 16 February 2023).

14 RF Wireless World (2023). STOMP Architecture basics | STOMP protocol in IoT. *RF Wireless World*. https://www.rfwireless-world.com/Terminology/STOMP-architecture.html (accessed 16 February 2023).

15 RF Wireless (2023). What is XMPP Protocol in IoT | XMPP Server | XMPP Client. *RF Wireless*. https://www.rfwireless-world.com/IoT/XMPP-protocol.html (accessed 16 February 2023).

16 Ha, M. and Lindh, T. (2018). Enabling dynamic and lightweight management of distributed Bluetooth low energy devices. *2018 International Conference on Computing, Networking and Communications (ICNC)*, Maui, HI (5–8 March 2018), pp. 620–624.

17 AVSYSTEM (2019). What is LwM2M? Lightweight M2M protocol overview. *AVSYSTEM*. https://www.avsystem.com/blog/lightweight-m2m-lwm2m-overview/ (accessed 16 February 2023).

18 Silabs (2019). AN988: Health Device Profile – iWRAP Application Note. *Silabs*. https://www.silabs.com/documents/public/application-notes/AN988.pdf (accessed 16 February 2023).

19 WS4D (2023). Web services for devices » Devices profile for web services. *WS4D*. http://ws4d.org/technology/dpws/ (accessed 16 February 2023).

20 Han, S.N., Park, S., Lee, G.M., and Crespi, N. (2015). Extending the devices profile for web services standard using a REST proxy. *IEEE Internet Computing* 19 (1): 10–17.

21 Narayanaswamy, S. and Kumar, A.V. (2019). Application layer security authentication protocols for the Internet of Things: a survey. *Advances in Science, Technology and Engineering Systems Journal* 4 (1): 317–328.

17

Popular Operating Systems of IoT

17.1 Introduction

An operating system is a set of software that manages hardware resources running on the computer and provides standard services for various application software. It provides communication between application programs and computer hardware. Operating systems such as Microsoft Windows, MAC OS X, GNU /Linux, BeOS, Android and IOS are popular operating systems used on computers and smartphones. Operating systems can be installed not only on computers, video game consoles, mobile phones, and Web Servers but also in cars and even white goods.

Similarly, operating systems are installed on network devices to make network connections, manage protocols, and perform input and output operations. Operating systems should be evaluated not by the breadth of their functions but by their ability to program the hardware for a specific purpose.

The essential functions of an operating system are listed as follows:

- Process management
- Job Accounting
- Memory management
- Secondary storage management
- File management
- Device management
- Security
- Input/output management
- Networking
- Device management

This section explains the most popular operating systems running on IoT devices.

Evolution of Wireless Communication Ecosystems, First Edition. Suat Seçgin.

Published 2023 by John Wiley & Sons, Inc.

17.2 OpenWSN

OpenWSN is a project created at the University of California, Berkeley, and extended at INRIA and the Open University of Catalonia (UOC), aiming to create an open, standards-based, and open-source implementation of a completely constrained network protocol stack for wireless sensor networks and IoT. The root of OpenWSN can be identified as a deterministic MAC layer that implements the IEEE 802.15.4e TSCH based on the concept of Time Slotted Channel Hopping (TSCH).

Combined with IoT standards such as IEEE 802.15.4e, 6LoWPAN, RPL, and CoAP, it provides ultralow-power and highly reliable mesh networks that are fully integrated into the Internet [1]. OpenWSN has been ported to several commercially available platforms, from legacy 16-bit microcontrollers to cutting-edge 32-bit Cortex-M architectures. The OpenWSN project provides a free and open-source implementation of a protocol stack. It provides debugging and integration tools to contribute to the overall goal of promoting the use of low-power wireless mesh networks [2] (Figure 17.1).

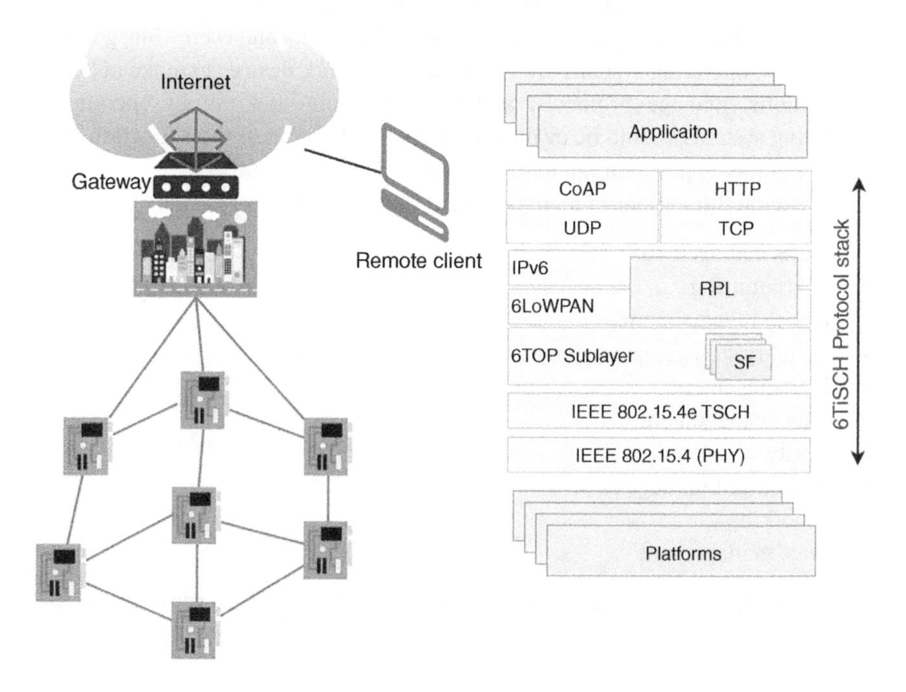

Figure 17.1 Open WSN protocol stack and architecture.

17.3 TinyOS

TinyOS is a free, open-source, BSD-licensed operating system designed for low-power embedded distributed wireless devices used in sensor networks. It is designed to support the intensive simultaneous operations required by networked sensors with minimal hardware requirements. TinyOS was developed by the University of California, Berkeley, Intel Research, and Crossbow Technology. It is written in the programming language nesC (Network Embedded Systems C), a version of C optimized to support components and concurrency. It also supports component-based, event-driven programming of applications for TinyOS [3] (Figure 17.2).

17.4 FreeRTOS

FreeRTOS is a real-time operating system (RTOS) kernel for embedded devices designed to be small and straightforward. It was ported to 35 microcontrollers and distributed under the GPL with one optional exception. The exception is keeping the kernel itself open source, allowing users' proprietary code to remain closed source, thus facilitating the use of free RTOS in private applications.

It is mainly written in C to make the code readable and easy to port and maintain (although some compilation functionality has been added to support architecture-specific timer routines). It provides methods for multiple threads or tasks, mutexes, semaphores, and software schedulers (Figure 17.3).

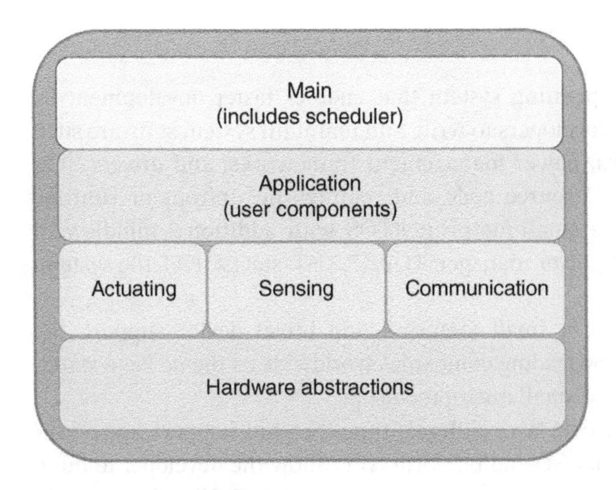

Figure 17.2 Tiny OS architecture.

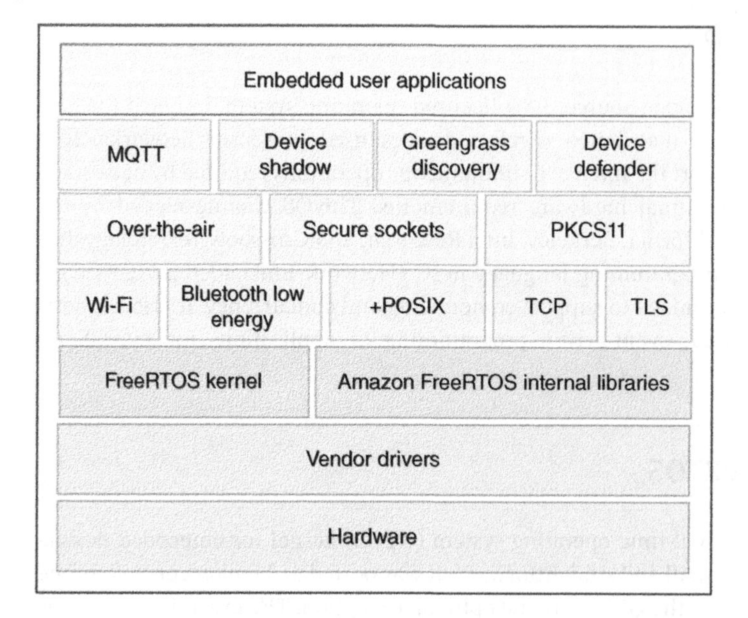

Figure 17.3 FreeRTOS.

Features such as a tiny, power-saving kernel, more than 40 architectural support, modular libraries, and Amazon Web Service (AWS) integrations provide essential advantages to FreeRTOS.

17.5 TI-RTOS

TI-RTOS is a real-time operating system that enables faster development by eliminating the need for developers to write and maintain system software such as timers, protocol stacks, power management frameworks, and drivers. It is supplied with complete C source code and requires no upfront or runtime license fees. TI-RTOS is a small-footprint RTOS with additional middleware components, including a power manager, TCP/IP, USB stacks, FAT file system, and device drivers.

With proven robustness, a small footprint, and broad device support, the FreeRTOS core is trusted by leading companies worldwide as the de facto standard for microcontrollers and small microprocessors.

Specifically, it relies on a core layer with real-time core, link support, and power management. On top of that, several platform APIs allow the developer to build custom applications. The operating system connects to TCP/UDP/IP network,

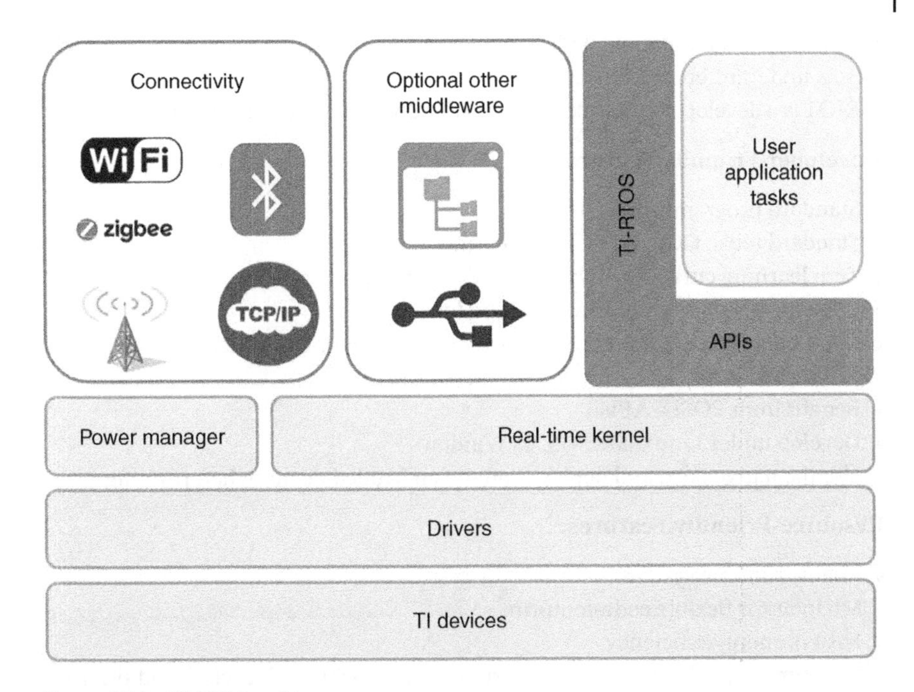

Figure 17.4 TI-RTOS architecture.

standard BSD socket interface, and HTTP, TFTP, Telnet, and DNS and provides a large set of ready-to-use libraries based on significant application layer protocols such as DHCP [4] (Figure 17.4).

17.6 RIOT

RIOT is a free, open-source operating system developed by a grassroots community scattered around the world, bringing together companies, academia, and hobbyists. RIOT supports most low-power IoT devices, microcontroller architectures (32-bit, 16-bit, and 8-bit), and external devices. RIOT aims to implement all relevant open standards that support a connected, secure, durable, and privacy-friendly Internet of Things (IoT).

RIOT supports DTLS transport layer security, IEEE 802.15.4 encryption, Secure Firmware Updates (SUIT), multiple cipher suites, and crypto-safe elements. RIOT is modular to adapt to application needs. We aim to support all common network technologies and Internet standards. RIOT is open to new developments and is often an early adapter in networking. Most of the software developed by the RIOT community is available under the terms of the GNU LGPLv2.1 published by the

Free Software Foundation. This provides an open Internet and allows building blocks under different licenses [5].

RIOT is a developer, resource, and IoT-friendly RTOS with the following features:

Developer-Friendly Features:

- Standard programming in C or C++
- Standard tools: GCC, GDB, and Valgrind
- Zero learning curve for embedded programming
- Code mostly without hardware dependence
- Code once, run on 8-bit (e.g. Arduino Mega 2560), 16-bit (e.g. MSP430), and 32-bit platforms
- Benefit from POSIX APIs
- Develop under Linux, Mac OS, or Windows
- Use the native port, and deploy it on the embedded device when running

Resource-Friendly Features:

- Robust runtime system
- Modular for flexible code-footprint
- Fosters energy-efficiency
- Real-time capable of limiting interrupt latency (~50 clock cycles) and priority-based scheduling
- Multithreading with ultralow overhead (<25 bytes per thread)

IoT-Friendly Features:

- 6LoWPAN, IPv6, RPL, UDP, TCP, and QUIC
- MQTT-SN, CoAP, and CBOR
- BLE, LoRaWAN, 802.15.4, WLAN, and CAN
- LwM2M client integration
- Static and dynamic memory allocation
- High-resolution and long-term timers
- Tools and utilities (System shell, Crypto primitives, . . .)
- Automated testing on various embedded hardware in the loop (Figure 17.5)

17.7 Contiki OS

Contiki is an operating system for networked, memory-constrained systems targeting low-power wireless IoT devices. Its main features are [2]:

- It is open source and is constantly being developed. Developers can work on custom applications and modify core operating system functions such as the TCP/IP stack and routing protocol, which are even less well-documented and less well-maintained than commercial operating systems.

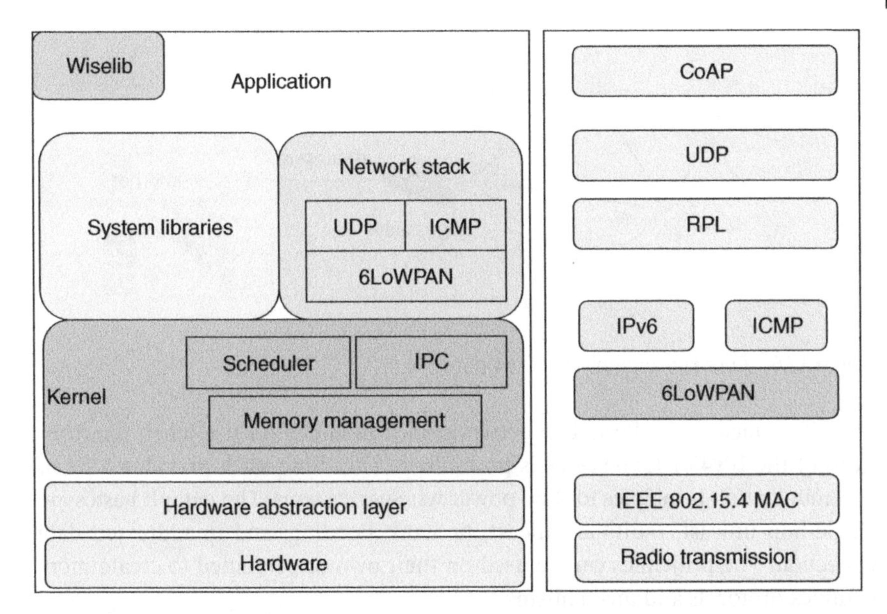

Figure 17.5 RIOT OS.

- It provides a full TCP/IPv6 stack using 6LoWPAN for the header. It creates LR-WPAN routes with RPL, the IPv6 routing protocol for compression and low-power and lossy networks.

Contiki was created by Adam Dunkels in 2002 and is further developed by a world-class developer team by Texas Instruments, Atmel, Cisco, ENEA, ETH Zurich, Redwire, RWTH Aachen University, Oxford University, SAP, Sensinode, Swedish Computer Science Institute, ST Microelectronics, Zolertia, and others.

Contiki is designed to operate in classes of hardware devices severely constrained in memory, power, processing power, and communication bandwidth. For example, in terms of memory, Contiki only needs about 10 kB of RAM and 30 kB of ROM, despite providing multitasking and a built-in TCP/IP stack. A typical Contiki system has kilobytes of memory, a milliwatt power budget, a processing speed measured in megahertz, and a communications bandwidth of hundreds of kilobits per second. This class includes embedded systems of various types and several legacy 8-bit computers (Figure 17.6).

Contiki provides three network mechanisms:

- Provides TCP/IP stack for IPv4 networking.
- Also provides the IPv6 stack, which enables IPv6 networking.
- Rime stack capability, a set of proprietary lightweight networking protocols explicitly designed for low-power wireless networks.

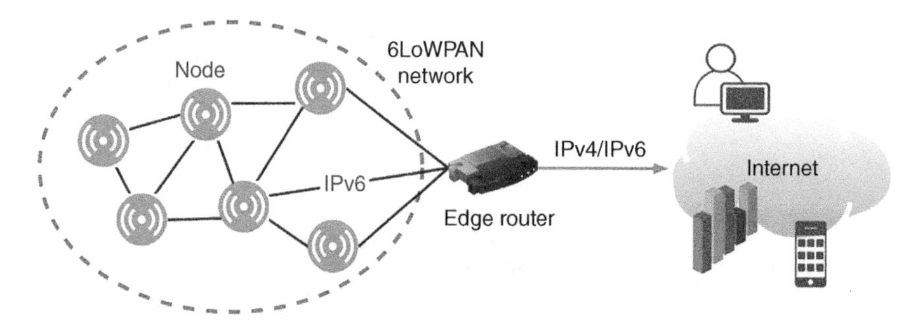

Figure 17.6 Contiki communication components.

The rime stack is an alternative network stack intended for use when the overhead of the IPv4 or IPv6 stacks is prohibitive. The Rime stack provides a set of communication principles for low-power wireless systems. The default basics are single-hop unicast, multitab unicast, network flooding, and unaddressed data collection. The principles can be used on their own or combined to create more complex protocols and mechanisms.

Contiki supports optional preemptive multithreading per process, interprocess communication using message-forwarding events, and an optional GUI subsystem with direct graphics support for locally attached terminals or virtual display networked via virtual network computing or Telnet.

A complete installation of Contiki includes the following features:

- Multitasking kernel
- Optional per-application pre-emptive multithreading
- Protothreads
- TCP/IP networking, including IPv6
- Windowing system and GUI
- Networked remote display using virtual network computing
- Web browser (claimed to be the world's smallest)

References

1 Watteyne, T., Vilajosana, X., Kerkez, B. et al. (2012). OpenWSN: a standards-based low-power wireless development environment. *Transactions on Emerging Telecommunications Technologies* 23 (5): 480–493.
2 Cirani, S., Ferrari, G., Picone, M., and Veltri, L. (2018). *Internet of Things*. Wiley.
3 TinyOS (2021). Home Page. http://www.tinyos.net/ (accessed 16 February 2023).
4 Texas Instruments (2017). TI-RTOS-MCU Operating system (OS). *TI.com*. https://www.ti.com/tool/TI-RTOS-MCU (accessed 16 February 2023).
5 RIOT (2023). RIOT – The friendly operating system for the Internet of Things. RIOT. https://www.riot-os.org/ (accessed 16 February 2023).

18

IoT Security

18.1 Introduction

The most critical issue in security is to use it at all levels, from the sensor to the communication system, from the router to the cloud. IoT security and privacy landscape should be created considering all of the IT and interconnection systems in the IoT ecosystem, which constitutes a large infrastructure starting from the most extreme sensor in the field to the cloud systems where the data is terminated and stored. As the system expands and its connectivity increases, many vulnerable points can occur. Security and privacy issues become significant as IoT enters critical facilities, our homes, our clothes, and our healthcare space.

To understand the importance of exploring security and privacy issues in the IoT space, it is necessary to look at the current state of IoT device deployments worldwide. A 2014 study by Hewlett-Packard [1] on commercialized IoT deployments found that 80% of such devices violate the privacy of personal information (e.g. name, date of birth, etc.). It was observed that 80% did not set passwords of sufficient complexity and length, 70% did not encrypt communication, and 60% had security vulnerabilities in their user interfaces [2].

Attacks on IoT devices are simple and easy to execute. Many events can be found that indicate the successful capture of smart things. The typical attack strategy is to compromise a device in the IoT network and perform fraudulent actions against another connected object by imitating the real one.

18.2 Limitations in IoT End Devices

Cyber attackers can organize thousands of security attacks through intelligent appliances such as home-networking routers, connected multimedia centers, smart TVs, refrigerators, and ovens. For example, the alarm system can be turned

Evolution of Wireless Communication Ecosystems, First Edition. Suat Seçgin.

Published 2023 by John Wiley & Sons, Inc.

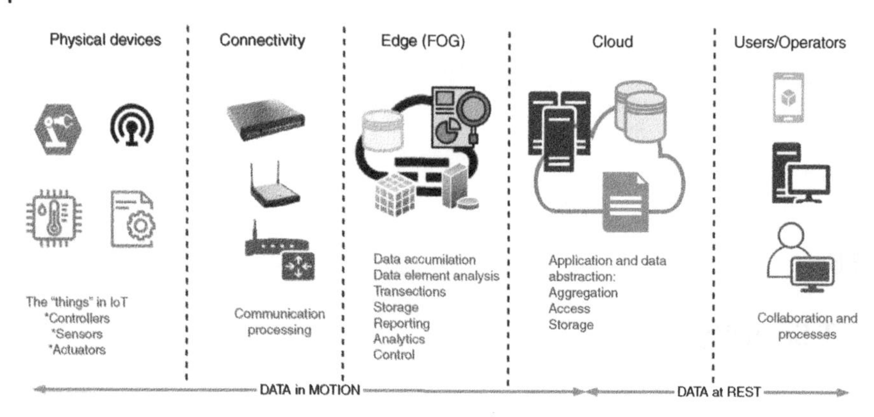

Figure 18.1 IoT hierarchy.

off by eavesdropping on the signals in a wireless security alarm system. Or, magnetic sensors used on the streets can be compromised by professional transmitters and antennas several miles away because only a few security protocols are used there.

The previous sections explained the elements of an IoT architecture in detail. The first important step before designing a security system is to have a good understanding of the infrastructure and architecture of the system. So, simply the layered architecture of an IoT system is given in Figure 18.1.

A secure system can be designed with a detailed analysis of the layers of this structure, the relationship between these layers, the connection types, and the device and software elements used as a whole.

There are some constraints and difficulties regarding a secure IoT system's hardware, software, and network design [2].

Hardware-based constraints:

- **Computation and energy constraints:** IoT devices are primarily designed with low-clock CPUs with low power consumption. Therefore, running computationally expensive security algorithms on such low-power devices may not be possible. The situation will be better understood when you consider the hardware capability of a heat sensor.
- **Memory constraint:** In addition to working with low battery power, IoT end devices (sensors, actuators, etc.) have limited RAM and flash memory as memory. These devices use Real-Time Operating System (RTOS) or a lightweight version of a General-Purpose Operating System (GPOS). They also contain system-specific software on them. For these reasons, traditional security algorithms cannot be run on these devices, as they have insufficient memory for security schemes.
- **Tamper-resistant packaging:** IoT devices can be deployed in remote areas and left unattended. An attacker can make changes to IoT devices through

device capture. They can extract cryptographic secrets, modify programs, or replace them with malicious nodes. Tamper-proof packaging is one way to protect against these attacks.

Software-based constraint:

- **Embedded software constraints:** Installing a dynamic security patch on IoT devices and mitigating potential vulnerabilities is no easy task. Remote reprogramming may not be possible for IoT devices, as the operating system or protocol stack may not be capable of acquiring and integrating new code or libraries.

Network-based constraints:

- **Mobility:** Mobility is one of the highlights of IoT devices where devices join a close network without prior configuration. This mobility structure increases the need to develop mobility-resistant security algorithms for IoT devices.
- **Scalability:** The number of IoT devices is increasing daily, and more and more devices are connecting to the global information network. Existing security schemes lack scalability; therefore, such schemes are not suitable for IoT devices.
- **Multiplicity of devices:** The variety of IoT devices within the IoT network ranges from full-fledged PCs to low-end RFID tags. Therefore, finding a single security scheme that accommodates even the simplest devices is complex.
- **Multiplicity of communication medium:** IoT devices connect to the local and public networks through various wireless connections. Therefore, it isn't easy to find a comprehensive security protocol when considering both wired and wireless media features.
- **Multiprotocol networking:** IoT devices can use a proprietary network protocol (e.g. non-IP protocol) for communication in nearby networks. It can also communicate with an IoT service provider over the IP network. These multi-protocol communication features make traditional security schemes unsuitable for IoT devices.
- **Dynamic network topology:** An IoT device can join or leave a network from anywhere at any time. The ability to add and remove temporal and spatial devices makes a network topology dynamic. The current security model for digital systems does not cope with sudden network topological changes. Therefore, such a model does not comply with the security of smart devices.

18.3 Security Requirements

There are several factors to consider when designing a security solution for IoT devices. The security requirements that IoT security schemes are expected to meet are given in Figure 18.2 [3].

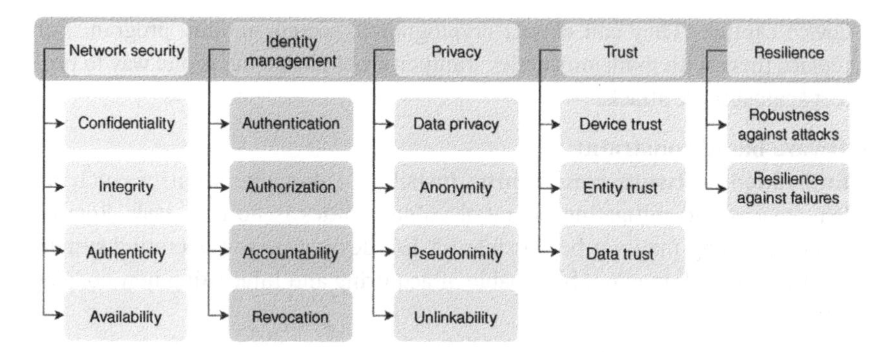

Figure 18.2 IoT security requirements.

Information Security Requirements

- **Confidentiality:** Confidentiality can ensure that data is only accessible to authorized users during the process and cannot be listened to or interfered with by unauthorized users. Confidentiality is an important security principle in IoT because many measuring devices (RFID, sensors, etc.) can be integrated into the IoT. It is, therefore, critical to ensure that data collected by a device does not provide secure information to neighboring devices [4]. Advanced techniques must be developed to achieve great confidentiality, including secure key management mechanisms and others.
- **Privacy:** Privacy can ensure that only the relevant user can control the data and that no other user can access or process the data. Unlike confidentiality, which aims to encrypt data without eavesdropping and interference by unauthorized users, privacy ensures that the user only has certain controls based on the data received and cannot extract other valuable information from the received data.
- **Trust:** Trust can ensure that the security mentioned here and privacy goals are achieved during interactions between different objects, IoT layers, and applications. The objectives of trust in IoT can be divided into trust between devices and between devices and applications [1]. With trust, security and privacy can be enforced.
- **Data integrity:** An adversary can alter data and compromise the integrity of an IoT system. Thus, integrity ensures that any received data is not modified during transmission. The data generated by IoT systems contain some secrets. This data is critical and should be protected from outsiders. In addition, this data should be kept confidential and stored for future use. Storage and other traditional centralized storage tools can be used and integrated with the IoT architecture. However, they suffer from inherent vulnerabilities. Also, due to

more devices having a central server model, multiple congestions, system scalability issues, and delayed responses can occur. In terms of data integrity, blockchain-based solutions can be developed to protect IoT data from deletion and pollution [5].

- **Information protection:** The secrecy and confidentiality of live and stored information must be strictly protected. It means limiting information access and disclosure to authorized IoT nodes and preventing unauthorized access or disclosure. For example, an IoT network should not display sensor readings to its neighbors if configured not to do so.
- **Anonymity:** Anonymity hides the source of the data. This security service helps with data security and privacy.
- **Nonrepudiation:** Nonrepudiation is the assurance that one cannot deny something one has done. An IoT node cannot refuse to send a message it has already sent.
- **Freshness:** It is necessary to ensure the freshness of each message. Freshness ensures that the data is very recent and no old messages are replayed.

Access-level security requirements:

- **Authentication and identification:** Authentication enables an IoT device to authenticate the peer it is communicating with (e.g. the receiver verifies whether the received data is coming from the correct source). It also requires providing valid users access to IoT devices and networks for administrative tasks (remote reprogramming or controlling IoT devices and networks). Identification can ensure that unauthorized devices or applications cannot connect to the IoT, authentication can ensure that data transmitted across networks is legitimate, and devices or applications requesting data are also fair.
- **Authorization:** Only authorized devices and users can access network services or resources.
- **Access control:** Access control ensures that an authenticated IoT node can only accesses what it is authorized to access.

Functional security requirements:

- **Exception handling:** Exception handling verifies that an IoT network is alive and continues to serve even in abnormal situations (node breach, node destruction, faulty hardware, software failures, and displacement of environmental hazards). Thus, it guarantees robustness.
- **Availability:** Availability ensures that IoT services can be made available to authorized parties when needed despite the denial of service (DoS) attacks. It also ensures that it can provide a minimum level of service in case of power cuts and malfunctions.
- **Resiliency:** A security plan should protect against attacks if several interconnected IoT devices are compromised.

- **Self-organization:** An IoT device can fail or run out of energy. The remaining device or collaborating devices must be capable of being refactored to maintain a certain level of security.

18.4 Attack Types and Points

Attacks can be launched at IoT layers (physical devices and sensors, connectivity equipment, edge/fog points, cloud, application, and local and public networks). Attack types will be classified and explained as physical, network, software, and data attacks as given here [6] (Figure 18.3).

Physical attacks:

- **Tampering devices:** In this attack, the attacker physically replaces the compromised node and can obtain sensitive information such as the encryption key. In other words, this type of attack is a type of physical attack where the attacker can manipulate the memory/computing, interact with the faulty device to gain additional information, and then attempt to break the security. IoT devices are attractive targets for the following reasons [7]:
  - They can be in private, remote, or unattended locations, so physical access is unlimited with no time constraints.
  - Some IoT devices may be small in size and therefore easily hidden in case of theft.
  - Most of such devices are technically limited. For example, limited processing capacity or power supply often means that security measures are similarly limited. This potentially facilitates the enforcement of a violation.
  - Due to the nature and location of a device, it is often clear that it can be used to attack a particular target when compromised.
  - When a device is compromised, it may be possible to compromise many similar devices more easily.
  - If multiple devices are compromised, a distributed DoS attack on another target may be possible.
  - Compromising an IoT device could give an attacker greater access to the IoT network.
  - Some devices can stay in a place for years. As technology advances, the security level may be low on an older device made from technology that is outdated over time.
- **Malicious code injection:** An attacker compromises a node by injecting it with malicious code that will allow it to physically access the IoT system; for example, imagine an attacker inserts a USB stick containing malware (i.e. virus) into the node. This means that the attacker can access the entire system.

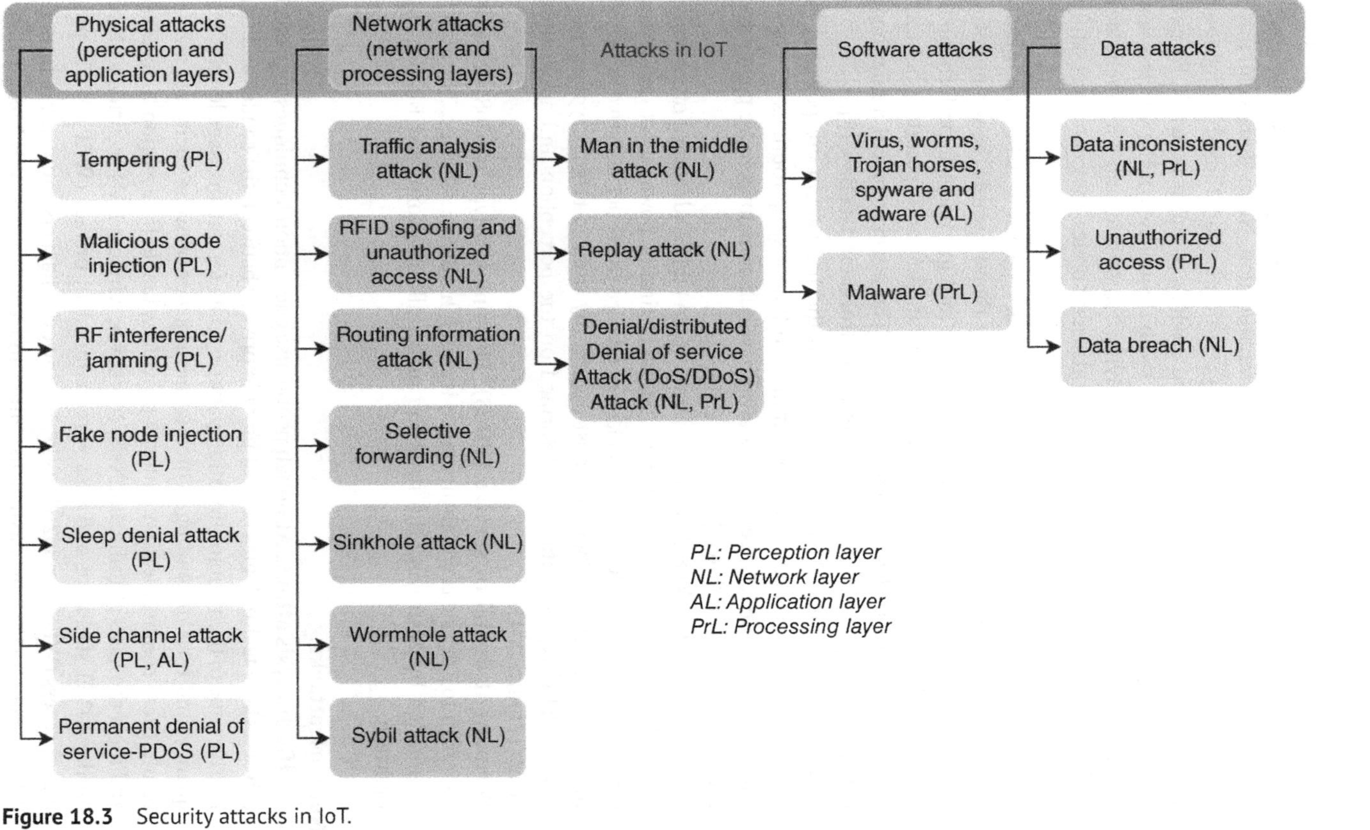

Figure 18.3 Security attacks in IoT.

- **RF interference/jamming:** In RF interference, a DoS attack can be performed on any RFID tag by generating and sending noise signals over the Radio Frequency signals that RFIDs use for communication. Noise signals can interfere with RFID signals that interfere with communication, thus shutting down the service. A jamming attack is based on a wireless sensor network. By jamming, the attacker can interfere with the radio frequencies of the wireless sensor nodes, corrupting the signals and denying communication with the nodes. If the attacker manages to block the critical sensor nodes, they can successfully disrupt the IoT service.
- **Fake\Malicious node injection:** In this attack type, the attacker can inject a malicious node between one or more nodes. It then forwards the wrong information to other notes. An attacker can also use multiple nodes to perform malicious node injections. This attack can control data flow and operations between IT and malicious nodes or connected links. The attacker places a fake node between two legitimate network nodes to control their data flow [8].
- **Sleep denial attack:** Most sensor nodes in the IoT system are powered by replaceable batteries and are programmed to follow sleep routines to extend battery life. This attack keeps nodes awake, which consumes more power and causes nodes to shut down [9].
- **Side-channel attack:** Side-channel attack (SCA) takes advantage of information leaks in the system. Leaks can be caused by timing, power, electromagnetic signals, sound, light, etc. These attacks can be used to extract sensitive information from the device. They are most commonly used to target cryptographic devices. This type of attack is performed on the perception and application layers (Figure 18.4).
- **Permanent denial of service (PDoS):** Phishing is a type of DoS attack in which an IoT device is wholly damaged through hardware sabotage. The attack is initiated by destroying the firmware or installing a corrupted BIOS using malware.

Network attacks:

- **Traffic analysis attack:** Although people use encryption techniques to protect privacy, these technologies cannot provide privacy between the sender and the receiver due to their communication patterns (data format, message length, frequency, and time of sending, which communicates between them, etc.). As a result, these communication patterns provide valuable information to hackers and form the basis of traffic analysis. Moreover, wireless LAN security is based on the 802.11 standards and uses wired equivalent privacy (WEP), Wi-Fi protected access (WPA), or 802.11 (WPA2). However, hackers can capture initialization vectors (IVs) by passive listening and try to crack WEP using these vectors. Hackers can register WPA's and WPA2's four-way handshakes and

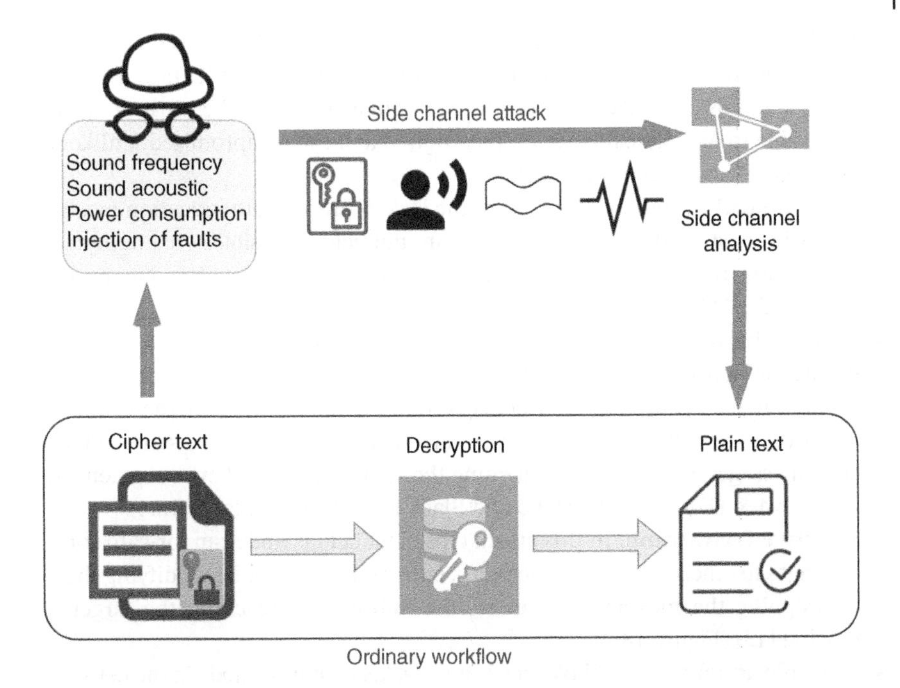

Figure 18.4 Side channel attack.

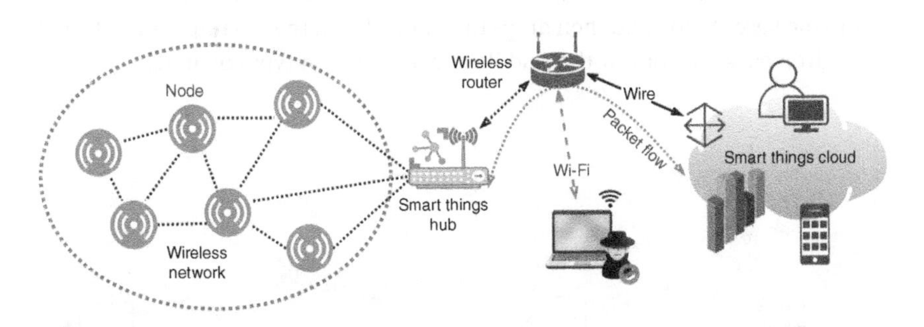

Figure 18.5 Traffic analysis attack.

organize dictionary attacks to discover passwords. Even if hackers cannot decrypt keys, they can monitor user data with the traffic analysis method [10] (Figure 18.5).

- **RFID spoofing:** An attacker emulates an RFID signal to read and record a data transmission from an RFID tag. Then the attacker can send his data containing the original tag ID to make it appear valid so the attacker can pretend to be the source and gain full access to the system [11].

- **RFID cloning:** The attacker clones an RFID tag by copying the data from the victim's RFID tag to another RFID tag. Although the two RFID tags have the same data, this method does not copy the original identity of the RFID. It makes it possible to distinguish between the original and the compromised, unlike in the event of an RFID fraud attack.
- **RFID unauthorized access:** Due to the lack of proper authentication mechanisms in most RFID systems, the tags are publicly accessible. This automatically means that the attacker can read, modify or even delete the data on the RFID nodes [12].
- **Routing information attack:** These are attacks in which the enemy can complicate the network by misleading, changing, or replaying routing information. In these attacks, traffic can be dropped as much as allowed. Again, these attacks are direct attacks where the adversary can create routing loops by sending false error messages, shortening or widening the source paths, and even segmenting the network. (For example. Hello and Black Hole Attack) (Figure 18.6).
- **Selective forwarding:** In this attack type, a malicious node can forward some messages to other nodes in the network, either selectively or by modifying them or dropping the message. Therefore, the information reaching the target is missing [13] (Figure 18.7).
- **Sinkhole attack:** In a sinkhole attack, an attacker captures a node in the network and executes the attack using that node. The compromised node sends fake routing information to its neighbor nodes where it has the minimum distance from the base station and then drops the traffic. It can then manipulate data and also drop packets. Man-in-the-middle attacks are such types of attacks.

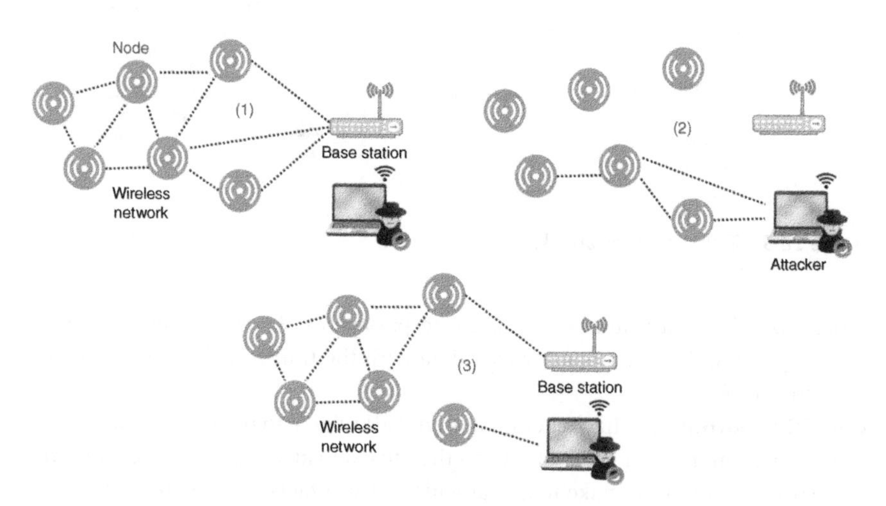

Figure 18.6 Hello flood attack.

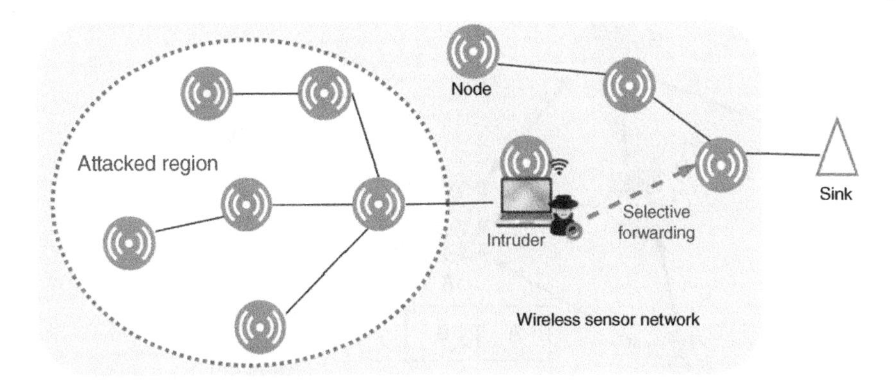

Figure 18.7 Sinkhole attack.

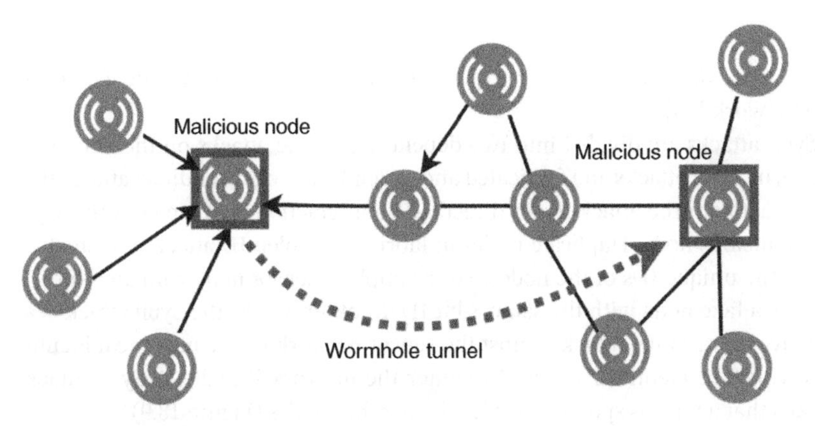

Figure 18.8 Wormhole attack.

- **Wormhole attack:** There is a bridge/tunnel between these malicious nodes. These nodes, which will attack the others, capture the packets from one point and tunnel them to an arbitrary point, and the attacking node there scatter it. This results in early or delayed arrival or, in some cases, the packet not reaching the appropriate node. The routing algorithms that depend on the path length between the nodes are performed because of these wormhole nodes [14] (Figure 18.8).
- **Sybil attack:** This is a malicious attack that shatters the network model. In this scenario, a node or a device receives many IDs, i.e. another node's ID from several other nodes leading to redundancy in the routing protocol. This results in a loss of data integrity, security, and resource usage. While encryption methods prevent an external attack on nodes, there may be an internal attack. Sybil node S mimics the other node N. Neighboring nodes receive messages from the Sybil

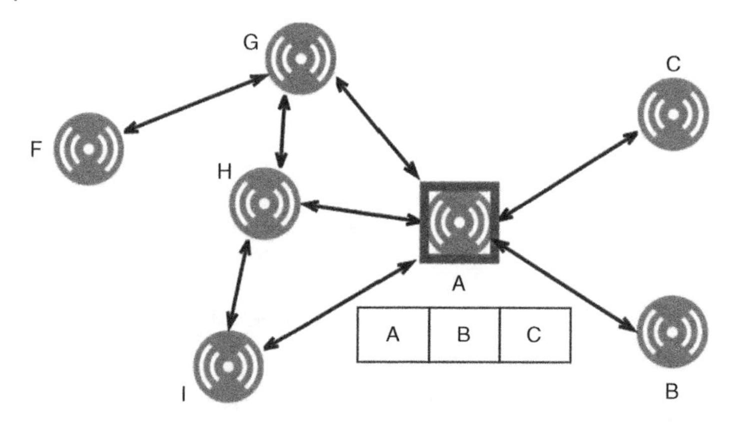

Figure 18.9 Sybil attack with multiple ID.

node with the ID of other nodes. This creates misbehavior and corrupts the network [15].

Sybil attacks are divided into two depending on the attacks on the network: direct/indirect attacks and fabricated and stolen ID attacks. In a direct attack, the nodes act with the attack nodes directly. The interaction is carried out through the malicious node. Duplicate nodes in fabricated/stolen ID attacks are created using the unique IDs of the nodes. For example, a sensor node with an 8-bit ID creates a fake node with the same 8-bit ID. In other words, the Sybil attack is a massive, devastating attack against the sensor network where many real identities with fake identities are used to enter the network illegally. A Sybil attack means that a node is spoofing its identity to other nodes (Figure 18.9).

- **Man-in-the-middle attack:** Over the Internet, the attacker interrupts the communication between the two nodes. They obtain sensitive information by eavesdropping.
- **Replay attack:** An attacker can intercept a signed packet and resend the packet to the destination multiple times, making the network busy and leading to a DoS attack (Figure 18.10).
- **Denial/Distributed denial of service (DoS/DDoS) attack:** An attacker can perform DoS or distributed denial-of-service DDoS attacks on the affected IoT network through the application layer and infect all users on the network. DoS is a cyber attack in which the perpetrator attempts to make a machine or network resource unavailable to its intended users by temporarily or indefinitely interrupting the services of a node connected to a network. A DDoS attack happens when multiple systems flood the bandwidth or resources of a targeted system. A DDoS attack uses multiple unique IP addresses or machines, often from thousands of malware-infected hosts (Figure 18.11).

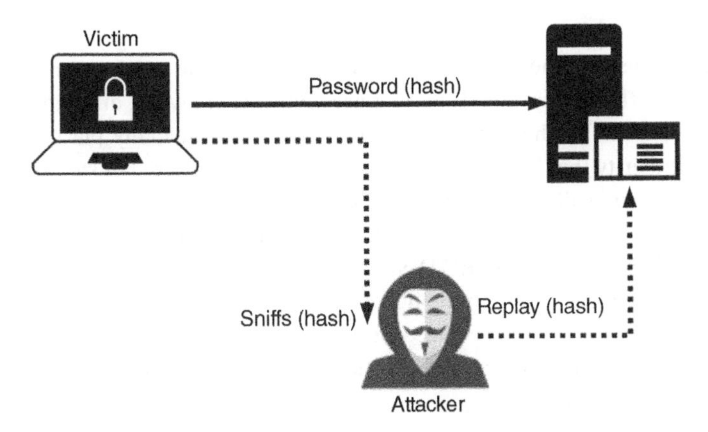

Figure 18.10 Replay attack.

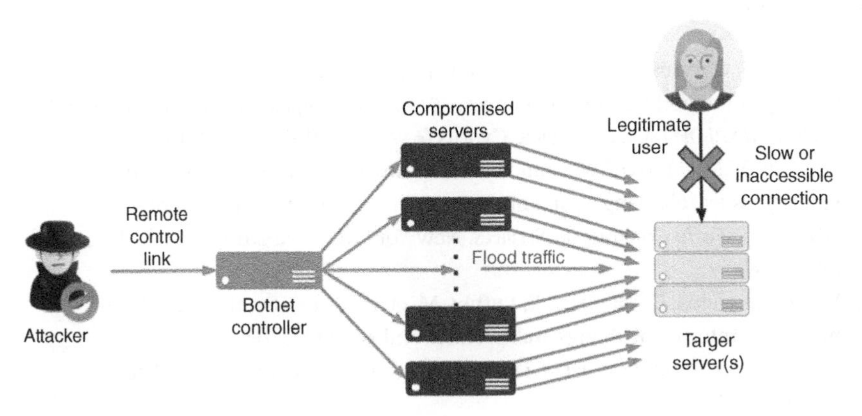

Figure 18.11 DDoS attack.

Software attacks:

- **Virus, worms, trojan horses, spyware, and adware:** Through this malware, an attacker can infect the system to falsify data, steal information, or even initiate DoS. These codes are spread by downloading files from the internet or via e-mail attachments. On the other hand, the worm can self-replicate without any human action and often locks up the network, causing the network resources to run out.
- **Malware:** The attacker can access the system by injecting a malicious script. Data located on IoT devices can be infected by malware that can contaminate the cloud or data centers. As you can see, this attack works in the processing layer of the IoT system.

Data attacks:

- **Data inconsistency:** It is the type of attack that occurs in the network and processing layers of the IoT system. Data inconsistency occurs when different and conflicting versions of the same information are found in many places. Attacks on data integrity that lead to inconsistency in data transferred or stored in a central database are called data inconsistency attacks.
- **Unauthorized access:** This attack is also on the processing layer. If the correct authentication is not provided in RFID systems, the attacker can observe, modify, or remove the information in the nodes.
- **Data breach:** It is a type of attack that occurs at the network layer. A data breach or memory leak refers to the unauthorized disclosure of personal, sensitive, or confidential data.

References

1 Andrea, I., Chrysostomou, C., and Hadjichristofi, G. (2015). Internet of Things: security vulnerabilities and challenges. *2015 IEEE Symposium on Computers and Communication (ISCC)*, Larnaca, Cyprus (6–9 July 2015), pp. 180–187.

2 Hossain, M.M., Fotouhi, M., and Hasan, R. (2015). Towards an analysis of security issues, challenges, and open problems in the Internet of Things. *2015 IEEE World Congress on Services*, New York (27 June 2015 to 2 July 2015), pp. 21–28.

3 Vasilomanolakis, E., Daubert, J., Luthra, M. et al. (2015). On the security and privacy of Internet of Things architectures and systems. *2015 International Workshop on Secure Internet of Things (SIoT)*, Vienna, Austria (21–25 September 2015), 49–57.

4 Lin, J., Yu, W., Zhang, N. et al. (2017). A survey on Internet of Things: architecture, enabling technologies, security and privacy, and applications. *IEEE Internet of Things Journal* 4 (5): 1125–1142.

5 Alferidah, D.K. and Jhanjhi, N.Z. (2020). A review on security and privacy issues and challenges in Internet of Things. *International Journal of Computer Science and Network Security IJCSNS* 20 (4): 263–286.

6 Sengupta, J., Ruj, S., and Das, B.S. (2020). A comprehensive survey on attacks, security issues and blockchain solutions for IoT and IIoT. *Journal of Network and Computer Applications* 149: 102481.

7 Secure IoT (2021). IoT Security Foundation. https://www.iotsecurityfoundation.org/best-practice-guidelines/ (accessed 16 February 2023).

8 Ahemd, M.M., Shah, M.A., and Wahid, A. (2017). IoT security: a layered approach for attacks & defenses. *2017 International Conference on Communication Technologies (ComTech)*, Rawalpindi, Pakistan (19–21 April 2017), pp. 104–110.

9 Alaba, F.A., Othman, M., Hashem, I.A.T., and Alotaibi, F. (2017). Internet of Things security: a survey. *Journal of Network and Computer Applications* 88: 10–28.

10 Yoshigoe, K., Dai, W., Abramson, M., and Jacobs, A. (2015). Overcoming invasion of privacy in smart home environment with synthetic packet injection. *2015 TRON Symposium (TRONSHOW)*, Tokyo, Japan (9–11 December 2015), 1–7.

11 Mitrokotsa, A., Rieback, M.R., and Tanenbaum, A.S. (2009). Classifying RFID attacks and defenses. *Information Systems Frontiers* 12 (5): 491–505.

12 Uttarkar, R. and Kulkarni, R. (2014). Internet of things: architecture and security. *International Journal of Computer Application* 3 (4): 12–19.

13 Varga, P., Plosz, S., Soos, G., and Hegedus, C. (2017). Security threats and issues in automation IoT. *2017 IEEE 13th International Workshop on Factory Communication Systems (WFCS)*, Trondheim, Norway (31 May 2017 to 2 June 2017), pp. 1–6.

14 Palacharla, S., Chandan, M., GnanaSuryaTeja, K., and Varshitha, G. (2018). Wormhole attack: a major security concern in Internet of Things (IoT). *International Journal of Engineering and Technology* 7 (3): 147–150.

15 Dhamodharan, U.S.R.K. and Vayanaperumal, R. (2015). Detecting and preventing sybil attacks in wireless sensor networks using message authentication and passing method. *The Scientific World Journal* 2015: 1–7.

19

IoT Applications

19.1 Introduction

This section is thought to develop the vision of the IoT system by giving sample applications to the readers. Since each application is based on a basic IoT architecture, it will also form the basis for other applications. The main differences between the applications are the access methods, the types of sensors used, the protocols, the data management system installed on the edge and cloud, and the various software tools.

19.2 Tactile Internet

The haptic Internet is a natural evolution of the Internet that went from a fixed and text-based Internet to the multimedia mobile Internet and later to the Internet of Things (IoT). The haptic Internet is for haptic communication over the network. Using the tactile Internet, a doctor can perform a remote surgical operation on a remote patient by sensing the real-time visual, auditory, and tactile senses of the distant environment. These new possibilities will revolutionize the set of applications and services provided by the Internet to date. They will take next-generation systems to an unprecedented level of human-like communication. The haptic Internet will revolutionize and unify machine-to-machine and human-to-machine interactions [1].

Tactile Internet applications such as Telesurgery often require ultra-low latency, high reliability, and security to operate accurately and securely. The latency requirements of haptic Internet applications may vary depending on the application type and the dynamics of the environment. More specifically, the

Evolution of Wireless Communication Ecosystems, First Edition. Suat Seçgin.

Published 2023 by John Wiley & Sons, Inc.

latency requirements of Haptic Internet applications can range from <10 ms to tens of milliseconds, while the Haptic Internet targets an ultra-low end-to-end round-trip latency of 1 ms. This requires keeping haptic applications reasonably close to the endpoints of applications (for example, keeping the robotic surgery application relatively close to the console of the surgeon and the console of the robot performing the surgery) [2] (Figure 19.1).

5G networks were initially known as promising candidates to enable Tactile Internet. Although 5G can provide a peak data rate of up to 20 Gbps, it may fall short of meeting the stringent requirements of Haptic Internet for the following reasons. First, emerging immersive Haptic Internet applications are expected to require Tbps-level data rates. For example, we note that a typical VR headset requires a communication link of multiple Gbps. In 5G networks, this can only be accomplished via a cable connection to a PC, posing severe limitations on users' mobility. These requirements may be even higher for Haptic Internet applications where there is a need to deliver multi-sensory content in real-time [3].

Second, it is clear that, like the previous generation of mobile networks, 5G is mainly driven by content and machine-centric design approaches to handle H2H and M2M traffic, respectively. In contrast, the Haptic Internet aims to realize human-machine interaction through Human to Machine Robot (H2M/R) communication, which differs in many ways from traditional Human to Human (H2H) and Machine to Machine M2M traffic.

Two Tactile Internet application examples are given below [4];

- **Remote Robotic Surgery (Telesurgery):** Remote robotic surgery will make surgery available anywhere, wherever surgeons and patients are. This will have many potential benefits for human society, including reducing the risks and delays associated with long-distance patient travel and providing surgery to patients living in underserved areas.
- **Autonomous Driving:** Autonomous driving is another potential application that the Haptic Internet could realize. It is expected to reduce traffic congestion, accidents, and greenhouse gas emissions. In general, an autonomous driving system can be divided into three main subsystems: perception, planning, and control. The sensing subsystem is responsible for collecting information from sensors about location, speeds, road conditions, nearby vehicles, and the environment and using this constantly updated information to create a dynamic environment model. The planning subsystem makes driving decisions based on this environment model and feedback from the control subsystem. The control subsystem then manipulates the vehicles according to their driving decisions – adjusting their speed, steering angle, acceleration, etc.

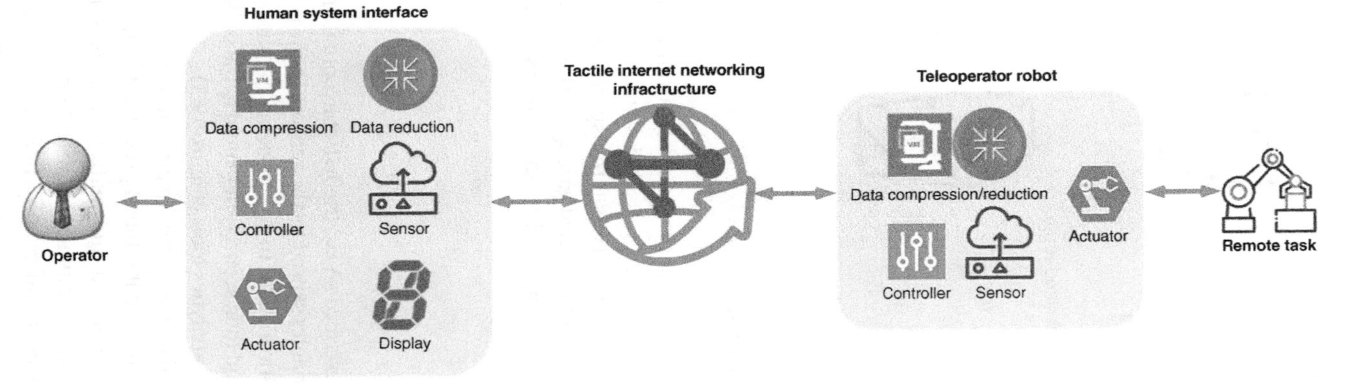

Figure 19.1 Tactile Internet operation.

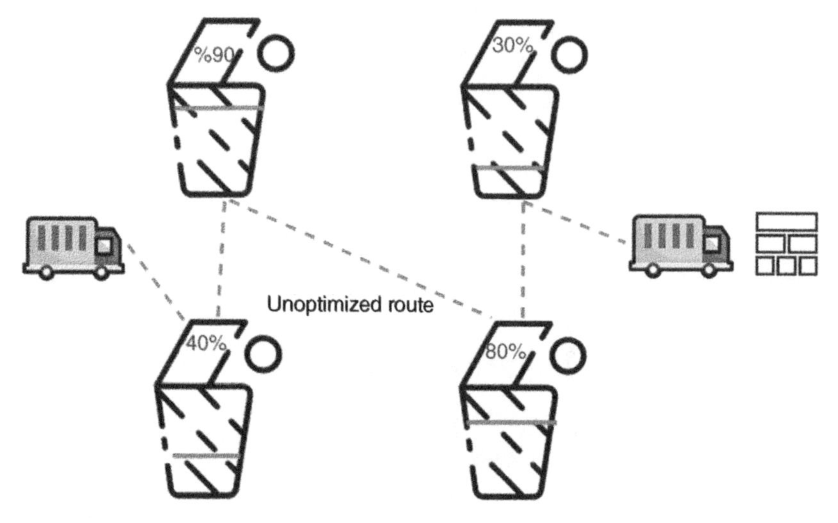

Figure 19.2 Waste management.

19.3 Waste Management

Waste management is one of the most critical environmental issues to be dealt with, especially in crowded cities. It is an issue that needs to be managed efficiently due to the cost of service (waste transporter trucks going around the garbage cans) and the storage of collected garbage. Information and communication technologies should be supported to save money and provide maximum benefit (Figure 19.2).

19.4 Healthcare

In the IoT architecture described so far, the IoT conceptual architecture does not change, except for the variables sensed by the sensors in the perception layer (data collection layer). The critical issue in IoT systems is that the collected data is aggregated on the cloud and then securely delivered to the evaluators (for example, doctors) with various data analytics tools (Figure 19.3).

Patient examination, diagnosis, and treatment can be made remotely by processing real-time and offline data (blood pressure, fever, heart rate, etc.).

We can classify cloud-connected information systems in healthcare applications as health information exchange, collaboration solutions for physicians, clinical information systems, office productivity solutions, and electronic medical records (EMR) storage. Mobile, web, etc., applications can monitor these systems.

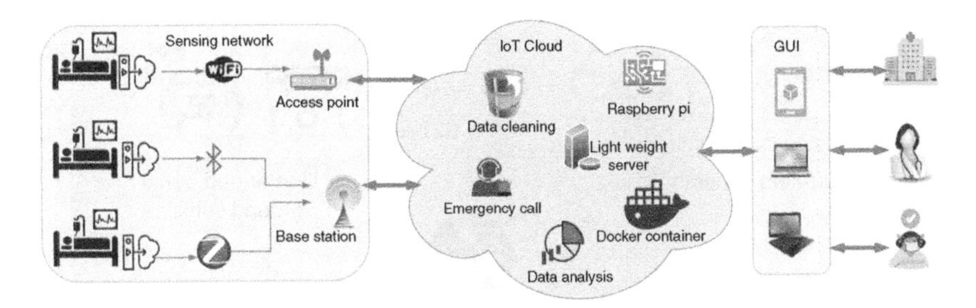

Figure 19.3 E-Health.

19.5 Smart Agriculture and Smart Water Supply

As in all other IoT applications, we can explain the IoT infrastructure in agricultural applications with a few basic architectures. At the extreme end, there are agriculture-specific reporting and feedback applications, starting with field sensors/actuators and unmanned aerial/ground vehicles for data collection, local and backbone access at the midpoint, and finally on the data stored with the cloud.

Of these mentioned layers, the most extreme is the data collection layer, where environmental/plant data is collected via sensors or remote sensing devices such as UAVs. Data such as soil moisture, pest status, fertilization need, etc., are collected and transmitted to the cloud via a gateway.

In the cloud layer, data is refined with data processing tools for reporting and operation feedback for the monitor and actuators. Necessary user and administrative applications are superimposed on this refined data. The system works in a closed loop. Considering certain factors such as soil condition, fertilization pattern, crop condition, weather and environmental conditions, and the like, the information obtained in the last step will be used for decision-making or creating area index maps for different purposes. This layer also provides data visualization and presentation services for user access (Figure 19.4).

Finally, the control and actions of agricultural operations such as cultivation, irrigation, fertilization, and harvesting are carried out at the top layer. Operations related to agricultural devices, machines, vehicles, an irrigation system based on decisions, or index maps created in the cloud in the final stage are performed. The Internet gateway will transmit the relevant control commands to the farming systems. Supported by GIS, these farming systems will fully process each atomic field so that optimum efficiency and productivity can be expected [5].

An irrigation system can be given as an example application. Information from sensors placed for end-point crop shade and air temperature measurements is transmitted to the gateway over a wireless sensor network (WSN). Wireless sensor network selection can be made by looking at the topics described in the previous sections. As ZigBee is seen as an alternative since it requires low bandwidth at

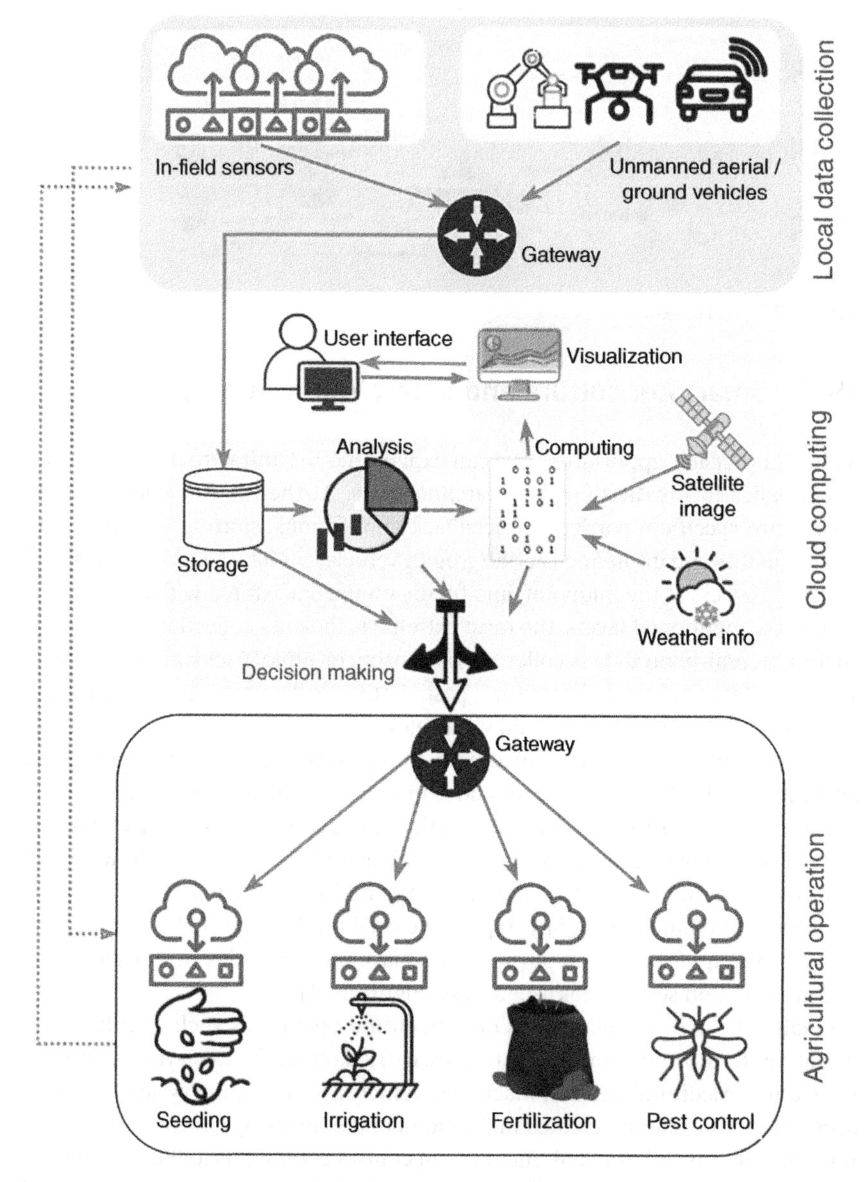

Figure 19.4 Smart agriculture application.

first glance, one of Lora or Sigfox technologies can be selected according to the width of the field and the measurement continuity requirement. It can determine irrigation times and amounts through the information obtained by various data analysis (machine learning, data science tools, etc.) methods of the data uploaded to the cloud via the network. Thus, we can reach the irrigation control system

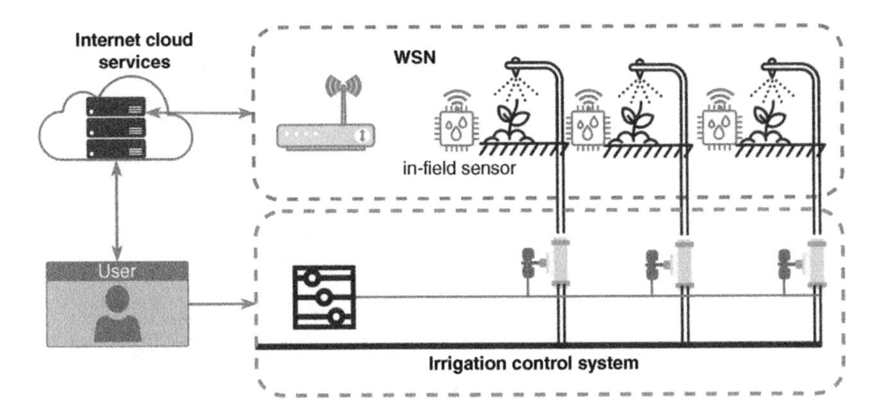

Figure 19.5 Irrigation system.

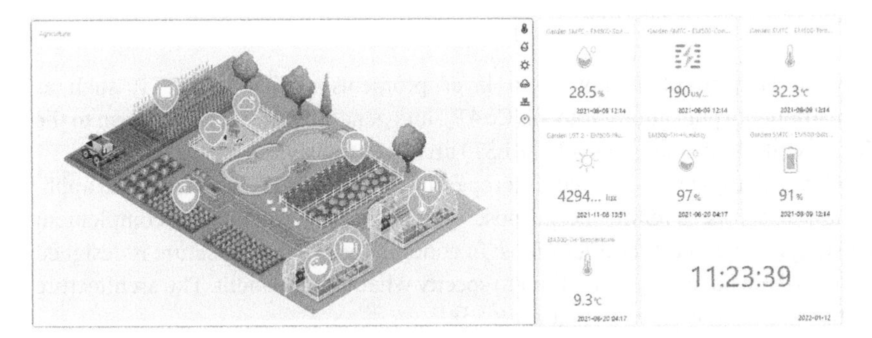

Figure 19.6 Smart farming dash/control board example.

through the terminals where we observe, monitor, and control the area, and start and stop irrigation (Figure 19.5).

Suppose these systems are installed in large areas. In that case, a dash/control board can be created by integrating the existing information with a geographic information system and measuring the fields' humidity, temperature, and shade conditions. Below is an example control panel of this type application [6] (Figure 19.6).

19.6 Web of Things (WoT)

To promote the development and dissemination of the IoT, applications are starting to be built around the well-known Web model, which gives rise to the Web of Things (WoT). The Web-based approach has allowed the driving force of the wide spread of the Internet to be used in the field of IoT as well [7]. WoT applications

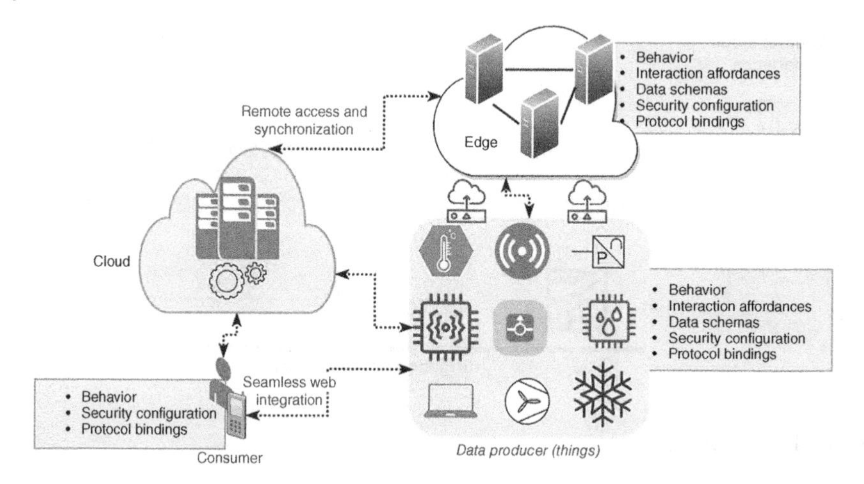

Figure 19.7 WoT architecture.

rely on Web-oriented application layer protocols similar to HTTP, such as Constrained Application Protocol (CoAP), and generally, protocols conform to the REpresentational State Transfer (REST) architectural style.

Web of Things (WoT) enables interoperability between IoT platforms and application domains. Overall, the purpose of WoT is to maintain and complement existing IoT standards and solutions. In general, the WoT architecture is designed to describe what exists rather than to specify what to implement. The architecture of the approach is given in Figure 19.7 [8].

References

1 Fettweis, G.P. (2014). The Tactile Internet: applications and challenges. *IEEE Vehicular Technology Magazine* 9 (1): 64–70.

2 Maier, M., Chowdhury, M., Rimal, B.P., and Van, D.P. (2016). The Tactile Internet: vision, recent progress, and open challenges. *IEEE Communications Magazine* 54 (5): 138–145.

3 Giordani, M., Polese, M., Mezzavilla, M. et al. (2020). Toward 6G networks: use cases and technologies. *IEEE Communications Magazine* 58 (3): 55–61.

4 Promwongsa, N., Ebrahimzadeh, A., Naboulsi, D. et al. (2021). A comprehensive survey of the Tactile Internet: state-of-the-art and research directions. *IEEE Communications Surveys and Tutorials* 23 (1): 472–523.

5 Zhang, L., Dabipi, I.K., and Brown, W.L. (2018). Internet of Things applications for agriculture. In: *Internet of Things A to Z* (ed. Q.F. Hassan), 507–528. Wiley.

6 Milesight (2022). IoT – LoRaWAN, 5G & AIoT. *Milesight*. https://www.milesight-iot.com/ (accessed 16 February 2023).
7 Want, R., Schilit, B.N., and Jenson, S. (2015). Enabling the Internet of Things. *Computer* 48 (1): 28–35.
8 W3C (2023). Web of Things (WoT) architecture. *W3C*. https://www.w3.org/TR/wot-architecture/ (accessed 16 February 2023).

Index

Evolution of Wireless Communication Ecosystems, First Edition. Suat Seçgin.

Published 2023 by John Wiley & Sons, Inc.

l

m

n

t

u

v

w

IEEE ComSoc
IEEE Communications Society

The ComSoc Guides to Communications Technologies
Nim K. Cheung, Series Editor

The ComSoc Guides to Communications Technologies

Nim K. Cheung, *Senior Editor*
Richard Lau, *Associate Editor*

The ComSoc Guide to Next Generation Optical Transport: SDH/SONET/OTN
Huub van Helvoort

The ComSoc Guide to Managing Telecommunications Projects
Celia Desmond

WiMAX Technology and Network Evolution
Kamran Etemad and Ming-Yee Lai

An Introduction to Network Modeling and Simulation for the Practicing Engineer
Jack Burbank, William Kasch, and Jon Ward

The ComSoc Guide to Passive Optical Networks: Enhancing the Last Mile Access
Stephen Weinstein, Yuanqiu Luo, and Ting Wang

Digital Terrestrial Television Broadcasting: Technology and System
Jian Song, Zhixing Yang, and Jun Wang

TV White Space: The First Step Towards Better Utilization of Frequency Spectrum
Ser Wah Oh, Yugang Ma, Edward Peh, and Ming-Hung Tao

Digital Services in the 21st Century: A Strategic and Business Perspective
Antonio Sanchez and Belen Carro

Toward 6G: A New Era of Convergence
Amin Ebrahimzadeh and Martin Maier

VCSEL Industry: Communication and Sensing
Babu Dayal Padullaparthi, Jim A. Tatum, and Kenichi Iga

Intelligent Reconfigurable Surfaces (IRS) for Prospective 6G Wireless Networks
Muhammad Ali Imran, Lina Mohjazi, Lina Bariah, Sami Muhaidat, Tie Jun Cui, and Qammer H. Abbasi

6G and Onward to Next G: The Road to the Multiverse
Martin Maier

Evolution of Wireless Communication Ecosystems
Suat Seçgin

Printed and bound by CPI Group (UK) Ltd, Croydon, CR0 4YY

16/04/2025

14658602-0002